About the Author

John L. Pollock is Professor of Philosophy and Cognitive Science at the University of Arizona. His previous publications in epistemology and philosophical logic include *An Introduction to Symbolic Logic* (1970), *Knowledge and Justification* (1974), *Subjunctive Reasoning* (1976), *Language and Thought* (1982), *The Foundations of Philosophical Semantics* (1984), and *Contemporary Theories of Knowledge* (1986).

Nomic Probability and the Foundations of Induction

NOMIC PROBABILITY AND THE FOUNDATIONS OF INDUCTION

John L. Pollock

New York   Oxford
OXFORD UNIVERSITY PRESS
1990

Oxford University Press

Oxford New York Toronto
Delhi Bombay Calcutta Madras Karachi
Petaling Jaya Singapore Hong Kong Tokyo
Nairobi Dar es Salaam Cape Town
Melbourne Auckland

and associated companies in
Berlin Ibadan

Copyright © 1990 by John L. Pollock

Published by Oxford University Press, Inc.,
200 Madison Avenue, New York, NY 10016

Oxford is a registered trademark of Oxford University Press

Library of Congress Cataloging-in-Publication Data
Pollock, John L.
Nomic probability and the foundations of induction / John L. Pollock.
p. cm. Bibliography: p. Includes index.
ISBN 0-19-506013-X
1. Probabilities. 2. Induction (Logic) I. Title.
QA273.P774 1989 519.2—dc20 89-31084

2 4 6 8 9 7 5 3 1

Printed in the United States of America
on acid-free paper

For Carole

PREFACE

My purpose in this book is to make general philosophical sense of objective probabilities and explore their relationship to the problem of induction. The species of probability under investigation is what I call 'nomic probability'. This is supposed to be the kind of probability involved in statistical laws of nature. The claim will be made that this is the most fundamental kind of objective probability, other important varieties of probability being definable in terms of it.

The theory of nomic probability is at base an epistemological theory. My main concern is to give a precise account of various kinds of probabilistic reasoning. I will make the claim that such a theory of probabilistic reasoning, all by itself, constitutes an analysis of the concept of nomic probability. No attempt will be made to give a *definition* of nomic probability. This contrasts sharply with most work in the philosophical foundations of probability theory. Virtually all such work has been centrally concerned with giving definitions. That work will be reviewed in Chapter 1. But my claim will be that no satisfactory definition of probability has been given, and it is unlikely that any satisfactory definition *can* be given. This does not make the concept of probability in any way unusual. If analytic philosophy has taught us anything, it should be that very few philosophically interesting concepts can be analyzed by giving definitions. This is a problem that I have addressed at length in my books *Knowledge and Justification* and *Contemporary Theories of Knowledge*. My general claim is that concepts are characterized by their role in reasoning. In exceptional cases, those roles make it possible to state informative necessary and sufficient conditions for the satisfaction of a concept, and in that case the concept can be defined. But that is usually impossible, in which case *the only way* to analyze the concept is to describe how it is used in reasoning. My proposal is that nomic probability should be analyzed in that way, and the purpose of the book is to provide such an analysis. On this view, the analysis of nomic probability becomes the same thing as the theory of reasoning about nomic probabilities.

The theory of nomic probability is essentially an epistemological theory, because it is a theory about reasoning. This theory differs from conventional probability theories mainly in the sophistication of the epistemological tools it employs. Traditional theories of probabilitistic reasoning employed sophisticated mathematics but crude epistemology. I have tried instead to couple the fairly standard mathematics of probability theory with state-of-the-art epistemology, and the result is the theory of nomic probability. The theory proceeds in terms of the technical paraphernalia of prima facie reasons and defeaters that are characteristic of all of my work on epistemology. This conceptual framework is now familiar to epistemologists, but many probability theorists will find it unfamiliar. Nevertheless, the framework is essential for the task at hand. If we are to get something new out of probability theory we must put something new into it, and my suggestion is that what is required is a more sophisticated epistemology.

I have referred to the mathematics of nomic probability as "fairly standard", but it is not entirely standard. One of the most important ingredients in the theory is the *principle of agreement*, which is a theorem of the calculus of nomic probabilities but is not a theorem of the classical probability calculus. Nomic probability involves a strengthening of the classical calculus, and that is the topic of Chapters 2 and 7. These chapters make use of an extravagant ontology of possible worlds and possible objects, and that will be objectionable to some readers. I confess that I am uncomfortable with it myself. It should be emphasized, however, that this ontology is used in only a small part of the book, and it is for the limited purpose of getting the mathematical structure of the calculus of nomic probability to come out right. For this purpose one can adopt an "as if" attitude, thinking of the ontology as just a heuristic model used in getting to the correct mathematical theory. If one is willing to accept the resulting calculus, on whatever grounds, then the rest of the theory of nomic probability follows and can be entirely insulated from the ontology. It is particularly to be emphasized that, casual appearances to the contrary, nomic probability is not *defined* in terms of sets of physically possible objects. In the development of the calculus, a principle is employed that *looks* like a definition, but it is really just a computational principle. No other theory of probability has ever been able to generate such theorems. This represents the meat of the theory, and I think it is also the strongest evidence for the correctness of the theory.

The most important part of the theory of nomic probability is the way it all hangs together mathematically. It does not consist of a bunch of unconnected principles governing different aspects of probabilistic reasoning but is a unified theory deriving accounts of important varieties of probabilistic reasoning from a very limited set of initial principles. For this kind of theory to be convincing, the treatment must be rigorous. This makes the book difficult, but to omit the technical details would make the book unbelievable. The theorems are sufficiently surprising that most philosophers are unwilling to believe them until they see that they have been proven. The importance of the theory is that the details actually work out, and the only way to demonstrate that they do is to actually work them out.

On the other hand, the reader need not read all the details in order to understand the theory. The details must be available for purposes of verification, but most readers will probably be willing to accept that various theorems can be proven without actually reading the proofs. The reader cannot entirely ignore the technical aspects of the theory, because they also play a role in guiding the construction of the theory. But I have tried to separate the technical details that are necessary for understanding the theory from those required only for a deep understanding of how to apply the theory to complex concrete situations. The latter have been put into Part II, and the general reader can ignore them. Insofar as the reader wishes to delve into the material in Part II, he can pick and choose what to read because the different chapters are largely independent of one another.

The reader's attention should be called to the appendix, which lists all named or numbered principles referenced at some distance from their original occurrence. This appendix is located unconventionally at the very end of the book, after the bibliography and index, in order to make it easy for the reader to find referenced principles quickly.

My overarching goal in epistemology is to construct a general theory of human rational architecture. A constraint that I want to impose upon all of this work is that the resulting theory of reasoning be sufficiently precise that it can be implemented in a computer program. The result will be an AI system that models human reasoning. The OSCAR project, currently underway at the University of Arizona, is devoted to the construction of this automated reasoning system. That project is the subject of my book *How to Build a Person*, which is appearing simultaneously with this book. The OSCAR project is an investigation in both epistemology and artificial intelligence. On the one hand, epistemological theory provides the theoretical basis for the automated reasoning system. On the other hand, implementing the epistemological theory on a computer is a way of testing the theory to see whether it works in reasonable ways. Experience indicates that the implementation almost invariably leads to the discovery of inadequacies in the theory and results in changes to the theory. The philosopher, sitting at home in his armchair and thinking about a theory, is limited in the complexity of the cases to which he can apply the theory. If the theory contains failings that are not too subtle, he will be able to find them by constructing fairly simple counterexamples. But if the difficulties are subtle, he may never find them in this way. On the other hand, a computer program implementing the theory can be applied to cases of arbitrary complexity. The computer becomes a kind of mechanical aid in the discovery of counterexamples. This is particularly valuable in an area like probabilistic reasoning where the counterexamples may be too complex to be easily evaluated by unaided reflection. Without trying this, one might question how useful such computer modeling actually is, but anyone who tries it discovers very quickly that the technique of computer modeling is an invaluable aid in constructing theories of reasoning. Theories evolve rapidly once they reach the stage of computer modeling. Researchers in AI have known this for years, and philosophers are now becoming aware of it.

Despite the power of computer modeling, the first step must still be a careful philosophical analysis producing the basic theory.

That theory can then be refined with the help of the computer. Much work in AI has proceeded in an *ad hoc* way without being based on careful philosophical analysis. This is largely because it has been carried out by researchers untrained in philosophy. What is required for success is a combination of philosophical analysis and the application of standard AI techniques to refine the analysis. I predict that in the future epistemology and artificial intelligence are going to become inextricably interwoven for just this reason.

The enterprise underaken by OSCAR project is an immense one, but the present work constitutes an important link in the overall structure. If we are to model our theory of probabilistic reasoning on a computer, we cannot be satisfied with giving the kind of vague untestable "picture", devoid of detail, that is characteristic of so much contemporary philosophy. Such a picture is not a theory. It is at best a proposal for where to start in constructing a theory. This problem is automatically avoided by insisting that a theory be capable of generating a computer program that models part of human reasoning. You cannot get a computer to do anything by waving your hands at it.

This book constitutes only the first step in the creation of an automated reasoning system that will perform probabilistic and inductive reasoning. This book provides the initial philosophical analysis from which an implementation attempt will begin. The actual implementation process will invariably turn up inadequacies in the theory, so this book represents just one stage of theory construction. The next stage is that of implementation and refinement. Implementation will not be explicitly discussed in the book, but the construction of the theory is guided throughout by the intent that it be implementable.

The general theory of nomic probability propounded in this book has important historical antecedents. The purely scholarly ones are detailed in Chapter 1. In a more personal vein, there are important connections to my own earlier work, and a very important connection to the work of Henry Kyburg. The theory is based upon the general epistemological theory that I have been developing throughout my philosophical career. That theory is

distinguished primarily by its treatment of epistemological problems in terms of prima facie reasons and defeaters. The most
recent statement of that theory occurs in my book *Contemporary
Theories of Knowledge*. Turning specifically to the epistemological
topics of the present book, I have always been interested in
induction, and the general formulation of inductive rules that is
defended here is an evolutionary development of the account I
proposed in *Knowledge and Justification*. The major difference is
that I now feel it possible to justify those rules on the basis of
more fundamental principles. The account of nomic generalizations, and the basic idea of nomic probability, can be traced to
Subjunctive Reasoning. However, I now regard my remarks about
probability in the latter book as somewhat of an embarrassment
and would not want anyone to take them very seriously. I feel
that some of the basic ideas were right, but I made little concrete
progress in thinking about probability until I began thinking about
Henry Kyburg's theory of direct inference. That theory contains
many important insights, and my own theory of direct inference
grew out of my thoughts about Kyburg's theory. On the surface,
my theory no longer looks much like Kyburg's, but the similarities
are still there, and I owe a profound intellectual debt to Kyburg.

Earlier versions of most of the material the theory of nomic
probability comprises have been published in various places, and
I want to thank the following publications for permission to
reprint parts of those papers in the present book. The papers
are: "A Theory of Direct Inference" (*Theory and Decision*, 1983),
"Epistemology and Probability" (*Synthese*, 1983), "Nomic Probability" (*Midwest Studies in Philosophy*, 1984), "A Solution to the
Problem of Induction" (*Nous*, 1984), "Foundations for Direct
Inference" (*Theory and Decision*, 1985), "The Paradox of the
Preface" (*Philosophy of Science*, 1986), "Probability and Proportions" (in *Theory and Decision: Essays in Honor of Werner
Leinfellner*, edited by H. Berghel and G. Eberlein and published
by D. Reidel, 1987), "Defeasible Reasoning" (*Cognitive Science*,
1987), and "The Theory of Nomic Probability" (*Synthese*, 1989).

I would also like to thank the National Science Foundation for two grants (Numbers SES-8306013 and SES-8510880) in partial support of this research and the University of Arizona for a sabbatical leave for the purpose of finishing this book.

Tucson   J. L. P.
March 1989

CONTENTS

PART I
THE GENERAL THEORY

1. Physical Probability
1.	Introduction	3
2.	The Laplacian Theory	5
3.	Logical Probability	8
4.	Frequency Theories	10
5.	Hypothetical Frequency Theories	15
6.	Empirical Theories	20
7.	Propensities	23
8.	Nomic Probability and Nomic Generalizations	32

2. Computational Principles
1.	Introduction	39
2.	The Logical Framework	39
3.	Nomic Generalizations	42
4.	Probabilities as Proportions	45
5.	The Boolean Theory of Proportions	48
6.	The Elementary Theory of Nomic Probability	53
7.	The Non-Boolean Theory of Proportions	60
8.	Non-Boolean Principles of Nomic Probability	65
9.	Exotic Principles	68
10.	Summary	72

3. Acceptance Rules
1.	Introduction	75
2.	An Acceptance Rule	75
	2.1 The Statistical Syllogism	75
	2.2 The Lottery Paradox and Collective Defeat	80
	2.3 Projectibility	81
	2.4 Defeaters	86
3.	The Epistemological Framework	87
	3.1 Justification and Warrant	87

3.2   Warrant and Ultimately
      Undefeated Arguments   89
3.3   The Structure of Arguments   92
4. Generalized Acceptance Rules   99
5. Derived Acceptance Rules   103
6. The Material Statistical Syllogism   104
7. Summary   106

4. Direct Inference and Definite Probabilities
1. Introduction   108
2. Classical Direct Inference   110
    2.1   Reichenbach's Rules   110
    2.2   Modifications to Reichenbach's Rules   115
    2.3   Frequencies vs. Nomic Probabilities   116
    2.4   Projectibility   123
3. Nonclassical Direct Inference   127
4. The Reduction of Classical Direct
    Inference to Nonclassical Direct Inference   132
5. Foundations for Direct Inference   140
6. Nonclassical Direct
    Inference for Frequencies   146
7. Summary   148

5. Induction
1. Introduction   149
2. A Generalized Acceptance Rule   155
3. The Statistical Induction Argument   170
4. Enumerative Induction   177
5. Material Induction   179
6. Concluding Remarks   183
7. Summary   185

6. Recapitulation
1. The Analysis of Nomic Probability   187
2. Main Consequences of the Theory   190
    2.1   Direct Inference   190
    2.2   Induction   191

3. Prospects: Other Forms of Induction   192
    3.1 Statistical Inference   192
    3.2 Curve Fitting   193
    3.3 Inference to the Best Explanation   195
4. The Problem of Projectibility   196

PART II
ADVANCED TOPICS

7. Exotic Computational Principles
1. Introduction   203
2. The Exotic Theory of Proportions   204
3. The Derivation of (*PFREQ*)   211
4. The Principle of Agreement   212
5. Summary   218

8. Advanced Topics: Defeasible Reasoning
1. Introduction   220
2. Reasons   220
3. Warrant   222
    3.1 An Analysis of Warrant   222
    3.2 Ultimately Undefeated Arguments   225
    3.3 Collective Defeat   226
    3.4 Two Paradoxes of Defeasible Reasoning   227
    3.5 Logical Properties of Warrant   231
    3.6 The Principle of Collective Defeat   232
    3.7 The Principle of Joint Defeat   235
4. Nonmonotonic Logic   238
5. Summary   243

9. Advanced Topics: Acceptance Rules
1. Introduction   244
2. The Gambler's Fallacy   245
3. The Paradox of the Preface   250
4. Defeasible Defeaters   253
5. Domination Defeaters   256
6. Summary   266

10. Advanced Topics: Direct Inference

1.	Propensities	267
	1.1 Counterfactual Probability	267
	1.2 Hard Propensities	270
	1.3 Weak Indefinite Propensities	274
	1.4 Strong Indefinite Propensities	276
2.	Joint Subset Defeaters	278
3.	Domination Defeaters	282
	3.1 Description of Domination Defeaters	282
	3.2 Defeaters for Domination Defeaters	288
	3.3 Extended Domination Defeaters	291
4.	Summary	295

11. Advanced Topics: Induction

1.	Projectibility	296
2.	Subsamples	303
3.	Fair Sample Defeaters for Statistical Induction	307
4.	Fair Sample Defeaters for Enumerative Induction	312
5.	Diversity of the Sample	315
6.	The Paradox of the Ravens	320
7.	Summary	322

Bibliography   323

Index   335

Appendix: Main Principles   339

PART I

THE GENERAL THEORY

CHAPTER 1
PHYSICAL PROBABILITY

1. Introduction

Probability theorists divide into two camps—the proponents of subjective probability and the proponents of objective probability. Opinion has it that subjective probability has carried the day, but I think that such a judgment is premature. I have argued elsewhere that there are deep incoherencies in the notion of subjective probability.[1] Accordingly, I find myself in the camp of objective probability. The consensus is, however, that the armies of objective probability are in even worse disarray. The purpose of this book is to construct a theory of objective probability that rectifies that. Such a theory must explain the meaning of objective probability, show how we can discover the values of objective probabilities, clarify their use in decision theory, and demonstrate how they can be used for epistemological purposes. The theory of nomic probability aims to do all that.

This book has two main objectives. First, it will propose a general theory of objective probability. Second, it will, in a sense to be explained, propose a solution to the problem of induction. These two goals are intimately connected. I will argue that a solution to the problem of induction is forthcoming, ultimately, from an analysis of probabilistic reasoning. Under some circumstances, probabilistic reasoning justifies us in drawing non-probabilistic conclusions, and this kind of reasoning underlies induction. Conversely, an essential part of understanding probability consists of providing an account of how we can ascertain the values of probabilities, and the most fundamental way of doing that is by using a species of induction. In *statistical induction* we observe the relative frequency (the proportion) of A's in a limited sample of B's, and then infer that the probability of a B being an A is approximately the same as that relative frequency. To provide philosophical foundations for probability

---

[1] Pollock [1986], 97ff.

we must, among other things, explain precisely how statistical induction works and what justifies it.

Probability is important both in and out of philosophy. Much of the reasoning of everyday life is probabilistic. We look at the clouds and judge whether it is going to rain by considering how often clouds like that have spawned rain in the past. We realize that we are going to be late to the train station, but we also think it likely that the train will be late, so we reflect upon whether it is worth risking a speeding ticket in order to get there on time. We read something extremely surprising in the newspaper, but we also know that this newspaper tends to be quite reliable, so we believe it. And so on. As Bishop Butler put it, "Probability is the very guide to life."[2] Probability is also one of the heavy weapons in the arsenal of contemporary philosophy. It is employed as a foundational tool in all branches of philosophy.

The term 'probability' encompasses two distinct families of ideas. On the one hand, we have *epistemic probability*, which is a measure (perhaps not quantitative) of how good our reasons are for various beliefs. I believe both that there are cacti in Tucson and that there are frozen methane clouds on Jupiter, but I regard myself as having much better reasons for the former. Accordingly I may assert it to be more probable that there are cacti in Tucson than it is that there are frozen methane clouds on Jupiter. Epistemic probability concerns "probable opinion". In contrast to this is *physical probability*, which is supposed to concern the structure of the physical world and be independent of knowledge or opinion. The laws of quantum mechanics concern physical probabilities. So too, it seems, do probability assessments in games of chance. The early history of probability theory is marked by a systematic confusion of epistemic and physical probability.[3] Perhaps Laplace was the first probabilist to become reasonably clear on the distinction. Laplace tended to use 'probability' in talking about epistemic probability, and 'degree of

---

[2] This is from the preface of Butler [1736]. This passage has probably been quoted by every probabilist since Keynes.

[3] This is well documented by Hacking [1975].

possibility' for physical probability.[4] Nowadays, philosophers are reasonably clear about the distinction between physical and epistemic probability. Most discussions of epistemic probability assimilate it to subjective probability, and there is a massive literature on that topic. There is a wider variety of competing theories of physical probability. The focus of this book will be on physical probability.

The main philosophical problems concerning physical probability are epistemological. We want to know how to reason with physical probabilities. We want to know both how to discover the values of physical probabilities and how to use physical probabilities to draw nonprobabilistic conclusions. Probabilists have tried to solve these problems using sophisticated mathematics but crude epistemology. Although some of the mathematics used in this book is novel, the main contribution of this book will consist of wedding a sophisticated epistemology to the reasonably familiar mathematics of probability theory. This will make it possible to propose new solutions to the epistemological problems of probability theory and also to propose logical analyses for a number of important probability concepts. One of the main theses of this book is that there is a central kind of physical probability—*nomic probability*—in terms of which a number of other important and useful probability concepts can be defined. The purpose of this chapter is to introduce the concept of nomic probability in an historical setting and give a brief sketch of the overall theory of nomic probability. I begin with a critical discussion of existing theories of physical probability.

2. The Laplacian Theory

It is not too much of an exaggeration to describe modern probability theory as beginning with Laplace.[5] The Laplacian

---

[4] See the discussion in Hacking [1975], Chapter 14.

[5] His most important works in probability are his [1795] and [1820]. This is not to deny the importance of earlier work, the most significant of which

theory has no contemporary defenders in its original form, but it has had a profound influence on contemporary probability theory. The Laplacian theory has two key elements. First is the definition of probability according to which probability is "the ratio of favorable to equally possible cases".[6] The paradigm example of this is the fair die having six faces, each face with an equal chance of showing after a roll. To compute the probability of an even number showing after a roll we note that there are three ways that can happen out of the six possibilities and so conclude that the probability is 1/2.

The Laplacian definition is framed in terms of "equally possible" cases, but what that amounts to is an appeal to equally *probable* cases. It is typically objected that this makes the definition circular. Laplace sought to avoid this circularity by adopting a principle for finding equally possible cases. Although both terms postdate Laplace, this has been called 'the principle of insufficient reason' and 'the principle of indifference'.[7] This principle has been stated in various ways, but the idea is that if we have no reason to favor one case over another they are to be regarded as equally probable. This has two construals depending upon whether we are talking about epistemic probability or physical probability, and Laplace seems to have endorsed both of

---

was Jacques Bernoulli [1713]. Ian Hacking's [1975] is an excellent survey of pre-Laplacian probability theory.

[6] Although this definition is most commonly associated with Laplace, Hacking [1975] (chapter 14) makes a strong case for Laplace's having gotten it from Leibniz.

[7] The first name originates with J. von Kries [1871], and the second name was coined by J. M. Keynes [1921]. The separation of the Laplacian theory into the definition and the principle of indifference is not so clear in Laplace as it is in his commentators. Laplace [1951] (6,7) writes, "The theory of chance consists in reducing all the events of the same kind to a certain number of cases equally possible, that is to say, to such as we may be equally undecided about in regard to their existence, and in determining the number of cases favorable to the event whose probability is sought. The ratio of this number to that of all the cases possible is the measure of this probability, which is thus simply a fraction whose numerator is the number of favorable cases and whose denominator is the number of all the cases possible."

them. The principle of indifference for physical probability is often illustrated by appeal to a die. If the faces are perfectly symmetrical, the center of mass is at the geometric center of the die, and so on, then there is nothing to favor one face over another, and so they are all equally possible.

As Richard von Mises observed, the Laplacian definition was repeated until about 1930 in nearly all textbooks on probability theory, and it is not uncommon even today. Despite this, there are two sorts of objections that are universally acknowledged to be fatal to the Laplacian theory. The first is that it assumes that in all applications of probability we can partition the situation into a finite number of equally probable alternatives. That may work for a fair die, but what about a skewed die? No actual die is ever exactly symmetrical, and other applications of probability (for instance, population statistics, birth and death rates, quantum mechanical phenomena, and so on) are even more difficult to cast into this mold.[8]

The second objection concerns the principle of indifference. This principle leads to probability assignments that are inconsistent with the probability calculus. Richard von Mises [1957] (p. 77) gives the following example:

> We are given a glass containing a mixture of water and wine. All that is known about the proportions of the liquids is that the mixture contains at least as much water as wine, and at most, twice as much water as wine. The range for our assumptions concerning the ratio of water to wine is thus the interval 1 to 2. Assuming that nothing more is known about the mixture, the indifference principle . . . tells us to assume that equal parts of this interval have equal probabilities. The probability of the ratio lying between 1 and 1.5 is thus 50%, and the other 50% corresponds to the probability of the range 1.5 to 2.
>
> But there is an alternative method of treating the same problem. Instead of the ratio water/wine, we consider the inverse ratio, wine/water; this we know lies between 1/2 and 1. We are again told to assume that the two halves of the total interval, i.e., the intervals 1/2 to 3/4 and 3/4 to 1, have equal probabilities

---

[8] This objection is raised by Richard von Mises [1957] (68ff), and Hans Reichenbach [1949] (68,69).

(50% each); yet, the wine/water ratio 3/4 is equal to the water/wine ratio 4/3. Thus, according to our second calculation, 50% probability corresponds to the water/wine range 1 to 4/3 and the remaining 50% to the range 4/3 to 2. According to the first calculation, the corresponding intervals were 1 to 3/2 and 3/2 to 2. The two results are obviously incompatible.[9]

Clearly, the same problem arises whenever we have two unknown quantities where one is the reciprocal of the other. This is known as 'Bertrand's paradox', and is due originally to Joseph Bertrand [1889].[10]

These objections to the original form of the Laplacian theory are decisive. Nevertheless, we will find that safer forms of some of the intuitive elements of the Laplacian theory have found their way into a number of contemporary theories.

3. Logical Probability

A theory in some ways reminiscent of the Laplacian theory is the theory of logical probability. Logical probabilities are conditional probabilities measuring what is often explained as a kind of "partial entailment". To say that P entails Q is to say that Q is true at every possible world at which P is true. But it might happen instead that Q is true at *most* of the worlds at which P is true, or 3/4 of the worlds at which P is true, and so forth. The *logical probability* $c(Q,P)$ of P given Q can be regarded as an *a priori* measure of the proportion of possible worlds making P true at which Q is also true. Logical probabilities are logical features of the world. If the logical probability $c(Q,P)$ is r, then it is necessarily true that $c(Q,P) = r$. This is taken to be characteristic of logical probability.[11] Contemporary work in logical proba-

---

[9] von Mises [1957], p. 77.

[10] The name 'Bertrand's paradox' is due to Poincare [1912].

[11] Henry Kyburg [1974] regards his theory as a theory of logical probability, but on my classification his theory is a theory of what I call below 'mixed physical/epistemic probability'. This is because his theory always assigns

bility began with John Maynard Keynes [1921], but the philosopher most prominently associated with logical probability is Rudolph Carnap (particularly [1950] and [1952]). Carnap's work has been further refined by other philosophers, the most notable of which is Jaakko Hintikka [1966].

Carnap's proposal was, in effect, to construct a measure m defined on sets of possible worlds, and then letting $|P|$ be the set of possible worlds at which a proposition P is true, he defined $c(P,Q) = m(|P\&Q|) \div m(|Q|)$.[12] This is an intuitively appealing idea, but it has proven very difficult to work out the details. In order to make sense of logical probability, we must determine what measure m to use. Various measures have been proposed,[13] but they are all *ad hoc*. Carnap [1952] showed how a whole continuum of probability measures could be constructed, but was unable to suggest any convincing way of choosing between them. In addition, all of his measures have undesirable features; for example, every universal generalization is assigned probability 0. Hintikka [1966] has suggested measures avoiding the latter defect, but his measures are still basically *ad hoc*. Furthermore, none of these measures is applicable to more than a very narrowly restricted class of propositions.[14]

Logical probability has usually been proposed as an analysis of epistemic probability, and Carnap in his later writings sought to combine logical and subjective probability.[15] ⌜c(P,Q)⌝ was supposed to be read ⌜the degree of confirmation of P by the evidence Q⌝. That puts it outside the scope of this book. I have

---

probabilities only relative to background knowledge that includes knowledge of relative frequencies, and different relative frequencies yield different probability assignments. If we consider the case of no background knowledge, his theory assigns the trivial intervals [0,1] as the probabilities of contingent propositions.

[12] Carnap did not actually talk about possible worlds. He talked instead about state descriptions, structure descriptions, etc., but his ideas can be recast in this way.

[13] See particularly Carnap [1950] and [1952], and Hintikka [1966].

[14] Specifically, propositions expressible in the monadic predicate calculus.

[15] See particularly Carnap [1962].

included this brief discussion of it anyway because of its connection with Laplacian probability.

I find logical probability an intriguing notion despite being unsure what to do with it. At this time, I do not think that logical probability can be either endorsed or dismissed.   It remains an interesting idea in need of further development.

4. Frequency Theories

I turn now to theories that are unabashedly theories of physical probability. With the demise of the principle of indifference, it became apparent that some other way was needed for discovering the values of physical probabilities. It seemed obvious to everyone that the way we actually do that is by examining relative frequencies. The relative frequency of A's in B's, symbolized ⌜freq[A/B]⌝, is the proportion of all B's that are A's. If there are m B's and n of them are A's, then freq[A/B] = n/m. We can talk about relative frequencies either in a small sample of B's, or relative frequencies among all the B's in the world.   Many probabilists were tempted by the idea that probability could be defined in terms of relative frequency. The simplest such theory identifies probabilities with relative frequencies.   Among the modern proponents of this theory are Bertrand Russell [1948], R. B. Braithwaite [1953], Henry Kyburg ([1961] and [1974]), and Lawrence Sklar ([1970] and [1973]).[16]

Theories identifying probability with relative frequency are generally considered inadequate on the grounds that we often want to talk about probabilities even in cases in which relative frequencies do not exist. In this connection, note that freq[A/B] does not exist if either there are no B's or there are infinitely

---

[16] William Kneale [1949] traces the frequency theory to R. L. Ellis, writing in the 1840's, John Venn [1888], and C. S. Peirce in the 1880's and 1890's. It is slightly inaccurate to include Kyburg in this camp, because he does not use the term 'probability' for relative frequencies, reserving it instead for definite probabilities (see following), but what plays the role of indefinite probability in his theory is relative frequency.

many. Applying this to quantum mechanics, for example, a certain configuration of particles may never have occurred, but we may still want to talk about the probability of the emission of a photon from such a configuration. Quantum mechanics even tells us how to calculate such probabilities. But as such a configuration has never occurred, there is no relative frequency with which the probability can be identified.

Kyburg [1976] responds to the so-called problem of the empty reference class, maintaining that there is really no problem for the frequency theory in the case in which there are no B's. His suggestion is that in such a case, our interest is not in the probability prob(A/B) at all, but rather in a probability prob(A/C) where C is some more general property entailed by B but of sufficient generality that there are C's. How else, he asks, could we arrive at a value for the probability? For example, suppose we have a coin of some description D, where D includes enough peculiar features of the coin to guarantee that there has never been another coin of that description. D might include the coin's being painted green, having certain scratches on its surface, etc. Despite its peculiar features, and despite the fact that the coin is destroyed without ever being tossed, we might judge that the coin is a fair coin; that is, the probability is 1/2 that a toss of this coin will land heads. But surely, says Kyburg, our reason for this judgment is that the coin satisfies some more general description D^* shared by many coins, and roughly half of the tosses of coins of description D^* have landed heads.

The difficulty with Kyburg's strategy for avoiding empty reference classes is that there are other examples of this phenomenon that cannot be handled so easily. There are more ways of evaluating probabilities than envisioned by Kyburg's response. For example, in quantum physics we might confirm a generalization of the form $\ulcorner(\forall r)[F(r) \supset \text{prob}(Axr/Bxr) = g(r)]\urcorner$ where r measures some physical parameter and g is a function of that parameter. For instance, this generalization might be about the behavior of particles in fields of varying strengths and shapes. For some values of r, $\{x \mid Bxr\}$ will be nonempty, but for infinitely many cases this set will be empty. By examining the nonempty cases we can confirm the generalization, and that tells us how to calculate probabilities in the empty cases as well. But in the empty cases, the probabilities cannot be identified with relative frequencies, and unlike the coin case there is no reason to think it will always be possible to subsume the empty cases under more general cases for which the probabilities are the same. Thus I am inclined to insist that the empty case constitutes an insuperable difficulty for anyone who wants to maintain that all of the probabilities of science can be identified with relative frequencies.

A more familiar difficulty for the identification of probabilities with relative frequencies concerns the case in which the reference class is infinite. The relative frequency does not exist in this case either. It is apt to seem that the infinite case is less problematic because we are rarely faced with infinite totalities. But that is a mistake. In talking about the probability of a *B* being an *A*, we are not usually referring to the probability of a *B* being an *A* *now*. Unless we explicitly relativize the properties and ask for the probability of a *B* *at time* t being an *A*, then we are typically asking about the probability at an arbitrary time. The relative frequency would then be the proportion of *B*'s that are *A*'s throughout all time, but in most cases there is no reason to think that there will be only finitely many *B*'s throughout the universe. For example, unless we adopt certain cosmological theories, there is no compelling reason to think that there will be only finitely many stars throughout the history of the universe, or even finitely many people. But then an identification of probabilities with relative frequencies will leave most probabilities undefined.

Theories identifying probabilities with actual relative frequencies in finite sets have come to be called *finite frequency theories*. In order to avoid the problem of the infinite reference class, most proponents of frequency theories have turned instead to *limiting frequency theories*. These theories identify probabilities with *limits of relative frequencies* rather than with relative frequencies simpliciter.[17] The idea is that if there are infinitely many *B*'s, then the

---

[17] This suggestion can be traced to John Venn [1888], p. 95. It has been defended in detail by von Mises [1957], Popper [1959], and Reichenbach [1949]. Kyburg [1974] makes passing concessions to limiting frequency

probability of a B being an A is defined as the limit to which the relative frequency approaches as we consider more and more B's. In the special case in which there are only finitely many B's, the probability is identified with the actual relative frequency. It is now generally recognized that limiting frequency theories face extreme difficulties. First, they are of no help at all in the case in which there are no B's. What is initially more surprising is that they do not work in the infinite case either. The difficulty is that the notion of the limit of relative frequencies of A's in B's makes no mathematical sense. If there are infinitely many B's, and infinitely many of them are A's, and infinitely many of them are non-A's, then you can obtain whatever limit you want by going through the B's in different orders. We might begin by enumerating the B's in some order $b_1,\ldots,b_n,\ldots$. Then suppose we construct a new ordering on the basis of the first: we begin by picking the first B that is an A; then we pick the first B that is not an A; then we pick the next two B's that are A's, followed by the next B that is not an A. Then we pick the next four B's that are A's, followed by the next B that is not an A; and so on. The limit of the relative frequency obtained in this way will be 1. If instead we reversed the above procedure and first picked B's that are not A's, we would arrive at a limit of relative frequencies of 0. And we can construct other orderings that will produce any limit between 0 and 1. The point is that there is no such thing as "the limit of relative frequencies of A's in B's". It only makes sense to talk about the limit relative to a particular ordering of the B's. Different orderings yield different limits.

For a while it was hoped that limiting frequency theories could be salvaged by placing some natural constraints on the ordering of the B's, and much effort went into the search for such constraints. The original idea goes back to von Mises [1957]. He proposed to define the probability of a B being an A in terms of *collectives*, where a collective is any ordering of the B's that is "random" in the sense that the same limiting frequency is exhibited by any infinite subsequence drawn from the whole

---

theories, but the spirit of his theory is finite frequentist.

sequence by "place selections". A place selection is an algorithm for choosing elements of the whole sequence in terms of their place in the sequence. If no restrictions are imposed on place selections then there are always place selections that generate subsequences with different limits.[18] Alonzo Church [1940] proposed to solve this problem by requiring place selections to be recursive functions, and Church's proposal was subsequently refined by other authors.[19] However, this whole approach seems misguided. Invariance over changes in place selections does not seem to have anything to do with the relationship between $prob(A/B)$ and the behavior of sequences of B's. Suppose, for example, that it is a physical law that B's occur sequentially, and every other B is an A. An example might be the temporal sequence of days and nights. This sequence is completely nonrandom and violates any criterion formulated in terms of place selections, but we would not hesitate to judge on the basis of this sequence that $prob(A/B) = 1/2$.

It is now generally acknowledged that the search for constraints on sequences has been in vain, and accordingly the limiting frequency theories have lost much of their popularity.[20] Reichenbach [1949] tried to avoid this difficulty by attaching probabilities directly to ordered sets rather than unordered sets, but such a ploy seems desperate. Reference to an ordering is out of place in normal attributions of probability. For example, if we inquire about the probability of a radium atom decaying during the interval of a year, we assume that this probability exists irrespective of whether there are infinitely many radium atoms, and to ask "The probability relative to which ordering?" is a confusion. The probability is not relative to an ordering. It is nonsense to suppose that the probability can be anything we want it to be if we just order the set of radium atoms appropriately.

---

[18] See Salmon [1977] for a discussion of the history of the development of this concept.

[19] See particularly Martin-Löf [1966] and [1969] and Schnorr [1971].

[20] For critical discussions of limiting frequency theories along this and other lines, see Russell [1948] (362ff), Braithwaite [1953] (p. 125), and Kneale [1949] (152ff).

5. Hypothetical Frequency Theories

The most common criticisms of frequency theories are those I discussed above, which focus on the cases of infinite or empty reference classes. But frequency theories would be inadequate even if we could ignore those cases. Even if there are finitely many B's, freq[A/B] need not be the same as prob(A/B). For example, suppose we have a coin of physical description D. It is the only coin of that description, and it is flipped only once. It is then melted down and no other coin of that description is ever constructed. We may be convinced that a coin of description D is a fair coin, i.e., the probability is 1/2 of a flip of such a coin landing heads. But as the coin was flipped only once, the relative frequency must be either 0 or 1. We have no inclination to suppose this to show that it was not a fair coin after all. Or suppose instead that the coin is flipped three times. If we regard the coin as fair we will not be surprised to discover that it landed heads twice and tails once, making the relative frequency 2/3 rather than 1/2. The relative frequency could not have been 1/2 if the coin was flipped three times. We do expect the probability and the relative frequency to be related to one another, but we do not expect them to be identical. If the coin were flipped a large number of times, we would be surprised if the relative frequency were *very* different from 1/2, but again we would not expect it to be exactly 1/2.

These considerations show that there is a kind of physical probability that cannot be identified with relative frequency. I think it would be overly hasty, however, to conclude that relative frequency is not also a kind of physical probability. Our pretheoretic use of the term 'probability' includes both relative frequencies and other kinds of physical probabilities. To illustrate this, consider our coin of description D that is tossed three times, landing heads twice and tails once, and then destroyed. One is inclined to say *both* that the probability of a toss of this coin landing heads is 1/2, and that it is 2/3. It would be natural to express the latter by saying that the probability of an *actual* toss of this coin landing heads is 2/3. Note that if we knew there were three actual tosses two of which landed heads and one tails, but we did not know which was which, and we were betting upon whether a specific one of those tosses landed heads, the appropriate probability to use would be 2/3 rather than 1/2. Our wanting to ascribe both the value 2/3 and the value 1/2 to the probability of a toss landing heads indicates that there are two kinds of probability here. We might say that the *material probability* is 2/3, and the non-material probability is 1/2. The material probability is only concerned with what is actually the case (with actual tosses of the coin).

If the relative frequency exists, then it must be the same as the material probability. But we are often interested in non-material probabilities. It cannot be denied that relative frequency is a much clearer notion than non-material probability. It would be very nice if we could focus exclusively on relative frequencies, ignoring non-material probabilities, but we have already seen that that is impossible. There are many purposes for which we require non-material probabilities. We noted above that in physics we often want to talk about the probability of some event in simplified circumstances that have never occurred. The relative frequency does not exist, but the non-material probability does, and that is what we are calculating. And in the case of the coin that is flipped only a few times, we are appealing to non-material probabilities in describing the coin as fair. As our investigation proceeds, additional reasons will emerge for wanting to make clear sense of non-material probability. It will turn out that a theory of direct inference (see below) can only be based upon non-material probability, and statistical induction leads to estimates of non-material probability.

If non-material probabilities cannot be identified with relative frequencies, how can we make sense of them? It was noted previously that if there are finitely many B's we would not expect prob(A/B) to be exactly the same as the relative frequency, but we would expect the relative frequency to approximate prob(A/B) more and more closely as the number of B's increases. This principle is known as *Bernoulli's theorem*.[21] It has suggested

---

[21] This is also known as the Weak Law of Large Numbers. Here is a

to some probabilists that prob(A/B) is the limit to which freq[A/B] *would* go *if* the number of B's increased without limit—it is what the relative frequency *would be* "in the long run". This is to explain prob(A/B) counterfactually—in terms of possible B's rather than actual B's. Karl Popper [1956] put this by saying that probabilities are limits of relative frequencies in "virtual sequences". More generally, I will call these *hypothetical frequency theories.*[22]

There is an important insight behind the hypothetical frequency theories. This is that prob(A/B) is not just about actual B's—it is about possible B's as well. For example, when we say that the probability is 1/2 of a toss of a coin of description D landing heads, we are not just talking about actual tosses (of which there may be none, or very few). We are also talking about possible tosses. We are saying that the proportion of possible tosses that would land heads is 1/2. Of course, 'possible' here does not mean 'logically possible'. It means 'really possible'—possible in light of the way the world is—or perhaps better, 'physically possible'.

---

more precise statement of the theorem: If prob(A/B) = r, then if we consider finite sequences of B's that are independent of one another and consider any finite interval around r, the probability of the relative frequency of A's (in a finite sequence) lying in that interval goes to 1 as we consider longer and longer sequences. This theorem is due to Jacques Bernoulli [1713], and is probably the most important theorem in probability theory. Virtually all theories of probability make use of this theorem in one way or another in trying to explain how we can know the values of probabilities.

[22] C. S. Peirce was perhaps the first to make a suggestion of this sort. Similarly, the statistician R. A. Fisher, regarded by many as "the father of modern statistics", identified probabilities with ratios in "a hypothetical infinite population, of which the actual data is regarded as constituting a random sample" ([1922], p. 311). Popper ([1956], [1957], and [1959]) endorsed a theory along these lines and called the resulting probabilities *propensities*. Henry Kyburg [1974a] was the first to construct a precise version of this theory (although he does not endorse the theory), and it is to him that we owe the name 'hypothetical frequency theories'. The most recent, and most detailed, hypothetical frequency theory is that of van Fraassen [1981]. Kyburg [1974a] insists that von Mises should also be considered a hypothetical frequentist.

I regard this insight as a major advance in understanding probability. There are obvious difficulties in spelling it out precisely, but in bold outline it is appealing. As we will see, the hypothetical frequency theory is just one way of trying to spell this out. That is important because, as formulated, the hypothetical frequency theory faces overwhelming difficulties. It assumes that as we suppose there to be more and more B's, the relative frequency will go to the probability in the limit. But that turns upon a mistaken interpretation of Bernoulli's theorem. Consider our fair coin of description D once more. As it is flipped more and more times we expect the relative frequency to approximate the probability more and more closely, but we never expect the relative frequency to be *exactly* the same as the probability. Indeed, if the flips are independent of one another, familiar calculations related to Bernoulli's theorem reveal that if the coin is flipped n times and n is an even number then the probability of the relative frequency being exactly 1/2 is just $n!/[(n/2)!(n/2)!2^n]$, which is approximately $(2/\pi n)^{1/2}$ (and of course, the probability is 0 if n is odd).[23] This probability goes to 0 as n increases without limit. Related calculations reveal that the probability of getting any *particular* relative frequency goes to 0 as n becomes indefinitely large. On the other hand, if n is large then it is probable that the relative frequency will approximate 1/2, and by making n large enough we can make it as probable as we like that the relative frequency will approximate 1/2 to any desired degree. But it never becomes probable that the relative frequency will be exactly 1/2. What this seems to indicate is that there is no number r such that freq[A/B] *would* go to r if the number of B's increased without limit. Rather there are many r's such that freq[A/B] *might* go to r if the number of B's increased without limit.[24] In fact, for each particular r the probability of

---

[23] This approximation is calculated using the Stirling approximation, according to which n! is approximately $e^{-n}n^n(2\pi n)^{1/2}$.

[24] Most contemporary theories of counterfactuals agree that $\ulcorner(P > Q) \vee (P > {\sim}Q)\urcorner$ ("disjunctive excluded middle") is invalid, and take $\ulcorner Q$ might be true if P were true$\urcorner$ to be analyzed as $\ulcorner{\sim}(P > {\sim}Q)\urcorner$. (For more about this, see section 2 of Chapter 2.) Thus to say that the frequency might be r is just

freq[A/B] going to r is the same—namely 0—and so it seems we must acknowledge that freq[A/B] might go to any r at all.[25] The most we can reasonably expect on the basis of Bernoulli's theorem is that it is extremely probable that if the number of B's increased without limit then freq[A/B] would go to a value arbitrarily close to r. It is actually a bit problematic how we could justify even this conclusion, but suppose we waive that objection for the moment. We could then characterize probabilities in terms of hypothetical frequencies as follows. Symbolizing ⌜the difference between r and s is less than δ⌝ as ⌜$r \approx_\delta s$⌝, we might have:

(5.1)   prob(A/B) $= r$ iff for any $\delta,\epsilon > 0$ there is a number n

such that prob$\Big($freq[A/X] $\approx_\delta r$ / X is a set of at least n

independent possible B's$\Big) \geq 1\text{-}\epsilon$.

Or we might have a slightly stronger principle. Where σ is an infinite sequence of B's, let limfreq[A/σ] be the limit of the relative frequency of A's in σ. We might then have:

(5.2)   prob(A/B) $= r$ iff for every $\delta > 0$, prob$\Big($limfreq[A/σ] $\approx_\delta r$

/ σ is an infinite sequence of possible B's$\Big) = 1$.

Informally, if prob(A/B) $= r$ then we expect (with probability 1) the limiting frequency in an infinite sequence of possible B's to be in each small interval around r, but this does not imply that the probability is 1 of its being in all of them at the same time (in fact, that probability is 0, because the limiting frequency is in all of them iff it equals r). Standard mathematics will not enable us to prove either (5.1) or (5.2), but they do seem intuitively

---

to deny that it definitely would not be r.

[25] van Fraassen [1981] (p. 192) is aware of this difficulty, but as far as I can see he does not propose an answer to it. His response is to talk about "idealizations", which seems to amount to no more than pretending that the difficulty is not there.

reasonable, so let us just suppose for now that one or the other is true.[26] The difficulty is that even given (5.1) or (5.2) we would not have an analysis of probabilities in terms of hypothetical frequencies. This is because 'prob' occurs in the analysans. This objection to hypothetical frequency theories seems to me to be conclusive. There is no nonprobabilistic way of relating probabilities to frequencies—even frequencies in hypothetical infinite populations.

6. Empirical Theories

It is simplistic to identify physical probabilities with either relative frequencies or limits of relative frequencies. Nevertheless, it appears that there must be important connections between physical probabilities and relative frequencies. At the very least there must be epistemic connections. It is undeniable that many of the presystematic probability judgments made by non-philosophers are based upon observations of relative frequencies. For example, the meteorologist assesses the probability of rain by considering how often it has rained under similar circumstances in the past. I will call any theory of physical probability that draws a tight connection between probabilities and relative frequencies an *empirical theory*. Frequency theories are empirical theories, but they are not the only possible candidates. The general difficulty in constructing a workable empirical theory lies in imagining how probabilities can be related to relative frequencies other than by being identical with them. The theory of nomic probability will be of the empirical genre, but its connection to

---

[26] Principles like this will be investigated more fully in Chapters 2 and 7. I endorsed (5.2) in Pollock [1984a], but it will not be endorsed in this book. It may be true, but I do not have a proof of it. Principle (5.1), on the other hand, is a theorem of the calculus of nomic probabilities.

relative frequencies will be less direct than that envisaged in conventional empirical theories.

Empirical theories as a group face two other difficulties. First, there is the problem of how to evaluate the probabilities. How can you discover, for example, that the probability of its raining under such-and-such circumstances is .3? It might seem that this problem would have a simple solution if we could identify probabilities with relative frequencies, but that is a mistake. We would still be faced with the problem of discovering the relative frequency. If the B's are widely extended in either space or time, there can be no question of actually counting the B's and seeing what proportion of them are A's. Instead, we use sampling procedures. We examine a random sample of B's and see what the relative frequency of A's is in the sample, and then we project that relative frequency (or a probability of that value) onto the class of all B's. This is a form of statistical induction. A workable empirical theory of probability must provide us with some account of statistical induction. Empirical probabilities are useless if there is no way to discover their values.

The second difficulty faced by empirical theories concerns a distinction between two kinds of probability. The probabilities at which we arrive by statistical induction do not concern specific individuals but classes of individuals or properties of individuals. We discover the probability of *a* smoker getting lung cancer, or the probability of its raining when we are confronted with a low pressure system of a certain sort. I will call such probabilities *indefinite probabilities*.[27] In the indefinite probability prob(F/G), I will sometimes call G the 'hypothesis' and at other times (following tradition) I will call G the 'reference property'. I will call F the 'consequent property'. To be contrasted with indefinite probabilities are the probabilities that particular propositions are true or particular states of affairs obtain. For example, we can talk about the probability that Jones will get cancer, or the probability that it will rain tomorrow. These are sometimes called 'single case probabilities', but I prefer to call them *definite*

---

[27] I take this term from Jackson and Pargetter [1973].

probabilities.[28]   Subjective and logical probability are species of definite probability.

We have roughly two kinds of reasons for being interested in probabilities—epistemic and practical.   The epistemic significance of probability is that we often take high probabilities to warrant belief. That is the topic of Chapter 3. Equally important is the role of probabilities in practical deliberation. The role of probability is obvious if we are betting on a dice game, but all of life is a gamble. I am engaged in decision-theoretic deliberations in deciding whether to take my umbrella with me when I go to the store.   It is awkward to carry it if it does not rain, but unpleasant to be caught without it if it does rain. In deciding what to do I make an estimate of how likely it is to rain and then try to maximize expectation values. The same thing is true in deciding how fast to drive on a winding road, or even what to have for dinner. Practical decisions like this are often based upon rough estimates of probability. The important thing to note is that the probabilities required for decision theory are definite probabilities. For example, it does little good to know that the probability of its raining under general circumstances of type C is

---

[28] The distinction between definite and indefinite probabilities is a fundamental one and, I should think, an obvious one. It is amazing then how often philosophers have overlooked it and become thoroughly muddled as a result. For example, virtually the entire literature on probabilistic causation conflates definite and indefinite probabilities. (One notable exception is Nancy Cartwright [1979].)

The definite/indefinite distinction is related to two other important distinctions, but is not the same as either. One of these distinctions is that between epistemic and physical probability. Epistemic probabilities are invariably definite probabilities, but as we will see, there can be both definite and indefinite physical probabilities. The second distinction is that of Carnap [1950] between probability$_1$ and probability$_2$. Carnap takes probability$_1$ to be degree of confirmation and probability$_2$ to be relative frequency. Degree of confirmation is the same thing as epistemic probability (although Carnap assumes that it can be assigned numerical values satisfying the probability calculus), and relative frequency is a variety of indefinite probability, but these categories are not exhaustive. There are other varieties of definite probability besides probability$_1$ and there are other varieties of indefinite probability besides probability$_2$.

3/4 unless I can relate that indefinite probability to my present situation and acquire some idea of the definite probability that it is going to rain in the next half hour. This is not a simple task. The mere fact that the present situation is of type C does not allow me to infer that the definite probability is 3/4, because the present situation may also be of type C^* where the indefinite probability of its raining under circumstances of type C^* is only 1/2.

Any empirical theory faces two problems concerning definite probabilities. The first is to explain the meaning of the definite probabilities. Even if we understand indefinite physical probabilities, we need a separate explanation of the definite probabilities to which we appeal in making practical decisions. Thus an empirical theory cannot be content with merely providing an analysis of indefinite probabilities. It must also provide an analysis of the related definite probabilities. The second problem the empirical theory must face is that of telling us how to discover the values of those definite probabilities. It appears that definite probabilities are somehow inferred from indefinite probabilities, and this kind of inference is called *direct inference*. An empirical theory must provide us with a theory of direct inference. That will be the topic of Chapters 4 and 10.

The definite probabilities at which we arrive by direct inference must have a strong epistemic element. For example, any decision on what odds to accept on a bet that Bluenose will win the next race must be based in part on what we know about Bluenose, and when our knowledge of Bluenose changes so will the odds we are willing to accept. It seems that these definite probabilities must combine epistemic and physical elements in a complicated way. The logical analysis of these probabilities will be a major constituent of any theory of empirical probability.

7. Propensities

Although there are some notable dissenters, there is a virtual consensus among recent probabilists that relative frequencies cannot do all the jobs required of physical probability. This led

Karl Popper to introduce propensities into philosophy, and most recent physical probability theories have been propensity theories of various sorts. Propensity theories take probabilities to be properties of individuals[29] and eschew any attempt to analyze probability in terms of relative frequencies. For example, when we say that a particular pair of dice are loaded, or that a certain coin is fair, we are talking about probabilistic properties of those dice and that coin. But despite their predominance in recent physical probability theory, most philosophers regard propensities as creatures of darkness. This is primarily because philosophers feel they do not understand what propensities are supposed to be or how they can be investigated empirically. Relative frequencies are clear. Beside them propensities are apt to seem like shadowy metaphysical creations, and propensity theories seem vague and muddled. But we must not lose sight of the fact that although they are clear, relative frequencies are inadequate for probability theory. To insist upon ignoring propensities because their analysis is more difficult is to be like the man who lost his watch in a dark alley but was observed looking for it under the street light. When asked why he was doing that he replied, "Because the light is better here."

Although there are a number of different propensity theories, there is little agreement about what propensities are supposed to be like. We can distinguish between several different kinds of propensities. First, there are *indefinite propensities* – propensities that are indefinite probabilities. Pondering a bent coin, I might conjecture that the probability of a toss of that coin landing heads is about 1/3. This is an indefinite probability, but it attaches to a particular object – the bent coin. There are also *definite propensities*. For example, I might want to know how probable it is that a particular radium atom will decay within the next half hour. A different kind of definite propensity is counterfactual: I may want to know how probable it is that I would be involved in a collision if I were to go through the red light I am

---

[29] The "individual" may be either an ordinary physical object or a "chance setup" – a physical-object-in-a-setting.

now approaching. I will call this *counterfactual probability*. No reasonable person could deny that these propensities make some kind of sense. All that can be in doubt is how they are to be analyzed.

Relative frequencies are extensional. That is:

If $(\forall x)(Ax \equiv Cx)$ and $(\forall x)(Bx \equiv Dx)$ then freq$[A/B]$ = freq$[C/D]$.

But indefinite propensities are intensional. For example, if the bent coin in the above example is tossed only once and lands heads on that occasion, then the set of tosses is the same as the set of tosses landing heads, but the propensity of a toss to land heads is not the same as the propensity of a toss to be a toss. There is some temptation to take intensionality to be the defining characteristic of indefinite propensities, but I prefer to draw the category a bit more narrowly. Within intensional indefinite probabilities, we can distinguish between those that are probabilistic laws and those that are not. For example, the probabilities we derive from the formulas of quantum mechanics are supposed to follow from laws of nature and do not represent physically contingent properties of objects. On the other hand, the propensity of our bent coin to land heads is a physically contingent property of that coin. We could alter the propensity by straightening the coin. I will reserve the term 'indefinite propensity' for those intensional indefinite probabilities that are not laws of nature (or do not follow from laws of nature, i.e., are not physically necessary). This is in keeping with the idea that propensities attach to individuals. Those intensional indefinite probabilities that are physically necessary will be called *nomic probabilities*. It may be doubted that nomic probabilities are actually a species of probability interestingly different from indefinite propensities, but I will argue that they are.

Now let us look a bit more closely at existing theories of propensities.[30] Popper introduced the term 'propensity', but he

---

[30] More recent propensity theorists include Ian Hacking [1965], Isaac Levi

was not clear about the distinction between definite and indefinite propensities. It seems that most propensity theorists have wanted to focus on definite propensities (although confusions remain).[31] The only clear example of an indefinite propensity theory is that of Ian Hacking [1965]. Hypothetical frequency theories *might* be regarded as theories of indefinite propensities, but my inclination is to interpret them instead as theories of nomic probability. At any rate, I claim to have disposed of those theories in the last section, so I will not discuss them further here.

Hacking does not try to give an analysis of indefinite propensities (he calls them 'chances'). He contents himself with giving enough examples to explain to his audience what he is talking about, and then he proceeds to the mathematical investigation of propensities. But other propensity theorists have tried to say just what propensities are. These theorists are all talking about definite propensities, and they take them to be dispositional properties of individuals. Propensity theorists differ regarding the nature of the purported disposition and regarding the kind of individual that is supposed to have a propensity. Sometimes propensities are attributed to ordinary physical objects like coins,[32] and at other times it is insisted that propensities are dispositions of "chance setups", for example, a coin together with an apparatus for tossing it.[33] I will not pursue this dispute here.[34] Concerning the nature of the disposition, Ronald Giere [1973] writes:

> Consider again a chance setup, CSU, with a finite set of outcomes and an associated probability function $P(E)$ with the usual formal

---

[1967], Ronald Giere ([1973], [1973a], [1976]), James Fetzer ([1971], [1977], [1981]), D. H. Mellor ([1969], [1971]), and Patrick Suppes [1973]. Ellory Eells' [1983] is a good general discussion of propensity theories.

[31] A possible exception is Bas van Fraassen [1981], although I think it is more reasonable to read him as talking about nomic probabilities rather than indefinite propensities.

[32] Mellor takes this line.

[33] This is the view of Fetzer, Giere, and Hacking.

[34] A good discussion of this can be found in Kyburg [1974a].

properties. The following is then a simple statement of the
desired physical interpretation of the statement $P(E) = r$:

The strength of the *propensity* of CSU to produce outcome E on
trial L is r.

This statement clearly refers to a particular trial. (p. 471)

Similarly, James Fetzer [1971] (p. 475) takes a propensity to be
"the strength of the dispositional tendency for an experimental
setup to produce a particular result on its singular trial." D. H.
Mellor takes propensities instead to be "dispositions of objects to
display chance distributions in certain circumstances."

There is a lot of infighting among propensity theorists
regarding which dispositional account of propensities is better.
Mellor objects to accounts like those of Giere and Fetzer on the
grounds that they misconstrue the nature of dispositions.
Considering the proposal that a propensity is simply a disposition
to produce a particular result on a trial, he compares this
purported disposition with the disposition of a fragile glass to
break when dropped:

The main point is that it cannot be the *result* of a chance trial
that is analogous to the breaking of a dropped glass. It is true
that the breaking may be regarded as the result of dropping the
glass, but to warrant the ascription of a disposition, it must be
supposed to be the *invariable* result. Other things being equal,
if a glass does not break when dropped, that suffices to show that
it is not fragile. But if propensity is to be analogous to fragility,
the result of a chance trial is clearly *not* the analogue of the glass
breaking since, in flat contrast to the latter, it must be supposed
not to be the invariable result of any trial on the set-up. If it
were so, the trial simply would not be a chance set-up at all.
Other things being equal, if a chance trial does not have any
given result, that does not suffice to show that the set-up lacks
the corresponding propensity.[35]

He goes on to conclude:

---

[35] Mellor [1969], p. 26.

> If propensity, then, is a disposition of a chance set-up, its display,
> analogous to the breaking of a fragile glass, is not the result, or
> any outcome, of a trial on the set-up. The display is the chance
> distribution over the results. . . .[36]

Virtually all commentators have objected that this makes propensities empirically vacuous because the chance distribution that constitutes the putative "display" is not experimentally accessible. In fact, it is hard to see what the difference is supposed to be between the chance distribution and the propensity itself.[37]

Mellor's position seems to be based upon a simplistic view of dispositions. It is just false that if something has a disposition to behave in a certain way under appropriate circumstances then it must always behave in that way under those circumstances. Consider abstemiousness, or kindness. These are *tendencies*. An abstemious person is one who *tends* not to eat too much, and a kind person is one who *tends* to behave kindly. Fetzer [1971] makes this same point in responding to Mellor that fragility is a universal disposition, whereas propensities are statistical dispositions. But what Fetzer in turn overlooks is that statistical dispositions, in the sense of tendencies, must be explained probabilistically. To say that Jones is kind is to say the probability of his behaving in a kindly manner is high. But that is just to say that he has a propensity to behave in that way.[38] For this reason, an explanation of propensities in terms of statistical dispositions is circular. For example, if we follow Fetzer and say that the propensity of a setup to produce A is "the strength of the dispositional tendency for an experimental set-up to produce a particular result on its singular trial", this gets cashed out as explaining the propensity by identifying it with "the strength of the propensity of the setup to produce A". Not much of an explanation!

It is for reasons like the above that most philosophers take a dim view of propensities. Propensity theorists have been unable

---

[36] Ibid.

[37] See Fetzer [1971], Sklar [1974], Burnor [1984], Giere [1973a].

[38] For more on dispositions and probability see Pollock [1976], Chapter 9.

to say anything about them that does much to clarify them. But as I insisted above, this should not lead us to condemn them utterly. Propensities do make sense even if we do not know how to analyze them.

Propensities might be philosophically useful even without an analysis. If we could develop enough of a theory of propensities, that might throw sufficient light on them to enable us to use them even if we cannot define the notion. Let us consider then just what role definite propensities play in probabilistic reasoning. Both Giere [1973] and Fetzer [1977] take the principal motivation for definite propensities to be that of solving the "problem of the single case":

> A single-case propensity interpretation, which makes no essential reference to any sequence (virtual or real), automatically avoids the problem of the single case and the problem of the reference class as well.[39]

The "problem of the single case" is, of course, the problem of recovering definite probabilities from indefinite probabilities or empirical observations. Empirical theories try to solve this problem with theories of direct inference. I indicated above just how important this problem is. It is definite probabilities that provide the basis for decisions regarding what to do in uncertain circumstances. But what Giere and Fetzer do not seem to realize is that the definite probabilities required for decision theoretic purposes are infected with an epistemic element. For example, suppose I am betting on whether a particular roll of a die ended with 3 showing. I may know that the propensity for this to happen is 1/6. But if I also have the collateral information that 5 was not showing, then the appropriate probability to use in betting on this case is 1/5—not 1/6. As I remarked above, the probabilities of use in decision theory combine physical and epistemic elements. Accordingly, they cannot be identified with propensities. The propensity theorist still has what is in effect a problem of direct inference—he must tell us how to get from

---

[39] Giere [1973], p. 473.

purely physical propensities to the mixed physical/epistemic probabilities we need.

Perhaps the most puzzling question about propensities concerns how we can ever discover their values. More than anything else, it is this epistemological problem that has made propensities seem mysterious to other philosophers. The problem is exacerbated by the fact that propensity theorists typically insist that definite propensities are the basic kind of probability. Most of them have no use for any kind of indefinite probability (even indefinite propensities). The reason this creates a problem is that it is much easier to see how indefinite probabilities are to be evaluated. Even if we cannot give the details it seems clear that we appeal to relative frequencies and use some kind of statistical induction. My own view will be that definite propensities are to be analyzed in terms of nomic probabilities, which are indefinite probabilities, and I will argue below that this makes it possible to infer their values from the values of nomic probabilities. But if instead we follow contemporary propensity theorists and insist that definite propensities are fundamental, how can we discover their values? Giere [1973] and Fetzer [1971] both give basically the same answer. They suggest that we can test propensities by using relative frequencies and Bernoulli's theorem. For example, if I am holding a coin in my hand and I want to know the propensity of this coin to land heads when tossed, I toss it a number of times. Ignoring details, Bernoulli's theorem tells us that the observed relative frequency in a long sequence of independent tosses will tend to approximate the propensity, and on that basis we can estimate the propensity. But note: This assumes that the propensity for landing heads does not change from toss to toss. How could we know that? It is surely an empirical question whether the propensity remains constant (it probably does not remain *quite* constant—the coin wears a little each time), and there seems to be no way we could know how the propensities on the different tosses compare without already knowing the propensities, which is the very matter at issue.[40] Fetzer [1971] (pp.

---

[40] Note that this is equally a problem if we are talking about indefinite

476-477) casts this as the objection that you cannot confirm propensity statements in this way because an individual event happens only once. He tries to meet it as follows:

> Although a single case probability statement surely does pertain to singular events, it is somewhat misleading to suppose that the events involved only occur a single time. A single case propensity statement, after all, pertains to every event of a certain kind, namely: to all those events characterized by a particular set of conditions. Such a statement asserts of all such events that they possess a specified propensity to generate a certain kind of outcome. Consequently, although the interpretation itself refers to a property of conditions possessed by single statements, such statements may actually be tested by reference to the class of all such singular occurrences—which is surely not restricted to a single such event. (p. 478)

But this is to confuse the definite propensity of *this* coin to land heads on the next toss with the indefinite probability of *a* coin of this sort and in this setting to land heads on *a* toss.

There is a further difficulty for the view that definite propensities constitute the basic kind of probability. This is that in a deterministic world all definite propensities would be either 1 or 0.[41] This constitutes an important difference between definite propensities and indefinite probabilities of various sorts.[42] The reason this creates a difficulty is that in many uses of probability, for instance, in statistics or in actuarial applications, it is assumed that probabilities can have values between 0 and 1, but do not seem to presuppose indeterminacy. Giere [1973] responds that our ignorance of the initial conditions makes it reasonable, in effect, to pretend that these phenomena are indeterministic and to make use of probability theory according-

---

propensities—they too can change from toss to toss.

[41] This is acknowledged by Giere [1973], p. 475.

[42] This point should be qualified. Definite propensities of the sort Giere, Fetzer, Mellor, etc., are talking about will be either 1 or 0 in a deterministic universe, but other kinds of definite propensities may not share this characteristic. An example is the counterfactual probability mentioned above.

ly.[43] But that strikes me as an implausible reconstruction of what is going on. A more reasonable stance is that although definite propensities exist, they do not constitute the fundamental kind of probability. In addition to definite propensities, there are also indefinite probabilities—either nomic probabilities or indefinite propensities—and it is to these we appeal in statistical and actuarial calculations.

To sum up, it must be granted that propensities make sense, but it must also be granted that existing propensity theories have not succeeded in clarifying them adequately. The inability of existing propensity theories to explain how we can discover the numerical values of definite propensities, coupled with the obvious need for indefinite probabilities in some contexts, suggests that definite propensities must be derived in some way from indefinite probabilities rather than being fundamental. That is the position defended in Chapter 4.

8. Nomic Probability and Nomic Generalizations

I have given reasons for thinking that no existing theory of physical probability is satisfactory. I propose to fill the gap with the theory of nomic probability. Recall that nomic probability is the kind of probability involved in statistical laws. Perhaps the best way to get an initial grasp of the concept of nomic probability is by looking first at nonstatistical laws. The logical positivists popularized the Humean view that there is no necessity in nature, and hence nonstatistical laws are just material generalizations of the form $(\forall x)(Fx \supset Gx)$. Such a view has profoundly influenced contemporary philosophy of science, but despite their Humean inclinations, philosophers of science have always known that there was a good reason for distinguishing between physical laws and material generalizations. Such a distinction is required by the possibility of accidental generalizations. For example, it might be true, purely by chance, that no one named 'Lisa' has ever been

---

[43] p. 481. See also the discussion of this point in Sklar [1974], p. 419.

stricken by Valley Fever. We would not regard such a true generalization as a law of nature. Laws entail material generalizations, but there must be something more to them than that. I call nonstatistical laws *nomic generalizations*.[44] This reflects the fact that laws are not just about actual objects—they are also about "physically possible objects". Nomic generalizations can be expressed in English using the locution ⌜Any F would be a G⌝. I will symbolize this nomic generalization as ⌜$F \Rightarrow G$⌝. What this means, roughly, is that any physically possible F would be G.

I propose that we think of nomic probabilities as analogous to nomic generalizations. Just as ⌜$F \Rightarrow G$⌝ tells that us any physically possible F would be G, for heuristic purposes we can think of the statistical law ⌜$\text{prob}(G/F) = r$⌝ as telling us that the proportion of physically possible F's that would be G's is r.[45] For instance, suppose it is a law of nature that at any given time, there are exactly as many electrons as protons. Then in every physically possible world, the proportion of electrons-or-protons that are electrons as 1/2. It is then reasonable to regard the probability of a particular particle being an electron, given that it is either an electron or a proton, as 1/2. Of course, in the general case, the proportion of G's that are F's will vary from one possible world to another. $\text{prob}(F/G)$ "averages" these proportions across all physically possible worlds. The mathematics of this averaging process is complex and is the subject of Chapters 2 and 7.

Nomic probability is illustrated by any of the examples discussed earlier that could not be handled by relative frequency theories—the fair coin of description D that is flipped only once, the simplified quantum mechanical situation that never occurs, and so forth. In all of these cases the nomic probability diverges from the relative frequency because the nomic probability pertains

---

[44] I introduced the concept of a nomic generalization in Pollock [1974a], and discussed it more fully in Pollock [1976]. I have previously called nomic generalizations 'subjunctive generalizations'.

[45] I first talked about nomic probability in Pollock [1976], where I called it 'subjunctive indefinite probability'. I would no longer endorse much of what I said there.

to possible situations of these types and not just to actual ones. I took this to be the fundamental insight underlying hypothetical frequency theories, although the latter theories do not successfully capture this insight. I argue that nomic probability is the most fundamental kind of physical probability, other probability concepts being defined in terms of it. We will later see how to define a number of different kinds of probability, including propensities and the mixed physical/epistemic probabilities required by decision theory, but the definitions all proceed in terms of nomic probability.

I think that it is helpful to characterize nomic generalizations and nomic probabilities in terms of physically possible objects, but we cannot rest content with such a characterization. Even ignoring ontological objections, it is ultimately circular, because to say that there could be a physically possible object of a certain sort is just to say that there being such an object is not precluded by laws of nature, and laws of nature are nomic generalizations and statements of nomic probability. Unfortunately, it is difficult to imagine what other kind of definition might be possible.

When faced with the task of clarifying a concept, the natural inclination of the philosopher is to seek a definition of it. Until as late as 1960, most attempts to solve philosophical problems proceeded by seeking definitions of concepts like *physical object*, *person*, or *red*, which would make our reasoning involving those concepts explicable. But virtually all such attempted definitions failed. The lesson to be learned from this is that philosophically interesting concepts are rarely *definable* in any interesting way. Thus our inability to define nomic generalizations and nomic probability does not distinguish these concepts from any other philosophically interesting concepts.

If a concept cannot be defined in terms of simpler concepts, then what kind of informative analysis can be given for it? The essence of a concept is the role it plays in thought. Concepts are categories for use in thinking about the world–they are the tools of rational thought. As such, they are characterized by their role in rational thought. The role a concept plays in rational thought is determined by two things: (1) what constitutes a good reason for thinking that something exemplifies the concept or its negation, and (2) what we can conclude from reasonably held beliefs involving the concept. In other words, the concept is characterized by its role in reasoning, or more simply, by its *conceptual role*.[46] I have defended this view of conceptual analysis at length over the last twenty years, and I now propose to apply it to nomic generalizations and to the analysis of probability.[47]

To illustrate my general point, consider the problem of perception. That is the problem of explaining how it is possible to acquire knowledge of the physical world on the basis of perception. Early-twentieth-century philosophers tried to solve this problem by defining physical object concepts like *red* in terms of concepts describing our perceptual states (like the concept of *looking red*). Any theory *defining* physical object concepts in terms of the contents of our perceptual states is a version of phenomenalism. But in the end, all phenomenalist theories collapsed in the face of the argument from perceptual relativity.[48] Once it is recognized that concepts can be characterized by their conceptual roles, the problem of perception has a relatively trivial solution. It is part of the conceptual role of the concept *red* that something's looking red to a person gives him a prima facie reason for thinking that it is red. The existence of this prima facie reason is a brute epistemic fact, partly constitutive of the concept *red* and not in need of being derived from anything more basic. Thus the problem of perception is solved (or perhaps "resolved") by giving a description of the structure of the reasoning involved in perceptual knowledge. It is a mistake to suppose that that reasoning must itself be justified on the basis of some-

---

[46] I used to make basically the same point by saying that concepts are characterized by their "justification conditions", but I identified justification conditions with just (1) above, overlooking (2).

[47] For a fuller discussion of this view of conceptual analysis, see Pollock [1986]. The general theory is applied to a wide range of philosophical problems in Pollock [1974].

[48] This discussion is short because, I take it, no one wants to defend phenomenalism nowadays. See Pollock [1974], Chapter 3, for a more careful discussion of all this.

thing deeper. There is nothing deeper. That reasoning is itself constitutive of our perceptual concepts.

My suggestion is now that nomic generalizations and nomic probability are best analyzed in terms of their conceptual role. Consider nomic generalizations first. Such an analysis will describe what are good reasons for believing nomic generalizations and what conclusions can be drawn from nomic generalizations. There are two kinds of reasons for believing nomic generalizations. The ultimate source of justification for believing nomic generalizations is enumerative induction. Enumerative induction carries us from observation of individual cases to nomic generalizations. Then once we have confirmed some nomic generalizations, we may be able to deduce others from them. For example, it seems reasonably clear that $\ulcorner(F \Rightarrow G) \& (F \Rightarrow H)\urcorner$ entails $\ulcorner F \Rightarrow (G\&H)\urcorner$. Thus a description of what are good reasons for believing nomic generalizations will consist of an account of enumerative induction together with an account of entailment relations between nomic generalizations. Turning to what inferences can be made *from* nomic generalizations, it appears that there are basically two kinds. The most obvious is that nomic generalizations entail material generalizations:

$$\ulcorner F \Rightarrow G\urcorner \text{ entails } \ulcorner(\forall x)(Fx \supset Gx)\urcorner.$$

In addition, nomic generalizations play an essential role in the analysis of counterfactual conditionals. All of this falls under the general topic of the "logic of nomic generalizations". I say more about this in Chapter 2. The upshot is that (1) analyzing nomic generalizations and (2) understanding enumerative induction and the logic of nomic generalizations become the same task.

Much the same conclusion can be drawn about nomic probabilities. The history of the philosophical theory of probability has consisted in large measure of philosophers attempting to construct philosophically adequate definitions of 'probability'. In the case of empirical theories, this has led to numerous attempts to define 'probability' in terms of relative frequencies. But none of the definitions that have been proposed have constituted reasonable analyses of the probability concepts employed either by the man in the street or by working scientists. The conclusion to draw from this is that we should be skeptical of the attempt to found the theory of probability on an informative *definition* of 'probability'.[49] A reasonable alternative is to seek characterizations of our probability concepts in terms of their conceptual roles, that is, in terms of their roles in reasoning. A conceptual role analysis of nomic probability will consist of three elements. First, it must provide an account of how we can ascertain the numerical values of nomic probabilities on the basis of relative frequencies. Second, it must contain an account of the various "computational" principles that enable us to infer the values of some nomic probabilities from others. Finally, it must provide an account of how we can use nomic probability to draw conclusions about other matters.

The first element of this analysis consists largely of an account of statistical induction. The second element consists of a calculus of nomic probabilities, and is the topic of Chapters 2 and 7. The final element of the analysis of nomic probability is an account of how conclusions not about nomic probabilities can be inferred from premises about nomic probability. This account has two parts.   First, nomic probabilities are indefinite probabilities, but we require definite probabilities for decision-theoretic purposes. Thus we need a theory of direct inference telling us how to obtain definite probabilities from nomic probabilities. Second, it seems clear that under some circumstances, knowing that certain probabilities are high can justify us in holding related nonprobabilistic beliefs. For example, I know it to be highly probable that the date appearing on a newspaper is the correct date of its publication. (I do not know that this is

---

[49] I often feel that I am crying out in the wilderness when I espouse this general view of conceptual analysis and urge that interesting concepts are rarely to be analyzed by giving definitions of them. But in probability theory I am not alone. Similar skepticism about defining probability was advocated even by such early notables as Emile Borel [1909] (p. 16) and Henri Poincare [1912] (p. 24). A more recent defender of a similar sort of view is R. B. Braithwaite [1953]. The unhealthy emphasis on definition in contemporary probability theory probably reflects, more than anything else, the fact that probability theory came of age in a time when logical positivism reigned supreme.

always the case—typographical errors do occur.) On this basis, I can arrive at a justified belief regarding today's date. The epistemic rules describing when high probability can justify belief are called *acceptance rules*. The acceptance rules endorsed by the theory of nomic probability will constitute the principal novelty of that theory. The other fundamental principles that will be adopted as primitive assumptions about nomic probability are all of a computational nature. They concern the logical and mathematical structure of nomic probability, and in effect amount to nothing more than an elaboration of the standard probability calculus. It is the acceptance rules that give the theory its unique flavor and constitute the main epistemological machinery making the theory run.

To summarize, the conceptual role analysis of nomic probability will consist of (1) a theory of statistical induction, (2) an account of the computational principles allowing some probabilities to be derived from others, (3) an account of acceptance rules, and (4) a theory of direct inference. These four elements are precisely the components that make up an empirical theory of probability. Consequently, the analysis and the theory are the same thing. To give an analysis of nomic probability is the same thing as constructing a theory of nomic probability accommodating statistical induction, direct inference, and acceptance rules.

CHAPTER 2
COMPUTATIONAL PRINCIPLES

1. Introduction

Much of the usefulness of probability derives from its rich logical and mathematical structure. That structure comprises the probability calculus. The classical probability calculus is familiar and well understood, but it will turn out that the calculus of nomic probabilities differs from the classical probability calculus in some interesting and important respects. The purpose of this chapter is to develop the calculus of nomic probabilities, and at the same time to investigate the logical and mathematical structure of nomic generalizations.

2. The Logical Framework

The mathematical theory of nomic probability is formulated in terms of possible worlds. Possible worlds can be regarded as maximally specific possible ways things could have been. This notion can be filled out in various ways, but the details are not important for present purposes.[1] I assume that a proposition is necessarily true iff it is true at all possible worlds, and I assume that the modal logic of necessary truth and necessary exemplification is a quantified version of *S5*.

States of affairs are things like *Mary's baking pies, 2 being the square root of 4, Martha's being smarter than John*, and the like. For present purposes, a state of affairs can be identified with the set of all possible worlds at which it obtains.[2] Thus if *P* is a state

---

[1] I have constructed a general theory of possible worlds in Pollock [1984], and the reader is referred to it for more details.

[2] For a more general theory of states of affairs, and an explanation for why they cannot always be identified with sets of possible worlds, see Pollock [1984].

of affairs and w is a possible world, P obtains at w iff $w \in P$. Similarly, we can regard monadic properties as sets of ordered pairs $\langle w,x \rangle$ of possible worlds and possible objects. For example, the property of being red is the set of all pairs $\langle w,x \rangle$ such that w is a possible world and x is red at w. More generally, an n-place property will be taken to be a set of $(n+1)$-tuples $\langle w,x_1,\ldots,x_n \rangle$. Given any n-place concept α, the corresponding property of exemplifying α is the set of $(n+1)$-tuples $\langle w,x_1,\ldots,x_n \rangle$ such that $x_1,\ldots,x_n$ exemplify α at the possible world w.

States of affairs and properties can be constructed out of one another using logical operators like conjunction, negation, quantification, and so on. Logical operators apply to states of affairs in the normal way; for example, if P and Q are states of affairs, $(P\&Q)$ is the state of affairs that is their conjunction. Recalling that P and Q are sets of possible worlds, $(P\&Q)$ is their intersection $P \cap Q$. We can also apply logical operators to properties to form their conjunctions, disjunctions, and so forth. Where F is a monadic property, $(\exists x)Fx$ is the state of affairs consisting of there being F's. I will usually suppress the variables and abbreviate this as $\ulcorner \exists F \urcorner$. I will use the same notation for the existential generalization of an n-adic property. Similarly, I will take $\forall F$ to be the state of affairs consisting of everything's having the property F.

There are important connections between nomic generalizations, nomic probabilities, and counterfactual conditionals. Counterfactuals can be taken to relate states of affairs. Contemporary analyses of counterfactuals analyze them in terms of the notion of a "nearest P-world", and have the following general form:

(2.1) $(P > Q)$ obtains at w iff Q obtains at every nearest P-world to w.

Of course, this must be supplemented by a definition of 'nearest P-world'. The most familiar analysis of this sort is that of David Lewis [1973], but I have argued in several places that that analysis

is inadequate.[3]   My own analysis can be loosely sketched as
follows. (See Pollock [1984] for precise definitions and a precise
formulation of the analysis.) The basic proposal is that counterfactuals are to be analyzed in terms of the notion of minimal
change. The nearest P-worlds to w are those worlds that, subject
to two constraints, are minimally changed from w to accommodate
P's obtaining. The first constraint is that of *legal conservatism*.
This requires that if P is consistent with the physical laws of w,
then the laws of the nearest P-worlds must be unchanged from
those of w, and if P is inconsistent with the laws of w, then in
constructing nearest P-worlds we begin by minimally changing the
laws of w to accommodate P's obtaining.[4]   The second constraint
is that of *undercutting*. P undercuts Q at w iff P is incompatible
with every "causal history" of Q at w. The undercutting constraint is that, having satisfied legal conservatism, in making
minimal changes to accommodate P we preserve states of affairs
that are not undercut in preference to those that are. The details
of the undercutting constraint will be irrelevant to the present
investigation, but legal conservatism will be important.
        This analysis of counterfactuals validates all of the theorems
of *modal SS*, which results from adding the following axioms to
the axioms and rules of $S5$:

(*SS*)   A1. $[(P > Q) \ \& \ (P > R)] \supset [P > (Q \& R)]$

          A2. $[(P > R) \ \& \ (Q > R)] \supset [(P \lor Q) > R]$

          A3. $[(P > Q) \ \& \ (P > R)] \supset [(P \& Q) > R]$

          A4. $(P \& Q) \supset (P > Q)$

          A5. $(P > Q) \supset (P \supset Q)$

---

[3] See Pollock [1976], [1981], and [1984]. The latter contains my currently
preferred version of the analysis, and contrasts it with David Lewis's somewhat
similar sounding analysis. Lewis does not actually endorse (2.1), preferring a
more complicated principle that accommodates the failure of the Limit
Assumption.

[4] David Lewis [1973] and [1979] vehemently denies this, but most
philosophers accept it, and I have defended it in Pollock [1984]. A good
discussion of this principle can be found in Jonathan Bennett [1984].

A6. $\Box(P \equiv Q) \supset [(P > R) \equiv (Q > R)]$

A7. $[(P > Q) \,\&\, \Box(Q \supset R)] \supset (P > R)$

A8. $\Box(P \supset Q) \supset (P > Q)$

3. Nomic Generalizations

Let us turn next to the logical structure of nomic generalizations. My purpose here is not to provide an analysis of this notion, but merely to get clear on its logical behavior. An obvious proposal is that nomic generalizations can be characterized in terms of physical necessity. We say that a proposition is *physically necessary* iff it is entailed by true physical laws:

(3.1) P is *physically necessary* (symbolized: $\ulcorner \Box_p P \urcorner$) iff P is entailed by the set of all true laws (both statistical and nonstatistical). P is *physically possible* (symbolized: $\ulcorner \Diamond_p P \urcorner$) iff $\sim\!\Box_p\!\sim\!P$.

These definitions of physical necessity and physical possibility entail that the logic of these operators is *S5*.

The most obvious initial suggestion for characterizing nomic generalizations in terms of physical necessity is that $\ulcorner F \Rightarrow G \urcorner$ is true iff any physically possible F would be a G, that is,

(3.2) $(F \Rightarrow G)$ iff $\Box_p \forall (F \supset G)$.

Of course, this could not be regarded as an analysis of '$\Rightarrow$', because '$\Box_p$' was defined in terms of '$\Rightarrow$', but it could be used to generate a simple characterization of the entailment relations between nomic generalizations. We know how '$\Box_p$' works—it is an *S5* modal operator—and the remaining operators on the right side of (3.2) are all familiar truth functional operators. Unfortunately, (3.2) must be false because it implies that all counterlegal nomic generalizations are vacuously true. Let me explain. We say that a property is *counterlegal* if it is physically impossible for anything to have that property. A counterlegal nomic generalization is one whose antecedent is counterlegal. It might seem either that no counterlegal nomic generalization could be true, or else that they are all vacuously true (in which case (3.2) is unobjectionable). But it is not difficult to find examples of counterlegal generalizations that are true but not vacuously so. For example, supposing that it is a physical law that electrons bear the charge they do (call it 'e'), and that this law is independent of the law of gravitation, it is surely a true (derived) law that any electrons bearing the charge $2e$ would still be subject to gravitational attractions as described by the law of gravitation. Furthermore, this law is not vacuously true, because it is not equally true that any electron bearing the charge $2e$ would fail to be subject to gravitational attractions as described by the law of gravitation.

It is fairly clear how counterlegal nomic generalizations work. We derive them from noncounterlegal generalizations. In evaluating a counterlegal generalization $\ulcorner F \Rightarrow G \urcorner$, we consider what the set of noncounterlegal generalizations would be if F were not counterlegal. To this end, we consider all of the different ways of minimally changing the set of noncounterlegal generalizations to make it consistent with $\ulcorner \exists F \urcorner$. We judge $\ulcorner F \Rightarrow G \urcorner$ to be true iff $\ulcorner \forall (F \supset G) \urcorner$ is entailed by every such minimally changed set of noncounterlegal generalizations. Such minimally changed sets of noncounterlegal generalizations are just the sets of noncounterlegal generalizations holding at different nearest $\ulcorner \diamond_p \exists F \urcorner$-worlds, and to say that $\ulcorner \forall (F \supset G) \urcorner$ is entailed by them is just to say that it is physically necessary at those worlds. Consequently, this account of nomic generalizations is equivalent to the following counterfactual characterization:

(3.3) $(F \Rightarrow G)$ iff $\diamond_p \exists F > \Box_p \forall (F \supset G)$.

As I have explained them, nomic generalizations relate properties to one another. Sometimes we want to talk about nomic connections between states of affairs. These are connections that follow from true nomic generalizations. Where P and Q are states of affairs, let us define:

(3.4) $(P \Rightarrow Q)$ is true iff $\diamond_p P > \square_p (P \supset Q)$.

(3.3) and (3.4) generate an account of entailment relations between nomic generalizations. We can use them to prove a number of simple theorems:

(3.5) $[(F \Rightarrow G) \ \& \ \square\forall(G \supset H)] \supset (F \Rightarrow H)$.

Proof: Suppose $(F \Rightarrow G) \ \& \ \square\forall(G \supset H)$. By (3.3), $\diamond_p \exists F > \square_p \forall (F \supset G)$, i.e., at any nearest $\ulcorner \diamond_p \exists F \urcorner$-world α, $\ulcorner \forall(F \supset G) \urcorner$ is entailed by the set of true nomic generalizations holding at α. As $\square\forall(G \supset H)$, it follows that $\ulcorner \forall(F \supset H) \urcorner$ is also entailed by the set of true nomic generalizations holding at α. Consequently, $\ulcorner \diamond_p \exists F > \square_p \forall(F \supset H) \urcorner$ is true, i.e., $\ulcorner F \Rightarrow H \urcorner$ is true.

The proofs of the following are analogous:

(3.6) $\square\forall(F \supset G) \supset (F \Rightarrow G)$.

(3.7) $[(F \Rightarrow G) \ \& \ (F \Rightarrow H)] \supset [F \Rightarrow (G \& H)]$.

(3.8) $[\square(F \equiv G) \ \& \ (F \Rightarrow H)] \supset (G \Rightarrow H)$.

(3.9) $[(F \Rightarrow H) \ \& \ (G \Rightarrow H)] \supset [(F \vee G) \Rightarrow H]$.

(3.10) $[(F \ \& \ {\sim}G) \Rightarrow H] \supset [F \Rightarrow (G \vee H)]$.

(3.11) $[(F \Rightarrow G) \ \& \ (F \Rightarrow H)] \supset [(F \& H) \Rightarrow G]$.

(3.12) $\{(F \Rightarrow G) \ \& \ [(F \& G) \Rightarrow H]\} \supset (F \Rightarrow H)$.

(3.13) $[\diamond_p \exists F \ \& \ (F \Rightarrow G) \ \& \ (G \Rightarrow H)] \supset (F \Rightarrow H)$.

(3.14) $[\diamond_p \exists {\sim}G \ \& \ (F \Rightarrow G)] \supset ({\sim}G \Rightarrow {\sim}F)$.

(3.15) $\diamond_p \exists F \supset [(F \Rightarrow G) \equiv \square_p \forall(F \supset G)]$.

(3.16) $[\exists F \ \& \ \square_p \forall(F \supset G)] \supset (F \Rightarrow G)$.

(3.17) $[\exists F \& \Box_p \forall (F \supset G)] \supset (F \Rightarrow G)$.

(3.18) $[\exists F \& \Box(P \equiv \Box_p P)] \supset \{[(F \Rightarrow G) \& P] \supset [(P \& F) \Rightarrow G]\}$.

(3.19) $[\exists G \& \Box(P \equiv \Box_p P)] \supset \{[(G \Rightarrow F) \& P] \supset [G \Rightarrow (P \& F)]\}$.

The preceding constitutes an account of entailment relations between nomic generalizations. For the purpose of providing a conceptual role analysis of nomic generalizations, all that remains is to give an account of enumerative induction.

4. Probabilities as Proportions

In developing the mathematical theory of nomic probability, the most natural approach would have us begin by adopting some intuitive axioms and would then proceed to derive theorems from those axioms. However, experience shows such a straightforward approach to be unreliable. Intuitions about probability frequently lead to inconsistent principles. The difficulty can be traced to two sources. First, nomic probabilities involve modalities in ways that are a bit tricky and easily confused. For example, the following two principles will be endorsed below:

(6.10) If $\Diamond \exists H$ and $[H \Rightarrow \sim(F \& G)]$ then $\text{prob}(F \lor G/H) = \text{prob}(F/H) + \text{prob}(G/H)$.

(6.11) If $\Diamond_p \exists H$ then $\text{prob}(F \& G/H) = \text{prob}(F/H) \cdot \text{prob}(G/F \& H)$.

The antecedent of (6.10) involves only logical necessity, whereas the antecedent of (6.11) involves physical necessity. Second, and more important, nomic probabilities are indefinite probabilities. Indefinite probabilities operate on properties, including relational properties of arbitrarily many places. This introduces logical relationships into the theory of nomic probability that are ignored in the classical probability calculus. One simple example is the *principle of individuals*:

$$(IND) \quad \text{prob}(Axy \ / \ Rxy \ \& \ y=b) = \text{prob}(Axb/Rxb).$$

This is an essentially relational principle and is not a theorem of the classical probability calculus. It might be wondered how there can be general truths regarding nomic probability that are not theorems of the classical probability calculus. The explanation is that, historically, the probability calculus was devised with definite probabilities in mind. The standard versions of the probability calculus originate with Kolmogoroff [1933] and are concerned with "events". The relationship between the calculus of indefinite probabilities and the calculus of definite probabilities is a bit like the relationship between the predicate calculus and the propositional calculus. Specifically, we will find that there are principles regarding relations and quantifiers that must be added to the classical probability calculus to obtain a reasonable calculus of nomic probabilities. But I have found direct intuition to be an unreliable guide in getting these principles right.

If we cannot trust our intuitions in developing the calculus of nomic probabilities, how can we proceed? I propose that we make further use of the heuristic model of nomic probability as measuring proportions among physically possible objects. A noncounterlegal nomic generalization $\ulcorner F \Rightarrow G \urcorner$ says that any physically possible F would be a G. Analogously, the noncounterlegal statistical law $\ulcorner \text{prob}(G/F) = r \urcorner$ can be regarded as telling us that the proportion of physically possible F's that would be G is r. This will not yield a logical analysis of nomic probability because in the end it will turn out that proportions must be explained in terms of nomic probabilities. Nevertheless, treating probabilities in terms of proportions proves to be a useful approach for investigating the logical and mathematical structure of nomic probability. The algebraic structure of proportions turns out to be somewhat simpler than that of probabilities, and our intuitions are correspondingly a bit clearer. Once we have an understanding of the algebraic structure of proportions we can use that to derive computational principles governing nomic probability.

Proportions operate on sets. Given any two sets A and B we can talk about the proportion of members of B that are also members of A. I will symbolize ⌜the proportion of members of B that are in A⌝ as ⌜$\wp(A/B)$⌝. For enhanced readability, I will sometimes write proportions in the form ⌜$\wp(\frac{A}{B})$⌝. The simplest case of proportions occurs when A and B are finite. In that case, $\wp(A/B)$ is just the relative frequency with which members of B are also in A, that is, the ratio of the cardinalities of the sets $A \cap B$ and B. But we can also talk about proportions when A and B are infinite. The concept of a proportion is a general measure-theoretic notion. It is quite natural to think of proportions as ratios of measures, taking $\wp(A/B)$ to be $\mu(A \cap B) \div \mu(B)$ for some measure μ. Such an approach is not entirely satisfactory, for reasons to be explained in section 6, but it does come close to capturing the measure-theoretic nature of proportions.

The theory of proportions will be developed in considerable detail. It will then be applied to the derivation of computational principles governing nomic probability. The derivation is accomplished by making more precise our model of nomic probability as measuring proportions among physically possible objects. Where F and G are properties and G is not counterlegal (so that there are physically possible G's), I have described the nomic probability prob(F/G) as the proportion of physically possible G's that would be F's. This suggests that we define:

(4.1)   If $\diamond_p \exists G$ then prob(F/G) = $\wp(\mathbf{F}/\mathbf{G})$

where $\mathbf{F}$ is the set of all physically possible F's and $\mathbf{G}$ is the set of all physically possible G's. This forces us to consider more carefully what we mean by 'a physically possible F'. We cannot mean just "a possible object that is F in some physically possible world", because the same object can be F in one physically possible world and non-F is another. Instead, I propose to understand a physically possible F to be an ordered pair $\langle w,x \rangle$ such that w is a physically possible world (i.e., one having the same physical laws as the actual world) and x is an F at w. We then define:

(4.2) $\mathbf{F} = \{\langle w,x \rangle \mid w$ is a physically possible world and x is F
             at $w\}$;

        $\mathbf{G} = \{\langle w,x \rangle \mid w$ is a physically possible world and x is G
             at $w\}$;

        etc.

With this understanding, we can regard noncounterlegal nomic
probabilities straightforwardly as in (4.1) as measuring proportions
between sets of physically possible objects. (4.1) will have to be
extended to cover the case of counterlegal probabilities, but it is
most convenient to postpone the discussion of this complication
until after we have investigated proportions.

      It deserves to be emphasized that this use of proportions
and sets of physically possible objects is just a way of getting the
mathematical structure of nomic probability right. It will play no
further role in the theory. Accordingly, if ontological scruples
prevent one from accepting this metaphysical foundation for
nomic probability, one can instead view it as merely a heuristic
model for arriving at an appropriate formal mathematical
structure. Once we have adopted that structure we can dispense
with the edifice of possible worlds and possible objects. This
formal mathematical structure will be coupled with epistemological
principles that are ontologically neutral, and then the theory of
nomic probability will be derived from that joint basis.

5. The Boolean Theory of Proportions

The simplest and least problematic talk of proportions concerns
finite sets. In that case proportions are just frequencies. Taking
$\#X$ to be the cardinality of a set X, relative frequencies are
defined as follows:

(5.1)   If X and Y are finite and Y is nonempty then freq$[X/Y]$
        $= \#(X \cap Y)/\#Y$.

We then have the *Frequency Principle*:

(5.2)   If X and Y are finite and Y is nonempty then $\wp(X/Y) =$ freq$[X/Y]$.

If Y is infinite and $\#(X \cap Y) < \#Y$, then $X \cap Y$ is infinitely smaller than Y, so we can also endorse the *Extended Frequency Principle*:

(5.3)   If Y is infinite and $\#(X \cap Y) < \#Y$ then $\wp(X/Y) = 0$.

But this is as far as we can get just talking about frequencies and cardinalities. We also want to talk about proportions among infinite sets of the same cardinality. The concept of a proportion in such a case is an extension of the concept of a frequency. The simplest laws governing such proportions are those contained in the classical probability calculus, which can be axiomatized as follows:

(5.4) $0 \leq \wp(X/Y) \leq 1$

(5.5)   If $Y \subseteq X$ then $\wp(X/Y) = 1$.

(5.6)   If $Z \neq \varnothing$ and $Z \cap X \cap Y = \varnothing$ then $\wp(X \cup Y/Z) = \wp(X/Z) + \wp(Y/Z)$.

(5.7) $\wp(X \cap Y/Z) = \wp(X/Z) \cdot \wp(Y/X \cap Z)$.

To proceed further it would be natural to take proportions to be ratios of measures of the sizes of sets. That is the way probability is usually studied in mathematical probability theory, and its generalization to proportions seems an obvious move. Rather than taking p as basic, it would be supposed that there is a real-valued function μ defined on some class $\mathbf{U}$ (typically a field) of sets such that if $X, Y \in \mathbf{U}$ then

(5.8) $\wp(X/Y) = \mu(X \cap Y)/\mu(Y)$.

μ would normally be supposed to be a finitely additive measure:

(5.9)   If $X \cap Y = 0$ then $\mu(X \cup Y) = \mu(X) + \mu(Y)$.[5]

The difficulty with this standard approach is that (5.8) leaves $\wp(X/Y)$ undefined when $\mu(Y) = 0$. This conflicts with the frequency principle, which we have already adopted. For example, suppose we are discussing proportions in sets of natural numbers. The frequency principle requires that $\wp(X/Y)$ exists whenever Y is finite and nonempty, and so it implies that every nonempty finite set of natural numbers receives a nonzero measure. The frequency principle also implies that two finite sets of the same cardinality must receive the same nonzero measure. But then (5.9) implies that any infinite set has a measure greater than every real number, which is impossible because μ is supposed to be real-valued. Consequently, we cannot both define proportions in terms of measures and endorse the frequency principle.[6] In case the reader is tempted to deny the frequency principle, remember that we are talking about proportions here—not probabilities. The frequency principle is surely *the* most obvious principle regarding proportions. The simple solution to this problem is to take proportions as basic rather than defining them in terms of measures. Proportions become a kind of "relative measure".[7]

---

[5] It would actually be more customary to suppose μ is countably additive, but that is inconsistent with the extended frequency principle: $\wp(\omega/\omega) = 1$ but for each $n \in \omega$, $\wp(\{n\}/\omega) = 0$. For the same reason countable additivity will fail for nomic probability. Such counterexamples to countable additivity are really quite obvious, and have been explicitly noted by some probabilists (e.g., Reichenbach [1949]). I find it a bit mysterious that most probabilists still assume that probability is countably additive. Mathematicians assume countable additivity in order to make things mathematically interesting (as one mathematician put it to me, "If you assume finite additivity, all you have is finite additivity"). But this is no justification for the philosopher, who means to be talking about real probability and not just mathematical abstractions.

[6] It is noteworthy that mathematicians do not endorse the frequency principle. Despite formal similarities, mathematical probability theory is not the same thing as the theory of proportions. Instead, the sets upon which mathematical probability operates are supposed to be abstractions representing sets of events or sets of states of affairs.

[7] It appears that Karl Popper ([1938] and [1959] (pp. 318-322)) was the first to suggest taking conditional probabilities as primitive rather than defining

There is a further possible advantage to taking proportions to be basic rather than defining them in terms of measures. For every set, there is some other set that is infinitely larger. If real-valued measures are to compare these sets, the smaller one must receive measure 0. It follows that a measure defined on all sets must assign measure 0 to every set, which makes it useless. Consequently, the measure-theoretic approach to proportions requires us to confine our attention to some restricted class **U** of sets, and take proportions to be defined only for members of **U**. On the other hand, if we take proportions to be basic, there appears to be no reason not to suppose that $\wp(X/Y)$ is defined for all sets. That is the assumption I will make here. It must be acknowledged, however, that this move is controversial. It is unclear whether there can be proportion functions of the sort described below that are defined for all sets. There are no known mathematical results that show there cannot be such functions, but there are unsolved mathematical problems that are related to this matter, and the existing evidence is equivocal.[8] It is worth

---

them in terms of nonconditional probabilities, and accordingly primitive nonconditional probability functions are sometimes called *Popper functions*. The mathematician Alfred Renyi [1955] was the first to apply this approach to the mathematical theory of probability. There are obvious similarities between Popper functions and proportions, but they should not be confused with one another. Probabilities are not the same things as proportions.

There is an alternative measure-theoretic approach that avoids the difficulties facing real-valued measures. That involves the use of measures whose values are the nonstandard reals. In fact, that is the way I first worked all of this out. On such an approach, the cross product principle (discussed in the following) requires that the measure of a finite set be its cardinality, and that suggests requiring that the measure always take nonstandard integers as its values. (Nonstandard integer-valued measures were, I believe, first studied by Bernstein and Wattenberg [1969].) The use of nonstandard measures makes possible the formulation of a more elegant and somewhat more powerful theory of proportions, but it also introduces a new source of possible error and makes our intuitions correspondingly less secure. Thus when I discovered that I could obtain the same basic theory of nomic probability without using nonstandard measures I opted to do so.

[8] A general discussion of matters related to this can be found in Bruckner and Ceder [1975]. They observe that the main obstacle to extending standard measures (e.g., Lebesgue measure) to broader classes of sets is the assumption

emphasizing then that I make this assumption mainly as a matter of convenience. It streamlines the theory somewhat, but at the expense of some additional complexity we could get basically the same theory without this assumption.

Although we cannot define proportions in terms of real-valued measure functions as in (5.8), the concept of a proportion is still basically a measure-theoretic notion. Although we cannot reduce proportions to any single measure μ, for each nonempty set Y there will be a finitely additive measure μ_Y defined on all subsets of Y as follows:

(5.10) If $X \subseteq Y$ then $\mu_Y(X) = \wp(X/Y)$.

It then follows by the multiplicative axiom (5.7) that:

(5.11) If $A \subseteq Y$ and $B \subseteq Y$ and $\wp(B/Y) \neq 0$ then $\wp(A/B) = \mu_Y(A \cap B)/\mu_Y(B)$.

This observation is helpful in mustering intuitions in support of certain principles regarding proportions. For example, suppose $X \subseteq Y$ and there is a $Z \subseteq Y$ such that $\mu_Y(Z) = r$, where $r < \mu_Y(X)$. Then there should be a $Z \subseteq X$ such that $\mu_Y(Z) = r$:

(5.12) If $X \subseteq Y$ and $r < \wp(X/Y)$ and $(\exists Z)[Z \subseteq Y \& \wp(Z/Y) = r]$ then $(\exists Z)[Z \subseteq X \subseteq Y \& \wp(Z/Y) = r]$.

I will adopt (5.12) as an axiom and call it *The Denseness Principle*. An immediate consequence of (5.12) is:

(5.13) If $X \subseteq Y$ and $r > \wp(X/Y)$ and $(\exists Z)[Z \subseteq Y \& \wp(Z/Y) = r]$ then $(\exists Z)[X \subseteq Z \subseteq Y \& \wp(Z/Y) = r]$.

---

of countable additivity, but I have already noted that proportions are not countably additive. Another troublesome principle is that stating that congruent sets receive the same measure, but that principle does not hold for proportions either (see the discussion of translation invariance, below).

Proof: By (5.12) there is a $W \subseteq Z$ such that $\wp(W/Y) = \wp(X/Y)$. By (5.6), $\wp(Z/Y)$ $= \wp((Z-W) \cup X \,/\, Y)$.

The probability calculus has a large number of useful theorems. A few familiar ones follow:

(5.14) If $Y \neq \emptyset$ then $\wp(Y-X/Y) = 1 - \wp(X/Y)$;

(5.15) If $X \cap Y \subseteq Z$ then $\wp(X/Y) \leq \wp(Z/Y)$;

(5.16) If $X \cap Z = Y \cap Z$ then $\wp(X/Z) = \wp(Y/Z)$;

(5.17) If $Y \neq \emptyset$ and $Z \neq \emptyset$ and $Y \cap Z = \emptyset$, then $\wp(X/Y \cup Z)$ $= \wp(X/Y) \cdot \wp(Y/Y \cup Z) + \wp(X/Z) \cdot \wp(Z/Y \cup Z)$.

(5.18) If $X \subseteq Y \subseteq Z$ then $\wp(X/Z) = \wp(X/Y) \cdot \wp(Y/Z)$.

(5.19) If $\wp(Z/Y) = 1$ then $\wp(X/Z \cap Y) = \wp(X/Y)$.

The theory of proportions resulting from the frequency principles, the classical probability calculus, and the denseness principle might be termed 'the Boolean theory of proportions', because it is only concerned with the Boolean operations on sets. This much of the theory of proportions must be regarded as beyond reasonable doubt. In the next section I turn to nomic probability and investigate what principles of nomic probability can be obtained from the Boolean theory of proportions. Then I return to the theory of proportions and investigate some non-Boolean principles and their consequences for nomic probability.

6. The Elementary Theory of Nomic Probability

I have explained the noncounterlegal nomic probability $\text{prob}(F/G)$ as the proportion of physically possible G's that would be F. More precisely:

(6.1) If $\Diamond_p \exists G$ then $\text{prob}(F/G) = \wp(\mathbf{F/G})$

where $\mathbf{F}$ is the set of all physically possible F's and $\mathbf{G}$ is the set of all physically possible G's. It was pointed out that (6.1) characterizes only noncounterlegal nomic probabilities. If G is counterlegal, prob(F/G) cannot be the proportion of physically possible G's that would be F, because there are no physically possible G's.[9] I propose instead that counterlegal nomic probabilities can be defined on analogy to counterlegal nomic generalizations. According to principle (3.3), if G is counterlegal then $(G \Rightarrow F)$ does not say that any physically possible G would be an F. Rather, it says that *if* it were physically possible for there to be G's *then* any physically possible G would be an F, that is, $\diamond_p \exists G > \square_p \forall (G \supset F)$. Similarly, if it is physically impossible for there to be G's, then rather than prob(F/G) being the proportion of physically possible G's that would be F, prob(F/G) should be the mathematical expectation of what that proportion *would be if* it were physically possible for there to be G's. For each possible world w, let prob$_w(F/G)$ be the value of prob(F/G) at w. If G is counterlegal at the actual world w then we should look at the nearest worlds w^* having laws consistent with $\exists G$. If two such worlds w^* and w^{**} have the same laws (both statistical and nonstatistical) then prob$_{w^*}(F/G)$ = prob$_{w^{**}}(F/G)$, because the value of prob(F/G) is one of those laws. Suppose just finitely many values $r_1,\ldots,r_k$ of prob$_{w^*}(F/G)$ occur at the different nearest $\diamond_p \exists G$ worlds. prob$_w(F/G)$ should then be a weighted average of the r_i's, the weight of each reflecting how likely it is to be the value of prob(F/G) given that one of the r_i's is. Where $\mathbf{N}$ is the set of nearest $\diamond_p \exists G$-worlds to w, this latter likelihood is given by the following proportion:

$$\wp(\{x|\ \text{prob}_x(F/G) = r_i\}\ /\ \mathbf{N}).$$

The proposal is then:

---

[9] It might be supposed that counterlegal probabilities are not really of much interest. It turns out, however, that counterlegal probabilities play an indispensable role in statistical induction. This is primarily because of the counterlegality of the principle (*PPROB*), following.

(6.2)   If $\sim\!\Diamond_p \exists G$ then

$$\text{prob}_w(F/G) = \sum_{i=1}^{k}\left[r_i \cdot \wp(\{x|\ \text{prob}_x(F/G) = r_i\}\ /\ \mathbf{N})\right].$$

This reduces counterlegal nomic probabilities to noncounterlegal nomic probabilities.

(6.2) assumes that there are just finitely many r_i's. Should that assumption prove false, then (6.2) must be replaced by its continuous generalization:

(6.3)   If $\sim\!\Diamond_p \exists G$ then

$$\text{prob}_w(F/G) = \int_0^1 \wp(\{x|\ \text{prob}_x(F/G) \geq r\}\ /\ \mathbf{N})dr.$$

(6.1), (6.2), and (6.3) reduce nomic probability to proportions among sets of physically possible objects. The calculus of proportions can now be used to generate a calculus of nomic probabilities. (6.1) has the immediate consequence that non-counterlegal nomic probabilities obey the probability calculus. That is, from (5.4)–(5.7) we obtain:

(6.4)   If $\Diamond_p \exists G$ then $0 \leq \text{prob}(F/G) \leq 1$.

(6.5)   If $\Diamond_p \exists G$ and $(G \Rightarrow F)$ then $\text{prob}(F/G) = 1$.

(6.6)   If $\Diamond_p \exists H$ and $[H \Rightarrow \sim(F\&G)]$ then $\text{prob}(F \vee G/H)$
        $= \text{prob}(F/H) + \text{prob}(F/H)$.

(6.7)   If $\Diamond_p \exists(F\&H)$ then $\text{prob}(F\&G/H)$
        $= \text{prob}(F/H) \cdot \text{prob}(G/F\&H)$.

The values of counterlegal nomic probabilities are determined by (6.2) or (6.3), according to which if $\sim\!\Diamond_p \exists G$ then $\text{prob}_w(F/G)$ is a weighted average of the $\text{prob}_x(F/G)$ for the different nearest $\Diamond_p \exists G$-worlds x. A weighted average of a constant value is equal to that value, and a weighted average of a sum is the sum of the weighted averages, so we can strengthen (6.4)–(6.6) as follows:

(6.8) $0 \leq \text{prob}(F/G) \leq 1$.

(6.9)   If $(G \Rightarrow F)$ then $\text{prob}(F/G) = 1$.

(6.10) If $\diamond \exists H$ and $[H \Rightarrow \sim(F\&G)]$ then $\text{prob}(F \vee G/H)$
        $= \text{prob}(F/H) + \text{prob}(G/H)$.

Principle (6.7), the multiplicative axiom, is not so easily generalized. If $\diamond_p \exists H$ but $\sim \diamond_p \exists (F\&H)$ then $(H \Rightarrow \sim F)$ and $[H \Rightarrow \sim(F\&G)]$. It then follows that $\text{prob}(F\&G/H) = 0 = \text{prob}(F/H)$. Thus (6.7) can be strengthened slightly to yield:

(6.11) If $\diamond_p \exists H$ then $\text{prob}(F\&G/H) = \text{prob}(F/H) \cdot \text{prob}(G/F\&H)$.

But we cannot generalize it any further. Specifically, it is not true in general that:

(6.12) $\text{prob}(F\&G/H) = \text{prob}(F/H) \cdot \text{prob}(G/F\&H)$.

The difficulty is that if H is counterlegal, then the set of nearest $\diamond_p \exists H$-worlds may be different from the set of nearest $\diamond_p \exists (F\&H)$-worlds, and hence there need be no connection between proportions in the one and proportions in the other. We can construct a concrete counterexample to (6.12) as follows. Suppose that H is counterlegal, and suppose that L_1 and L_2 are the sets of laws that would be true in the different nearest worlds at which $\diamond_p \exists H$ is true. Let $\mathbf{L}_1$ be the set of all nearest worlds at which L_1 is the set of true physical laws, and let $\mathbf{L}_2$ be the set of all nearest worlds at which L_2 is the set of true physical laws. Suppose these are the same laws as would be true at the nearest $\diamond_p \exists (F\&H)$ worlds. Then suppose we have the following probabilities:

$\text{prob}(\mathbf{L}_1/\diamond_p \exists H) = \text{prob}(\mathbf{L}_2/\diamond_p \exists H) = 1/2$;
If $x \in \mathbf{L}_1$ then $\text{prob}_x(F/H) = \text{prob}_x(G/F\&H) = 1/2$;
If $x \in \mathbf{L}_2$ then $\text{prob}_x(F/H) = \text{prob}_x(G/F\&H) = 1/4$.

Then by (6.11),

If $x \in L_1$ then $\text{prob}_x(F \& G/H) = 1/4$;
If $x \in L_2$ then $\text{prob}_x(F \& G/H) = 1/16$.

By (6.2),

$\text{prob}(F \& G/H) = 1/2 \cdot 1/4 + 1/2 \cdot 1/16 = 5/32$;
$\text{prob}(F/H) = 1/2 \cdot 1/2 + 1/2 \cdot 1/4 = 3/8$;
$\text{prob}(G/F \& H) = 1/2 \cdot 1/2 + 1/2 \cdot 1/4 = 3/8$;

and hence $\text{prob}(F/H) \cdot \text{prob}(G/F \& H) = 9/64 \neq 10/64 = \text{prob}(F \& G/H)$.

The following is a useful generalization of (6.11):

(6.13)   If $\ulcorner \lozenge_p \exists (G \& F_1 \& \ldots \& F_n) \urcorner$ then $\text{prob}(F_1 \& \ldots \& F_n/G)$
      $= \text{prob}(F_1/F_2 \& \ldots \& F_n \& G) \cdot \text{prob}(F_2/F_3 \& \ldots \& F_n \& G)$
      $\cdot \ldots \cdot \text{prob}(F_n/G)$.

Proof: $\text{prob}(F_1 \& \ldots \& F_n/G)$
      $= \text{prob}(F_1/F_2 \& \ldots \& F_n \& G) \cdot \text{prob}(F_2 \& \ldots \& F_n/G)$
      $= \text{prob}(F_1/F_2 \& \ldots \& F_n \& G) \cdot \text{prob}(F_2/F_3 \& \ldots \& F_n \& G) \cdot \text{prob}(F_3 \& \ldots \& F_n/G)$
      $\vdots$

      $= \text{prob}(F_1/F_2 \& \ldots \& F_n \& G) \cdot \text{prob}(F_2/F_3 \& \ldots \& F_n \& G) \cdot \ldots \cdot \text{prob}(F_n/G)$.

Although nomic probabilities do not satisfy the unrestricted multiplicative axiom, (6.8)–(6.11) comprise an only slightly weakened version of the classical probability calculus. Some familiar consequences of these axioms are as follows:

(6.14)   If $\lozenge \exists G$ then $\text{prob}(\sim F/G) = 1 - \text{prob}(F/G)$;

(6.15)   If $[(F \& G) \Rightarrow H]$ then $\text{prob}(F/G) \leq \text{prob}(H/G)$.

(6.16)   If $[H \Rightarrow (F \equiv G)]$ then $\text{prob}(F/H) = \text{prob}(G/H)$.

(6.17) If $(G \Leftrightarrow H)$ then $\text{prob}(F/G) = \text{prob}(F/H)$.

(6.18) If $\text{prob}(H/G) = 1$ then $\text{prob}(F/G\&H) = \text{prob}(F/G)$.

A more general consequence of either (6.2) or (6.3) is:

(6.19) If $(\diamond_p \exists G > \diamond_p \exists B)$ and $(\diamond_p \exists B > \diamond_p \exists G)$ and $[\diamond_p \exists G > \text{prob}(F/G) = \text{prob}(A/B)]$ then $\text{prob}(F/G) = \text{prob}(A/B)$.

A term r is a *rigid designator of a real number* iff r designates the same real number at every possible world. Rigid designators include terms like '2.718' and 'π'. Rigid designators play a special role in nomic probability, as is exemplified by the next theorem:

(6.20) If r is a rigid designator of a real number and $\diamond \exists G$ and $[\diamond_p \exists G > \text{prob}(F/G) = r]$ then $\text{prob}(F/G) = r$.

Proof: Suppose $\diamond \exists G$. Let N be the set of nearest $\diamond_p \exists G$-worlds. If w is the actual world and $[\diamond_p \exists G > \text{prob}(F/G) = r]$ holds at w, then

$$\Box(\forall x)[x \in N \supset \text{prob}_x(F/G) = r].$$

Thus by (6.2) or (6.3), $\text{prob}(F/G)$ is a weighted average of terms all of which are equal to r, and hence $\text{prob}(F/G) = r$.

Notice that the last step of this proof would be fallacious if r were not a rigid designator. If r designated different numbers at different worlds, then we would not be taking a weighted average of a constant value.

The following principle will be very useful:

(6.21) $(\diamond_p \exists G \& \Box_p P) \supset \text{prob}(F/G\&P) = \text{prob}(F/G)$.

Proof: Suppose $\diamond_p \exists G$ and $\Box_p P$. Then $[G \Leftrightarrow (G\&P)]$, so by (6.17), $\text{prob}(F/G\&P) = \text{prob}(F/G)$.

With the help of (6.21), we can prove a theorem that will play a pivotal role in the derivation of principles of induction. This theorem concerns probabilities of probabilities (hence the name '(*PPROB*)'). On many theories, there are difficulties making sense of probabilities of probabilities, but there are no such problems within the theory of nomic probability. 'prob' can relate any two properties, including properties defined in terms of nomic probabilities. Our theorem is:

(*PPROB*) If r is a rigid designator of a real number and $\diamond[\exists G$
& $\text{prob}(F/G) = r]$ then $\text{prob}\big(F \,/\, G \,\&\, \text{prob}(F/G) = r\big) = r.$

Proof: Suppose r is a real number and $\diamond[\exists G \,\&\, \text{prob}(F/G) = r]$. Then $\diamond\exists[G$ & $\text{prob}(F/G) = r]$. Given that r is a rigid designator, if $\ulcorner\text{prob}(F/G) = r\urcorner$ is true then it is a physical law, and hence it is physically necessary. In other words,

(a) $\Box[\text{prob}(F/G) = r \supset \Box_p\text{prob}(F/G) = r].$

If r is some value other than the actual value of $\text{prob}(F/G)$, then $\ulcorner\text{prob}(F/G) = r\urcorner$ is incompatible with true physical laws, so if the latter is physically possible then it is true:

(b) $\Box[\diamond_p\text{prob}(F/G) = r \supset \text{prob}(F/G) = r].$

Clearly

(c) $\Box\big[\diamond_p\exists(G \,\&\, \text{prob}(F/G) = r) \supset \diamond_p\text{prob}(F/G) = r\big].$

By (a), (b), and (c),

(d) $\Box\big[\diamond_p\exists(G \,\&\, \text{prob}(F/G) = r) \supset \Box_p\text{prob}(F/G) = r\big]$

and hence by the logic of counterfactuals

(e) $\diamond_p\exists[G \,\&\, \text{prob}(F/G) = r] > \Box_p\text{prob}(F/G) = r.$

Therefore,

(f) $\diamond_p\exists[G \,\&\, \text{prob}(F/G) = r] > [\diamond_p\exists G \,\&\, \Box_p\text{prob}(F/G) = r],$

so by (6.21)

(g) $\diamond_p \exists [G \,\&\, \mathrm{prob}(F/G) = r] > \mathrm{prob}\big(F \,/\, G \,\&\, \mathrm{prob}(F/G) = r\big) = \mathrm{prob}(F/G)$
       $= r.$

Therefore, by (6.20)

(h) $\mathrm{prob}\big(F \,/\, G \,\&\, \mathrm{prob}(F/G) = r\big) = r.$

Note that if $\mathrm{prob}(F/G) \neq r$ then $\mathrm{prob}\big(F \,/\, G \,\&\, \mathrm{prob}(F/G) = r\big)$ is a counterlegal probability.

7. The Non-Boolean Theory of Proportions

The Boolean theory of proportions is only concerned with the Boolean operations on sets. In this respect, it is analogous to the propositional calculus. However, in $\wp(X/Y)$, X and Y might be sets of ordered pairs, that is, relations. There are a number of principles that ought to hold in that case but are not contained in the Boolean theory of proportions. The classical probability calculus takes no notice of relations, and to that extent it is seriously inadequate. For example, the following *Cross Product Principle* would seem to be true:

(7.1)   If $A \subseteq C$ and $B \subseteq D$ and $C \neq \varnothing$ and $D \neq \varnothing$ then
        $\wp(A \times B / C \times D) = \wp(A/C) \cdot \wp(B/D).$

In the special case in which A, B, C and D are finite, the cross product principle follows from the frequency principle because

$$\begin{aligned}
\mathrm{freq}[A \times B / C \times D] &= \#(A \times B)/\#(C \times D) \\
&= (\#A \cdot \#B)/(\#C \cdot \#D) = (\#A \,/\, \#C) \cdot (\#B \,/\, \#D) \\
&= \mathrm{freq}[A/C] \cdot \mathrm{freq}[B/D].
\end{aligned}$$

(7.1) is a generalization of this to infinite sets. It seems clearly true, but it is not a consequence of the classical probability calculus. It follows from the axioms we have already adopted that $\wp(A/C) = \wp(A \cap C / C)$, so the cross product principle can be simplified to read:

(7.2)   If $C \neq \emptyset$ and $D \neq \emptyset$ then $\wp(A{\times}B/C{\times}D) = \wp(A/C){\cdot}\wp(B/D)$.

I have found that experienced probabilists tend to get confused at this point and raise two sorts of spurious objections to the cross product principle. The first is that it is not a new principle—"It can be found in every text on probability theory under the heading of 'product spaces'." That is quite true, but irrelevant to the point I am making. My point is simply that this is a true principle regarding proportions that is not a theorem of the classical probability calculus. The second spurious objection acknowledges that the cross product principle is not a theorem of the probability calculus but goes on to insist that it should not be, because it is false. It is "explained" that the cross product principle does not hold in general because it assumes the statistical independence of the members of C and D. This objection is based upon a confusion, and it is important to get clear on this confusion because it will affect one's entire understanding of the theory of proportions. The confusion consists of not distinguishing between probabilities and proportions. These are two quite different things. What the probabilist is thinking is that we should not endorse the following principle regarding *probabilities*:

$$\mathrm{prob}(Ax\&By/Cx\&Dy) = \mathrm{prob}(Ax/Cx){\cdot}\mathrm{prob}(Bx/Dx)$$

because the C's and the D's need not be independent of one another. This is quite right, but it pertains to probabilities—not proportions. The cross product principle for proportions does not imply the above principle regarding probabilities.[10] Proportions

---

[10] Examining the definition of nomic probability reveals that what would be required to get the cross product principle for probabilities is not (7.2), but rather

$$\wp(\{\langle w,x,y \rangle \,|\, Awx \ \& \ Bwy\} \,/\, \{\langle w,x,y \rangle \,|\, Cwy \ \& \ Dwy\})$$
$$= \wp(\{\langle w,x \rangle \,|\, Awx\} \,/\, \{\langle w,x \rangle \,|\, Cwy\}){\cdot}\wp(\{\langle w,y \rangle \,|\, Bwy\} \,/\, \{\langle w,y \rangle \,|\, Dwy\}),$$

but this principle has nothing to recommend it.

are simply relative measures of the sizes of sets. What the cross product principle tells us is that the relative measure of $A \times B$ is the product of the relative measures of A and B, and this principle is undeniable. As noted above, when A, B, C, and D are finite this principle is an immediate consequence of the fact that if A has n members and B has m members then $A \times B$ has $n \cdot m$ members. Talk of independence makes no sense when we are talking about proportions.

Proportions are "relative measures". They only measure the size of a set relative to a larger set. But we can also use them to make absolute comparisons. We can compare any two sets X and Y by comparing $\wp(X/X \cup Y)$ and $\wp(Y/X \cup Y)$. X and Y "are the same size" iff $\wp(X/X \cup Y) = \wp(Y/X \cup Y)$. Let us abbreviate this as $\ulcorner X \eqsim Y \urcorner$. The Boolean axioms imply:

(7.3)   If $A \eqsim C$ & $B \eqsim D$ & $A \subseteq B$ & $C \subseteq D$ then $\wp(A/B) = \wp(C/D)$.

There are a number of simple set-theoretic operations that should not alter the size of a set. These are described by "invariance principles". One simple invariance principle results from observing that the definition of 'ordered triple' is to a certain extent arbitrary. For the sake of specificity, I will define:

$$\langle x,y,z \rangle = \langle x, \langle y,z \rangle \rangle.$$

(More generally, I will define ordered $(n+1)$-tuples in terms of ordered n-tuples as follows: $\langle x_1, \ldots, x_{n+1} \rangle = \langle x_1, \langle x_2, \ldots, x_{n+1} \rangle \rangle$.) But we could just as well have defined $\langle x,y,z \rangle$ to be $\langle \langle x,y \rangle, z \rangle$. Which definition we adopt should not make any difference. Accordingly, we should have the following *Associative Principle*:

(7.4)   If R is a set of ordered triples then $R \eqsim \{ \langle \langle x,y \rangle, z \rangle \mid \langle x, \langle y,z \rangle \rangle \in R \}$.

I will take this as an axiom. It immediately implies:

(7.5)   If S is a set of ordered triples and $R \subseteq S$ then
$$\wp\big(\{\langle\langle x,y\rangle,z\rangle \mid \langle x,\langle y,z\rangle\rangle \in R\} / \{\langle\langle x,y\rangle,z\rangle \mid \langle x,\langle y,z\rangle\rangle \in S\}\big) =$$
$$\wp(R/S).$$

We can derive a number of important principles from the cross product and associative principles. Let us define the *concatenation* of an ordered n-tuple and an ordered m-tuple as follows:

(7.6) $\langle x_1,\dots,x_n\rangle \,^\frown\, \langle y_1,\dots,y_m\rangle = \langle x_1,\dots,x_n,y_1,\dots,y_m\rangle.$

An n-tuple is also an m-tuple for every $m < n$. In constructing concatenations it is to be understood that an ordered set is always to be treated as an n-tuple for the largest possible n. It is also convenient to define the notion of the concatenation of a finite ordered set with an object that is not a finite ordered set:

(7.7)   If y is not a finite ordered set then $\langle x_1,\dots,x_n\rangle \,^\frown\, y = \langle x_1,\dots,x_n,y\rangle$ and $y \,^\frown\, \langle x_1,\dots,x_n\rangle = \langle y,x_1,\dots,x_n\rangle.$

The concatenation of two relations or of a relation and a nonrelational set (a set containing no finite ordered sets) is the set of all concatenations of their members:

(7.8)   If either R and S are both relations or one is a relation and the other is a nonrelational set then $R \,^\frown\, S = \{x \,^\frown\, y \mid x \in R \,\&\, y \in S\}.$

The cross product principle (7.2) and associative principle (7.5) imply the *Concatenation Principle*:

(7.9)   If $A \subseteq R$, $B \subseteq S$, $R \neq \emptyset$, $S \neq \emptyset$, and either R and S are both relations or one is a relation and the other is a nonrelational set then $\wp\big(A \,^\frown\, B \,/\, R \,^\frown\, S\big) = \wp(A/R) \cdot \wp(B/S).$

A second invariance principle that will be adopted as an axiom is:

(7.10) If K is a set of sets and b is not in any member of K then
$K \doteq \{X \cup \{b\} \mid X \in K\}$.

For lack of a better name I will refer to (7.10) as *The Addition Principle*. Numerous other invariance principles suggest themselves, but I have not found any general account that seems to capture them all. To that extent the present axiomatization of proportions is incomplete.

Most of the non-Boolean principles in the theory of proportions make essential reference to relations. This makes it convenient to adopt the use of variables in expressing proportions. Variables serve as a bookkeeping device enabling us to keep track of how the argument places in different relations are being connected in a given proportion. Let us arbitrarily choose some ordering of the variables, to be called the 'alphabetical ordering', and then define:

(7.11) If φ and θ are open formulas containing the free variables $x_1,\ldots,x_n$ (ordered alphabetically) then $\wp(\varphi/\theta) = \wp\big(\{\langle x_1,\ldots,x_n\rangle \mid \varphi\} / \{\langle x_1,\ldots,x_n\rangle \mid \theta\}\big).$

Thus p becomes a variable-binding operator, binding the variables free in φ and θ. (7.11) requires that the same variables occur free in both the antecedent and consequent of a proportion, but it is often convenient to allow fewer free variables in the consequent than in the antecedent. That can be accomplished as follows:

(7.12) If fewer variables occur free in φ than in θ then $\wp(\varphi/\theta) = \wp(\varphi \& \theta/\theta).$

Using our variable notation, we obtain the following theorem from (7.10):

(7.13) If $b \in B$ then $\wp\big(b \in X / X \subseteq B\big) = 1/2.$

Proof: Let $K = \{X \mid X \subseteq B \ \& \ b \notin X\}$ and $K_b = \{X \mid X \subseteq B \ \& \ b \in X\}$. By

(7.10), $K \doteq K_b$, and K and K_b are disjoint, so $\wp(K/K\cup K_b) = \wp(K_b/K\cup K_b) = 1/2$. Let $\wp(B)$ be the power set of B. Then $\wp(B) = K\cup K_b$. $\wp(b\in X / X \subseteq B) = \wp(X\in K_b / X\in\wp(B)) = \wp(K_b/K\cup K_b) = 1/2$.

Using the multiplicative axiom we can obtain the following generalization of (7.13), which will be of use later:

(7.14) If B is infinite, $b_1,\ldots,b_n\in B$ and $b_1,\ldots,b_n$ are distinct, $C \subseteq B$ and $b_1,\ldots,b_n\notin C$, then $\wp\big(b_1,\ldots,b_n\in X / C \subseteq X \subseteq B\ \&\ X$ infinite$\big) = 1/2^n$.

8. Non-Boolean Principles of Nomic Probability

There are some important principles regarding nomic probabilities that are not forthcoming from the Boolean theory of proportions but are consequences of the non-Boolean theory. The concatenation principle has the following consequence for nomic probability:

(8.1)   If $\diamond_p(\exists x)(\exists y)(Bx\&Cy)$ and $\Box_p(\forall y)(Cy \supset \Box_p Cy)$ then
prob$(Ax/Bx\&Cy) =$ prob(A/B).

Proof: Assume the antecedent, and let **W** be the set of all physically possible worlds. Let us symbolize $\ulcorner x$ has the property A at $w\urcorner$ by $\ulcorner A_w x\urcorner$. Then

prob$(Ax/Bx\&Cy) =$

$$\wp\left[\frac{\{\langle w,\langle x,y\rangle\rangle \mid\ w\in W\ \&\ A_w x\ \&\ B_w x\ \&\ C_w y\}}{\{\langle w,\langle x,y\rangle\rangle\mid\ w\in W\ \&\ B_w x\ \&\ C_w y\}}\right]$$

By the associative principle (7.5) and the concatenation principle (7.9):

$$\wp\left[\frac{\{\langle w,\langle x,y\rangle\rangle\mid\ w\in W\ \&\ A_w x\ \&\ B_w x\ \&\ C_w y\}}{\{\langle w,\langle x,y\rangle\rangle\mid\ w\in W\ \&\ B_w x\ \&\ C_w y\}}\right]$$

$$= \wp\left[\frac{\{\langle w,\langle x,y\rangle\rangle\mid\ w\in W\ \&\ A_w x\ \&\ B_w x\ \&\ \diamond_p Cy\}}{\{\langle w,\langle x,y\rangle\rangle\mid\ w\in W\ \&\ B_w x\ \&\ \diamond_p Cy\}}\right]$$

$$= \wp\left[\frac{\{\langle\langle w,x\rangle,y\rangle \mid w\in W \;\&\; A_w x \;\&\; B_w x \;\&\; \diamond_p Cy\}}{\{\langle\langle w,x\rangle,y\rangle \mid w\in W \;\&\; B_w x \;\&\; \diamond_p Cy\}}\right]$$

$$= \wp\left[\frac{\{\langle w,x\rangle \mid w\in W \;\&\; A_w x \;\&\; B_w x\}^\frown\{y\mid \diamond_p Cy\}}{\{\langle w,x\rangle \mid w\in W \;\&\; B_w x\}^\frown\{y\mid \diamond_p Cy\}}\right]$$

$$= \wp(A\cap B/B) = \mathrm{prob}(A/B).$$

An important consequence of (8.1) is a principle concerning the role of individuals in nomic probability:

(8.2)   If $\diamond_p(\exists x)Rxb$ then $\mathrm{prob}\big(Axy \;/\; Rxy \;\&\; y=b\big)$
        $= \mathrm{prob}(Axb/Rxb)$.

Proof: By (6.16) and (6.17), $\mathrm{prob}(Axy \;/\; Rxy \;\&\; y=b) = \mathrm{prob}(Axb \;/\; Rxb \;\&\; y=b)$. If $\diamond_p(\exists x)Rxb$ then $\diamond_p(\exists x)(\exists y)(Rxb \;\&\; y=b)$, and $\Box_p(\forall y)(y=b \supset \Box_p y=b)$, so by (8.1), $\mathrm{prob}(Axb \;/\; Rxb \;\&\; y=b) = \mathrm{prob}(Axb/Rxb)$.

The antecedent of (8.2) can be deleted to yield:

(*IND*) $\mathrm{prob}(Axy \;/\; Rxy \;\&\; y=b) = \mathrm{prob}(Axb/Rxb)$.

Proof: If $\sim\diamond(\exists x)Rxb$ then by (6.9), $\mathrm{prob}(Axy \;/\; Rxy \;\&\; y=b) = \mathrm{prob}(Axb \;/\; Rxb \;\&\; y=b) = 1$. If $\diamond(\exists x)Rxb$ then it follows from (8.2) and (6.19) that $\mathrm{prob}(Axy \;/\; Rxy \;\&\; y=b) = \mathrm{prob}(Axb/Rxb)$.

I have only stated (*IND*) for binary relations, but it holds in general for relations of arbitrarily many places.[11]

Suppose φ and θ are open formulas whose free variables are $x_1,\ldots,x_n$. It is not usually the case that $\mathrm{prob}(\varphi/\theta) = \wp(\varphi/\theta)$. By definition,

---

[11] In Pollock [1983], I erroneously assumed the false principle $\ulcorner\mathrm{prob}(Ax/Rxy) = \mathrm{prob}(Ax \;/\; (\exists y)Rxy)\urcorner$. Fortunately, most of the purposes for which I used that principle can be served equally by (8.1) and (*IND*).

$$\wp(\varphi/\theta) = \wp(\{\langle x_1, \ldots, x_n \rangle \mid \varphi\} \,/\, \{\langle x_1, \ldots, x_n \rangle \mid \theta\}).$$

The latter is the proportion of θ's that are φ's in the actual world, whereas prob(φ/θ) is the proportion of θ's that are φ's in all physically possible worlds. For example, if $\{\langle x_1, \ldots, x_n \rangle \mid \theta\}$ is finite, then it follows by the frequency principle that $\wp(\varphi/\theta) = \text{freq}[\varphi/\theta]$, but it is not generally true that prob$(\varphi/\theta) = \text{freq}[\varphi/\theta]$ even when the latter frequency exists. There is one important class of cases, however, in which proportions and nomic probabilities agree:

(8.3)   If $\diamond_p \exists \theta$ and $\square_p \forall(\varphi \supset \square_p\varphi)$ and $\square_p \forall(\theta \supset \square_p\theta)$ then $\wp(\varphi/\theta) = \text{prob}(\varphi/\theta)$.

Proof: Suppose $\diamond_p \exists \theta$ and $\square_p \forall(\varphi \supset \square_p\varphi)$ and $\square_p \forall(\theta \supset \square_p\theta)$. Then prob$(\varphi/\theta) =$

$$\wp\left[\frac{\{\langle w, x_1, \ldots, x_n \rangle \mid w \text{ is a physically possible world } \& \langle x_1, \ldots, x_n \rangle \text{ satisfies } \varphi \text{ at } w\}}{\{\langle w, x_1, \ldots, x_n \rangle \mid w \text{ is a physically possible world } \& \langle x_1, \ldots, x_n \rangle \text{ satisfies } \theta \text{ at } w\}}\right]$$

As $\square_p \forall(\varphi \supset \square_p\varphi)$ and $\square_p \forall(\theta \supset \square_p\theta)$, it follows that if $\langle x_1, \ldots, x_n \rangle$ satisfies $(\varphi \& \theta)$ at one physically possible world then $\langle x_1, \ldots, x_n \rangle$ satisfies $(\varphi \& \theta)$ at every physically possible world, and hence if $\mathbf{W}$ is the set of all physically possible worlds and α is the actual world then

$\{\langle w, x_1, \ldots, x_n \rangle \mid w \text{ is a physically possible world } \& \langle x_1, \ldots, x_n \rangle \text{ satisfies } (\varphi \& \theta) \text{ at } w\} = \mathbf{W}^\wedge \{\langle x_1, \ldots, x_n \rangle \mid \langle x_1, \ldots, x_n \rangle \text{ satisfies } \varphi \text{ at } \alpha\}.$

Similarly,

$\{\langle w, x_1, \ldots, x_n \rangle \mid w \text{ is a physically possible world } \& \langle x_1, \ldots, x_n \rangle \text{ satisfies } \theta \text{ at } w\} = \mathbf{W}^\wedge \{\langle x_1, \ldots, x_n \rangle \mid \langle x_1, \ldots, x_n \rangle \text{ satisfies } \theta \text{ at } \alpha\}.$

Hence

$$\text{prob}(\varphi/\theta) = \wp\left[\frac{\mathbf{W}^\wedge \{\langle x_1, \ldots, x_n \rangle \mid \langle x_1, \ldots, x_n \rangle \text{ satisfies } (\varphi \& \theta) \text{ at } \alpha\}}{\mathbf{W}^\wedge \{\langle x_1, \ldots, x_n \rangle \mid \langle x_1, \ldots, x_n \rangle \text{ satisfies } \theta \text{ at } \alpha\}}\right]$$

and hence by the concatenation principle:

$\text{prob}(\varphi/\theta)$

$$= \wp \left[\frac{\{\langle x_1,\ldots,x_n \rangle \text{ satisfies } (\varphi \& \theta) \text{ at } \alpha\}}{\{\langle x_1,\ldots,x_n \rangle \text{ satisfies } \theta \text{ at } \alpha\}} \right]$$

$$= \wp(\varphi/\theta).$$

A special case of theorem (8.3) is of particular interest. I have explained nomic probability in terms of proportions, thus introducing a new primitive concept into the theory. But the next theorem shows that proportions constitute nothing but a special case of nomic probability:

(8.4) $\wp(X/Y) = \text{prob}(x \in X/x \in Y)$.

Proof: I assume that sets have their members essentially, i.e., $\Box(\forall x)(x \in X \supset \Box x \in X)$.[12] Consequently, if $X = \emptyset$ then $\sim \Diamond(\exists x) x \in X$. Then by (6.9), $\text{prob}(x \in X/x \in Y) = 1$, and by (5.5), $\wp(X/Y) = 1$. Suppose instead that $Y \neq \emptyset$. In that case, $(\exists x) x \in Y$, so $\Diamond_p (\exists x) x \in Y$. Then by (8.3), $\text{prob}(x \in X / x \in Y) = \wp(x \in X / x \in Y)$. But by definition, $\wp(x \in X / x \in Y) = \wp(\{x| \ x \in X\} / \{x| \ x \in Y\}) = \wp(X/Y)$.

In light of this theorem, we can regard our axioms about proportions as being axioms about nomic probability. What we have done is to reduce all nomic probabilities to the special case of nomic probabilities concerning properties of the form $\lceil x \in Y \rceil$.

9. Exotic Principles

Although the non-Boolean theories of proportions and nomic probabilities contain some principles that are novel, they do not seem at all problematic. Accordingly, their status seems as secure as that of the Boolean principles. There remain some important principles of nomic probability that cannot be derived from the theory of proportions in its present form. Stronger principles can be added to the theory of proportions that will enable the

---

[12] For a defense of this, see Pollock [1984], Chapter 3.

derivation of these principles of nomic probability, but there is reason to be at least somewhat suspicious of the stronger principles regarding proportions. The difficulty is not that the principles are unintuitive. Each by itself is quite intuitive. The problem is that there are a number of principles that are all intuitive, but they are jointly inconsistent, and it is hard to choose between them.

If we look at the principles of nomic probability that we would like to derive from a strengthened theory of proportions, the former seem more secure than the latter. I will call these *exotic principles* of nomic probability and proportions respectively. The exotic principles of nomic probability can be derived from exotic principles of proportions, but I am inclined to regard the derivations as having more of an explanatory role than a justificatory role. The derivations explain why, within the general theory of probabilities as proportions, the exotic principles of nomic probability turn out to be true. But the explanations would not be convincing if we were not already convinced that the principles are true. This is because, as far as I can see, the only way to choose between different mutually inconsistent candidates for exotic principles of proportions is by choosing those that allow us to prove what we want to prove about nomic probabilities.

In this section I state the exotic principles of nomic probability that will be presupposed by the rest of the theory of nomic probability. I postpone the discussion of the exotic theory of proportions until Chapter 7. This is partly because of its mathematical complexity, and partly because it accomplishes less than the weaker theory of proportions already endorsed.

Two exotic principles will be required. The first relates probabilities and frequencies. Suppose the property G is such that, necessarily, freq$[F/G] = r$ (where r designates the same real number at every physically possible world). prob(F/G) is supposed to be the proportion of physically possible G's that would be F's. As the proportion of actual G's that are F is r in every possible world, it seems that averaging over all physically possible worlds should yield r as the proportion of *all* physically possible G's that would be F's. Consequently, r should also be the value of prob(F/G). Precisely:

(9.1)   If r designates the same real number at every physically
         possible world and $\Diamond_p \exists G$ and $\Box[\exists G \supset \text{freq}[F/G] = r]$ then
         $\text{prob}(F/G) = r$.

In the above reasoning, r need not designate the same real
number at every logically possible world, only at every physically
possible world, because we only average the value of r over
physically possible worlds. Let us call such terms *nomically rigid
designators*.   Notice that terms of the form 'prob(A/B)' are
nomically rigid designators. Principle (9.1) cannot be derived from
our current theory of proportions, because it tells us something
about how proportions among physically possible objects are
computed, namely, by averaging the proportions that hold *within*
different physically possible worlds. Still, it seems quite unproblematic, so I will assume it.

An immediate consequence of (6.20), (9.1), and the "centering principle" (*A4*) for counterfactuals, is:

(*PFREQ*)   If r is a nomically rigid designator of a real number
             and $\Diamond_p[\exists G \ \& \ \text{freq}[F/G] = r]$ then
             $\text{prob}(F \ / \ G \ \& \ \text{freq}[F/G] = r) = r$.

This principle, relating probabilities and frequencies, will play an
important role in the theory of direct inference.

The second exotic principle is the *Principle of Agreement*. To
motivate this principle, let us begin by noting a rather surprising
combinatorial fact (at least, surprising to the uninitiated in
probability theory).   Consider the proportion of members of a
finite set B that are in some subset A of B. Subsets of B need
not exhibit the same proportion of A's, but it is a striking fact of
set theory that subsets of B *tend* to exhibit *approximately* the same
proportion of A's as B, and both the strength of the tendency and
the degree of approximation improve as the size of B increases.
More precisely, where $\ulcorner x \approx_\delta y \urcorner$ means $\ulcorner x$ is approximately equal
to y, the difference being at most $\delta \urcorner$, the following is a theorem
of set theory:

(9.2) For every $\delta, \gamma > 0$, there is an n such that if B is a finite set containing at least n members then

$$\text{freq}\big[\text{freq}[A/X] \approx_\delta \text{freq}[A/B] \ / \ X \subseteq B\big] > 1\text{-}\gamma.$$

It seems inescapable that when B becomes infinite, the proportion of subsets agreeing with B to any given degree of approximation should become 1. This is *The Principle of Agreement*:

(9.3) If B is infinite and $\wp(A/B) = p$ then for every $\delta > 0$:

$$\wp\big(\wp(A/X) \approx_\delta p \ / \ X \subseteq B\big) = 1.$$

This principle seems undeniable. It is simply a generalization of (9.2) to infinite sets. But it cannot be proven within standard probability theory, or within our existing theory of proportions.[13] It is shown in Chapter 7 that (9.3) can be derived from suitably strengthened axioms regarding proportions, but those axioms seem considerably less certain than (9.3) itself. My strategy here is to assume (9.3) without worrying about just why it is true.

The importance of the Principle of Agreement for Propor-

---

[13] There is a strong intuitive connection between the principle of agreement and the Weak Law of Large Numbers of classical probability theory. But although the connection is quite intuitive, it is not a trivial one. I have taken an amazing amount of abuse for touting the importance of this principle. It is gratifying that no one seems to want to deny that it is true, but when presented with this principle a surprising number of probabilists have the reaction that it is nothing new. I repeatedly hear the allegation that it is a trivial consequence of the Weak Law of Large Numbers. One referee for NSF accused me of "reinventing the wheel." I wish these people were right, because that would make it possible to greatly simplify the theory of nomic probability, but they are wrong. In order even to formulate the principle of agreement within the standard framework of mathematical probability theory you need two separate measures—one defined for subsets of B and the other defined for sets of subsets of B. In order to prove the principle of agreement you need some way to connect these two measures with one another, but classical probability theory provides no way of doing that. Some powerful additional assumptions are required, and that is just what the exotic theory of proportions is all about.

tions will be that it implies a Principle of Agreement for Nomic Probabilities. Let us define:

(9.4) H is a *subproperty of* G (relative to the actual world) iff H is a property such that $\diamond_p \exists H$ & $\square_p \forall (H \supset G)$.

(9.5) H is a *strict subproperty of* G iff (1) H is a subproperty of G and (2) if $\langle w, x_1, \ldots, x_n \rangle \in H$ then w is a physically possible world.

Strict subproperties are subproperties that are restricted to the set **W** of physically possible worlds. In effect, they result from chopping off those parts of properties that pertain to physically impossible worlds. The following is an easy consequence of the Principle of Agreement for Proportions and principle (8.3):

(*AGREE*)   If F and G are properties and there are infinitely many physically possible G's and $\text{prob}(F/G) = p$ (where p is a nomically rigid designator) then for every $\delta > 0$, $\text{prob}\big(\text{prob}(F/X) \approx_\delta p \:/\: X$ is a strict subproperty of $G\big) = 1$.

This is *The Principle of Agreement for Probabilities*. (*AGREE*) is the single most important computational principle of nomic probability. It will lie at the heart of the theory of direct inference, and that in turn will make it fundamental to the theory of statistical and enumerative induction.

10. Summary

1. The purpose of this chapter has been to investigate the computational structure of nomic probability. $\text{prob}(F/G)$ is taken to measure the proportion of physically possible G's that would be F's. Then principles of nomic probability are derived from principles regarding proportions among sets. This provides us with a characterization of the formal structure

of nomic probability, but not an analysis. This is because physical possibility is defined in terms of nomic generalizations and nomic probabilities.

2. The proportion function ρ is characterized by a set of axioms, but that does not characterize it uniquely. If any function satisfies those axioms then infinitely many functions satisfy them. The identity of p is actually determined by its relationship to nomic probability, which is given by theorem (8.4), according to which $\rho(X/Y) = \text{prob}(x \in X/x \in Y)$. This defines the proportion function in terms of nomic probability, but it also means that to give further content to both nomic probability and proportions they must be anchored to the world in some other way. That will be done by the acceptance rules discussed in Chapter 3.

3. The axiomatization of proportions and characterization of nomic probabilities in terms of proportions can be viewed as one big axiom regarding nomic probabilities. It can be formulated as follows:

(*COMP*) Where ρ is the real-valued set function defined by:

$$\rho(X/Y) = \text{prob}(x \in X/x \in Y),$$

ρ satisfies the following axioms:

(a) If X and Y are finite and Y is nonempty then $\rho(X/Y) = \text{freq}[X/Y]$;

(b) If Y is infinite and $\#(X \cap Y) < \#Y$ then $\rho(X/Y) = 0$;

(c) $0 \leq \rho(X/Y) \leq 1$;

(d) If $Y \subseteq X$ then $\rho(X/Y) = 1$;

(e) If $Z \neq \emptyset$ and $Z \cap X \cap Y = \emptyset$ then $\rho(X \cup Y/Z) = \rho(X/Z) + \rho(Y/Z)$;

(f) $\rho(X \cap Y/Z) = \rho(X/Z) \cdot \rho(Y/X \cap Z)$;

(g) If $X \subseteq Y$ and $r < \rho(X/Y)$ and $(\exists Z)[Z \subseteq Y \& \rho(Z/Y) = r]$ then $(\exists Z)[Z \subseteq X \subseteq Y \& \rho(Z/Y) = r]$;

(h) If $C \neq \emptyset$, $D \neq \emptyset$, $A \subseteq C$, $B \subseteq D$, then $\rho(A \times B/C \times D) = \rho(A/C) \cdot \rho(B/D)$;

(i) If R is a set of ordered triples then $R \doteq \{\langle\langle x,y\rangle,z\rangle\rangle \mid \langle x,\langle y,z\rangle\rangle \in R\}$;

(j) If K is a set of sets and b is not in any member of K then $K \doteq \{X \cup \{b\} \mid X \in K\}$;

and prob is related to $\wp$ by the following condition:

If $\Diamond_p \exists G$ then $\text{prob}(F/G) = \wp(\mathbf{F}/\mathbf{G})$; otherwise, where $\mathbf{N}$ is the set of nearest $\Diamond_p \exists G$-worlds to w:

$$\text{prob}_w(F/G) = \int_0^1 \wp(\{x \mid \text{prob}_x(F/G) \geq r\} \,/\, \mathbf{N})dr.$$

4. The theory includes two exotic principles of nomic probability, (*PFREQ*) and (*AGREE*). (*AGREE*) will be the most important formal principle regarding nomic probability.

5. The calculus resulting from these axioms is more powerful than the standard probability calculus, but the basic principles are not different in kind. What we have is still just a calculus of probabilities. It will be rather remarkable, then, that we need no assumptions specifically about probability beyond these axioms and the acceptance rules developed in the next chapter to construct the entire theory of nomic probability and derive principles of direct inference and principles of induction.

6. A profligate ontology of sets of possible objects and possible worlds underlies the constructions of this chapter. I would be happier if I could do all of this with a less extravagant ontology, but at this time I do not see how to accomplish that. It is important to realize, however, that the ontology does not affect the rest of the theory. The ontology only plays a role in getting the mathematical structure right. If we are so inclined we can regard the ontology as nothing but a heuristic model employed in guiding our intuitions toward a correct description of the formal structure of nomic probability.

CHAPTER 3
ACCEPTANCE RULES

1. Introduction

I have urged that nomic probability be analyzed in terms of its conceptual role. The conceptual role analysis of nomic probability has four parts: (1) an account of statistical induction; (2) an account of the computational principles that allow some nomic probabilities to be derived from others; (3) an account of acceptance rules; and (4) an account of direct inference. The purpose of the present chapter is to develop and defend the acceptance rules that will play a central role in the theory of nomic probability. The theories of direct inference and statistical induction will then be derived from the acceptance rules and the computational principles defended in the last chapter. Although some of the computational principles are novel, they still amount to little more than an embellishment of the classical probability calculus. The main philosophical weight of the theory of nomic probability will be borne by the acceptance rules.

A simple acceptance rule will be described and defended in section 2. The epistemological framework presupposed by the rule will be discussed and refined in section 3. Sections 4 and 5 will demonstrate that more powerful rules can be derived from the simple acceptance rule described in section 2.

2. An Acceptance Rule

2.1 The Statistical Syllogism

The philosophical literature contains numerous proposals for probabilistic acceptance rules. For instance, the following "Simple Rule" has had a number of proponents:

Belief in *P* is justified iff *P* is probable.

Note, however, that this rule is formulated in terms of *definite* probabilities. This is true of most candidate acceptance rules. However, nomic probability is an indefinite probability. It would make no sense to propose a rule like the Simple Rule for nomic probability. Nevertheless, there is an obvious candidate for an acceptance rule formulated in terms of nomic probability. This is the *Statistical Syllogism*, whose traditional formulation is something like the following:

> Most A's are B's.
> This is an A.
> _____
> Therefore, this is a B.

It seems clear that we often reason in roughly this way. For instance, on what basis do I believe what I read in the newspaper? Certainly not that everything printed in the newspaper is true. No one believes that. But I do believe that *most* of what is published in the newspaper is true, and that justifies me in believing individual newspaper reports. Similarly, I do not believe that every time a piece of chalk is dropped, it falls to the ground. Various circumstances can prevent that. It might be snatched in midair by a wayward hawk, or suspended in air by Brownian movement, or hoisted aloft by a sudden wind. None of these are at all likely, but they are possible. Consequently, the most of which I can be confident is that chalk, when dropped, will almost always fall to the ground. Nevertheless, when I drop a particular piece of chalk, I expect it to fall to the ground.

⌈Most A's are B's⌉ can have different interpretations. It may mean simply that most actual A's are B's. If I report that most of the apples on my tree are green, I am talking only about the apples that are actually there right now. Such a 'most' statement can be cashed out in terms of relative frequencies — the frequency of B's among actual A's is high. But 'most' statements need not have this interpretation. Sometimes they are about more than just actual A's. For instance, in asserting that most pieces of chalk will fall to the ground when dropped, I am not confining the scope of 'most' to actual pieces of chalk. If by some statistical fluke it happened that at this instant there was just one piece of chalk in the world, all others having just been used up and no new ones manufactured for the next few seconds, and that single remaining piece of chalk was in the process of being dropped and snatched in midair by a bird, I would not regard that as falsifying my assertion about most pieces of chalk.

At least sometimes, 'most' statements can be cashed out as statements of nomic probability. On that construal, ⌜Most A's are B's⌝ means ⌜prob(B/A) is high⌝. I think this is the most natural construal of my claim about the behavior of pieces of chalk. This suggests the following acceptance rule, which can be regarded as a more precise version of the statistical syllogism:

> prob(B/A) is high.
> Ac
> _____
>
> Therefore, Bc.

It is perhaps more common for ⌜Most A's are B's⌝ to mean that most actual A's are B's, that is, that freq[B/A] is high. This suggests an alternative version of the statistical syllogism, which I will call *the material statistical syllogism*:

> freq[B/A] is high.
> Ac
> _____
>
> Therefore, Bc.

It is shown in section 6 that a version of this principle can be derived from the statistical syllogism construed in terms of nomic probabilities. Consequently, these need not be regarded as competing interpretations of the statistical syllogism. They are both correct, but the version that proceeds in terms of nomic probability is the more basic, because the material version can be derived from it.

We must ponder the significance of the 'therefore' in the statistical syllogism. Clearly, the conclusion ⌜Bc⌝ does not follow deductively from the premises. Furthermore, although the premises may often make it reasonable to accept the conclusion, that is not always the case. For instance, I may know that most ravens are black, and Josey is a raven, but I may also know that Josey is an albino raven and hence is not black. The premises of the statistical syllogism can at most create a presumption in favor of the conclusion, and that presumption can be defeated by contrary information. In other words, the inference licensed by this rule must be a *defeasible* inference: The inference is a reasonable one in the absence of conflicting information, but it is possible to have conflicting information in the face of which the inference becomes unreasonable.

Defeasible reasoning has been the subject of much discussion in recent epistemology.[1] As recently as the 1960's it was often supposed that in order for P to be a reason for believing Q, P must logically entail Q. But such a view of reasons leads inexorably to skepticism. Our knowledge of the world cannot be reconstructed in terms of logically conclusive reasons.[2] It is now generally granted in epistemology that there are two kinds of reasons–defeasible and nondefeasible. Nondefeasible reasons are always conclusive–they logically entail the conclusions for which they are reasons. Defeasible reasons are what I have called 'prima facie reasons'.[3] These are reasons providing justification only when unaccompanied by defeaters. For example, something's looking red to me is a reason for me to think that it is red. This is a defeasible reason. Despite the fact that an object looks red to me, I may have reasons for believing that the object is not red. For instance, I may be assured of this by someone I regard as reliable, or I may know that it was orange just a minute ago when I examined it in better light and it is unlikely to have changed color in the interim.

Considerations making it unreasonable to accept the con-

---

[1] This stems mainly from the work of Roderick Chisholm and myself. See Chisholm [1957], [1966], [1977], [1981], and Pollock [1967], [1968], [1971], [1974], [1986]. See also Kyburg [1974] and [1983].

[2] For a defense of this, see Pollock [1974] and [1986].

[3] In Pollock [1974].

clusion of a prima facie reason are *defeaters*. If *P* is a prima facie reason for *Q*, there can be two kinds of defeaters for *P*. *Rebutting defeaters* are reasons for denying *Q* in the face of *P*. To be contrasted with rebutting defeaters are *undercutting defeaters*, which attack the connection between the prima facie reason and its conclusion rather than attacking the conclusion itself.[4] For example, if I know that *x* is illuminated by red lights and such illumination often makes things look red when they are not, then I cannot be justified in believing that *x* is red on the basis of it looking red to me. What I know about the illumination constitutes an undercutting defeater for my prima facie reason. An undercutting defeater for *P* as a prima facie reason for believing *Q* is a reason for denying that *P* would not be true unless Q were true. To illustrate, knowing about the peculiar illumination gives me a reason for denying that *x* would not look red to me unless it were red. $\ulcorner P$ would not be true unless Q were true$\urcorner$ can be written as a conditional $\ulcorner (P \geqslant Q) \urcorner$. It is not clear how this conditional is to be analyzed, but we can remain neutral on that topic for now.[5] Defeaters are defeaters by virtue of being reasons either for $\sim Q$ or for $\sim (P \geqslant Q)$. They may be only defeasible reasons for these propositions, in which case we can have defeater defeaters, and defeater defeater defeaters, and so forth.[6]

---

[4] In Pollock [1974] I called rebutting and undercutting defeaters 'type I' and 'type II' defeaters, respectively.

[5] In Pollock [1986] I suggested that $\ulcorner (P \geqslant Q) \urcorner$ could be analyzed as being equivalent to the counterfactual conditional $\ulcorner (\sim Q > \sim P) \urcorner$, but I am no longer sure that that is correct.

[6] It is interesting that researchers in artificial intelligence became interested in defeasible reasoning at about the same time as philosophers. In AI this led to the current theories of 'nonmonotonic reasoning'. This is represented by the work of Doyle [1979], Etherington [1983], Hanks and McDermott [1985], [1987], Lifschitz [1984], Loui [1987], McCarthy [1980], [1984], McDermott [1982], McDermott and Doyle [1980], Moore [1983], Nute [1986], [1987], [1988], Poole [1985], Reiter [1980], and others. In both philosophy and AI, undercutting defeaters have generally been overlooked, leading to accounts that do not capture the full logical structure of defeasible reasoning.

2.2 The Lottery Paradox and Collective Defeat

It appears that the principle of statistical syllogism should be formulated in terms of prima facie reasons. A plausible initial formulation is as follows:

(2.1)   If $r > .5$ then $\ulcorner Ac$ and $\mathrm{prob}(B/A) \geq r \urcorner$ is a prima facie reason for $\ulcorner Ac \urcorner$, the strength of the reason depending upon the value of r.

Because the reason provided by the statistical syllogism is only a prima facie reason, we automatically avoid a difficulty that has plagued many proposed acceptance rules. This is the lottery paradox.[7] Suppose you hold one ticket in a fair lottery consisting of 1 million tickets, and suppose it is known that one and only one ticket will win. Observing that the probability is only .000001 of a ticket being drawn given that it is a ticket in the lottery, it seems reasonable to accept the conclusion that your ticket will not win. But by the same reasoning, it will be reasonable to conclude for each ticket that it will not win. However, these conclusions conflict jointly with something else we are justified in believing, namely, that some ticket will win. Assuming that we cannot be justified in believing each member of an explicitly contradictory set of propositions, it follows that we are not warranted in believing of each ticket that it will not win.[8] But this is no problem for our rule of statistical syllogism as long as it provides only a prima facie reason. What is happening in the lottery paradox is that that prima facie reason is defeated.

The lottery paradox is a case in which we have prima facie reasons for a number of conclusions but they jointly defeat one another. This illustrates the *principle of collective defeat*. This

---

[7] The lottery paradox is due to Kyburg [1961].

[8] Kyburg [1970] draws a different conclusion, namely, that we can be justified in holding inconsistent sets of beliefs, and that it is not automatically reasonable to adopt the conjunction of beliefs one justifiably holds (i.e., adjunction fails). I am not going to discuss this position here. I will just assume that adjunction is a correct rule of reasoning. For a discussion of Kyburg's position, see my [1986], pp. 105-112.

principle will turn out to be of considerable importance in probability theory, so I will say a bit more about it. Starting from propositions we are justified in believing, we may be able to construct arguments supporting some further propositions. But that does not automatically make those further propositions justified, because some propositions supported in that way may be defeaters for steps of some of the other arguments. That is what happens in cases of collective defeat. Suppose we are warranted in believing some proposition R and we have equally good prima facie reasons for each of $P_1,\ldots,P_n$, where $\{P_1,\ldots,P_n\}$ is a minimal set of propositions deductively inconsistent with R (i.e., it is a set deductively inconsistent with R and has no proper subset that is deductively inconsistent with R). Then for each i, the conjunction $\ulcorner R \,\&\, P_1 \,\&\, \ldots \,\&\, P_{i\text{-}1} \,\&\, P_{i+1} \,\&\, \ldots \,\&\, P_n \urcorner$ entails $\sim P_i$. Thus by combining this entailment with the arguments for R and $P_1,\ldots,P_{i\text{-}1},P_{i+1},\ldots,P_n$ we obtain an argument for $\sim P_i$ that is as good as the argument for P_i. Accordingly, we have equally strong support for both P_i and $\sim P_i$, and hence we could not reasonably believe either on this basis. This holds for each i, so none of the P_i is justified. They collectively defeat one another. The simplest version of the principle of collective defeat can be formulated as follows:

(2.2)   If we are justified in believing R and we have equally good independent prima facie reasons for each member of a minimal set of propositions deductively inconsistent with R, and none of these prima facie reasons is defeated in any other way, then none of the propositions in the set is justified on the basis of these prima facie reasons.

This principle need not be viewed as a primitive epistemic principle. It can be derived from a general theory of the operation of prima facie reasons. This will be explored in section 3.

2.3 Projectibility

It is the principle of collective defeat that allows the principle (2.1) of statistical syllogism to escape the lottery paradox.

But it turns out that the very fact that (2.1) can handle the lottery paradox by appealing to collective defeat shows that it cannot be correct. The difficulty is that every case of high probability can be recast in a form that makes it similar to the lottery paradox. We need only assume that $\text{prob}(B/A) < 1$. Pick the smallest integer n such that $\text{prob}(B/A) < 1 - 1/2^n$. Now consider n fair coins, unrelated to each other and unrelated to c's being A or B. Let T_i be ⌜is a toss of coin i⌝ and let H be ⌜is a toss that lands heads⌝. There are 2^n *Boolean conjunctions* of the form ⌜$(\sim)Hx_1$ & ... & $(\sim)Hx_n$⌝ where each tilde in parentheses can be either present or absent. For each Boolean conjunction $\beta_j x_1,\ldots,x_n$,

$$\text{prob}(\beta_j x_1,\ldots,x_n / T_1 x_1 \& \ldots \& T_n x_n) = 2^{-n}.$$

Consequently, because the coins were chosen to be unrelated to A and B,

$$\text{prob}(\sim\beta_j x_1,\ldots,x_n / Ax \& T_1 x_1 \& \ldots \& T_n x_n) = 1 - 2^{-n}.$$

By the probability calculus, a disjunction is at least as probable as its disjuncts, so

$$\text{prob}(\sim Bx \vee \sim\beta_j x_1,\ldots,x_n / Ax \& T_1 x_1 \& \ldots \& T_n x_n)$$
$$\geq 1 - 2^{-n} > \text{prob}(Bx/Ax).$$

Let $t_1,\ldots,t_n$ be a sequence consisting of one toss of each coin. As we know ⌜$Ac \& T_1 t_1 \& \ldots \& T_n t_n$⌝, (2.1) gives us a prima facie reason for believing each disjunction of the form

$$\sim Bc \vee \sim\beta_j t_1,\ldots,t_n.$$

By the propositional calculus, the set of all these disjunctions is equivalent to, and hence entails, $\sim Bc$. Thus we can construct an argument for $\sim Bc$ in which the only defeasible steps involve the use of (2.1) in connection with probabilities at least as great as that used in defending Bc. Hence, we have a situation formally identical to the lottery paradox. Therefore, the principle of collective defeat has the consequence that if prob(B/A) has any probability less than 1, we cannot use statistical syllogism to draw any justified conclusion from this high probability.[9]

At this point it is tempting to conclude that the traditional view is just wrong and that high nomic probability never gives us a reason for believing anything. Perhaps all we are justified in believing on probabilistic grounds is that various propositions are *probable*. But it seems to me that there are clear cases in which we must be justified in believing propositions on probabilistic grounds. A great deal of what we believe and regard ourselves as justified in believing comes to us through the testimony of others. We do not regard such testimony as infallible. That is, we do not believe the contents of such testimony on the basis of inductive generalizations of the form ⌜All testimony of such-and-such a kind is true⌝. Rather, we believe that *most* testimony of certain kinds is true. This seems an unmistakable case in which we reason from ⌜Most B's are A's⌝ to ⌜This B is an A⌝. Such reasoning must sometimes be permissible. Thus we are led to the conclusion that ⌜Bc & prob(A/B) ≥ r⌝ is sometimes a prima facie reason for ⌜Ac⌝, but not always. There must be some restrictions on its applicability. Discovering the nature of those restrictions is the central problem for constructing a correct acceptance rule.

A clue to the nature of the desired constraint emerges from thinking about the difference between the lottery case, in which we want (2.1) to be applicable, and the contrived disjunctions employed above to show that instances of (2.1) are always subject to collective defeat. If we examine the behavior of disjunctions in probabilities, we find that they create general difficulties for the application of (2.1). For instance, it is a theorem of the probability calculus that prob($F/G \lor H$) ≥ prob(F/G)·prob($G/G \lor H$). Consequently, if prob(F/G) and prob($G/G \lor H$) are sufficiently large, it follows that prob($F/G \lor H$) ≥ r. To illustrate, because the vast majority of birds can fly and because there are many more birds than giant sea tortoises, it follows that most things that are

---

[9] I was led to this construction by thinking about Keith Lehrer's racehorse paradox, in Lehrer [1981].

either birds or giant sea tortoises can fly. If Herman is a giant sea tortoise, (2.1) would give us a reason for thinking that Herman can fly, but notice that this is based simply on the fact that most birds can fly, which should be irrelevant to whether Herman can fly. This indicates that arbitrary disjunctions cannot be substituted for B in (2.1).

Nor can arbitrary disjunctions be substituted for A in (2.1). By the probability calculus, $\text{prob}(F \vee G/H) \geq \text{prob}(F/H)$. Therefore, if $\text{prob}(F/H)$ is high, so is $\text{prob}(F \vee G/H)$. Thus, because most birds can fly, it is also true that most birds can either fly or swim the English Channel. By (2.1), this should be a reason for thinking that a starling with a broken wing can swim the English Channel, but obviously it is not.

There must be restrictions on the properties A and B in (2.1). To have a convenient label, let us say that B is *projectible with respect to A* iff (2.1) holds. What we have seen is that projectibility is not closed under disjunction; that is, neither of the following hold:

> If C is projectible with respect to both A and B, then C is projectible with respect to $(A \vee B)$.

> If A and B are both projectible with respect to C, then $(A \vee B)$ is projectible with respect to C.

On the other hand, it seems fairly clear that projectibility is closed under conjunction. More precisely, the following two principles hold:

(2.3)   If A and B are projectible with respect to C, then $(A \& B)$ is projectible with respect to C.

(2.4)   If A is projectible with respect to both B and C, then A is projectible with respect to $(B \& C)$.

We can give what amounts to a proof of (2.3). Suppose A and B are projectible with respect to C, and suppose $\ulcorner Cc \ \& \ \text{prob}(A \& B/C) \geq r \urcorner$. It then follows by the probability calculus that $\text{prob}(A/C)$

$\geq r$ and prob(B/C) $\geq r$. Hence we have prima facie reasons for $\ulcorner Ac \urcorner$ and $\ulcorner Bc \urcorner$, and can then infer $\ulcorner Ac \ \& \ Bc \urcorner$ deductively. Therefore, inferences resulting from (2.3) are reasonable.

We cannot establish (2.4) so conclusively, but consideration of actual scientific methodology provides a fairly compelling reason for thinking that (2.4) must be true. Suppose A is projectible with respect to B. Suppose we want to know whether $\ulcorner Ac \urcorner$ is true, and we know that $\ulcorner Bc \urcorner$ is true. But suppose prob(A/B) is not large enough to warrant an inference in accordance with (2.1). Then we will typically look for other properties of c that we can conjoin with B to raise the probability. In this way we may find a conjunction $B_1 \& \ldots \& B_n$ of properties possessed by c such that prob($A/B_1 \& \ldots \& B_n$) $\geq r$, and then we will infer $\ulcorner Ac \urcorner$. This can only be legitimate if (2.4) is true.

It seems that the best we can do in formulating the principle of statistical syllogism is to build in an explicit projectibility constraint:

($A1$)   If F is projectible with respect to G and $r > .5$, then $\ulcorner Gc$ & prob(F/G) $\geq r \urcorner$ is a prima facie reason for believing $\ulcorner Fc \urcorner$, the strength of the reason depending upon the value of r.

Of course, if we define projectibility in terms of (2.1), ($A1$) becomes a mere tautology, but the intended interpretation of ($A1$) is that *there is* a relation of projectibility between properties, holding in important cases, such that $\ulcorner Gc$ & prob(F/G) $\geq r \urcorner$ is a prima facie reason for $\ulcorner Fc \urcorner$ when F is projectible with respect to G. To have a fully adequate theory we must augment ($A1$) with an account of projectibility, but that proves very difficult and I have no account to propose. At best, our conclusions about closure conditions provide a partial account. Because projectibility is closed under conjunction but not under disjunction, it follows that it is not closed under negation. Similar considerations establish that it is not closed under the formation of conditionals or biconditionals. It is not clear how it behaves with quantifiers. Although projectibility is not closed under negation, it seems likely that negations of "simple" projectible properties are projectible.

For instance, both 'red' and 'nonred' are projectible with respect to 'robin'. A reasonable hypothesis is that there is a large class **P** containing most properties that are "logically simple", and whenever A and B are conjunctions of members of **P** and negations of members of **P**, A is projectible with respect to B. This is at best a sufficient condition for projectibility, however, because we will find numerous cases of projectibility involving properties that are logically more complex than this.

2.4 Defeaters

The reason provided by $(A1)$ is only a prima facie reason, and as such it is defeasible. As with any prima facie reason, it can be defeated by having a reason for denying the conclusion. The reason for denying the conclusion constitutes a rebutting defeater. But there is also an important kind of undercutting defeater for $(A1)$. In $(A1)$, we infer the truth of $\ulcorner Fc \urcorner$ on the basis of probabilities conditional on a limited set of facts about c (i.e., the facts expressed by $\ulcorner Gc \urcorner$). But if we know additional facts about c that alter the probability, that defeats the prima facie reason:

$(D1)$ If F is projectible with respect to H then $\ulcorner Hc$ & $\mathrm{prob}(F/G\&H) \neq \mathrm{prob}(F/G) \urcorner$ is an undercutting defeater for $(A1)$.

I will refer to these as *subproperty defeaters*. $(D1)$ amounts to a kind of "total evidence requirement". It requires us to make our inference on the basis of the most comprehensive facts regarding which we know the requisite probabilities. Notice that according to $(D1)$, the use of $(A1)$ is defeated even if $\mathrm{prob}(F/G\&H)$ is still high. That might seem unreasonable. But this does not mean that you cannot use $(A1)$ to infer $\ulcorner Fc \urcorner$ – it just means that you cannot infer $\ulcorner Fc \urcorner$ from $\ulcorner Gc$ & $\mathrm{prob}(F/G) \geq r \urcorner$. If $\mathrm{prob}(F/G\&H)$ is also high but different from $\mathrm{prob}(F/G)$, then you can infer $\ulcorner Fc \urcorner$ from $\ulcorner Gc$ & Hc & $\mathrm{prob}(F/G\&H) \geq r \urcorner$.[10]

---

[10] I first proposed this account of defeat in Pollock [1983]. Similar

It will turn out that the defeater (*D1*) is itself defeasible, but the complications that involves are best postponed until we have a better understanding of the logic of defeasibility. That will be taken up in section 3, and then the matter of acceptance rules is reopened in section 4.

3. The Epistemological Framework

3.1 Justification and Warrant

A distinction can be made between several different concepts of epistemic justification. The basic concept is that of justification *simpliciter*—a person is justified in believing *P* just in case he has adequate reason to believe *P* and he does not have any defeaters at his immediate disposal. This of course is not a definition—only a rough characterization whose primary purpose is to distinguish between justification simpliciter and the other concepts I will now define. A person might be justified in believing *P* even though further reasoning is in principle available to him that would lead to the discovery of a defeater for *P*. Justification simpliciter requires only that the availability of such reasoning not be obvious. This suggests a stronger concept of justification—a person is *objectively justified* in believing *P* iff he is justified in believing *P* and his justification could not be defeated "in the long run" by any amount of reasoning proceeding exclusively from other propositions he is justified in believing. This formulation is intended to allow that the justification could be defeated "in the short run" but then reinstated by further reasoning. A third concept is that of a "justifiable" proposition, or as I will say, a *warranted* proposition. *P* is warranted for *S* iff *S* could become objectively justified in believing *P* through reasoning proceeding exclusively from the propositions he is objectively justified in believing. A warranted proposition is one that *S* would become justified in believing if he were an "ideal reasoner".

---

proposals were made in AI by Touretzky [1984], Nute [1988], Poole [1985], and Loui [1987].

These three concepts of epistemic justification have importantly different logical properties. Let us say that a proposition P is a *deductive consequence* of a set Γ of propositions, and Γ *implies* P, iff there exists a deductive argument leading from members of Γ to the conclusion P. I will say that a set of propositions is *deductively consistent* iff it does not have an explicit contradiction as a deductive consequence. P is a *logical consequence* of Γ iff, necessarily, P is true if all the members of Γ are true (i.e., Γ entails P). Logical consequence and deductive consequence may or may not coincide, depending upon whether all necessary truths are demonstrable (i.e., provable *a priori*). If they are not, there will be logical consequences that are not deductive consequences, and deductively consistent sets of propositions that are not logically consistent. Only deductive consequences can be relevant to reasoning.

There is no reason to expect a person's set of justified beliefs to be either deductively consistent or closed under deductive consequence. For example, prior to the discovery of the set-theoretic antinomies, people were presumably justified in believing the (demonstrably inconsistent) axiom of comprehension in set theory. Every proposition is a deductive consequence of that axiom, but clearly people were not thereby justified in believing everything.

In contrast to this, a person's set of objectively justified beliefs must be deductively consistent. If a contradiction could be derived from it, then reasoning from some objectively justified beliefs would lead to the denial (and hence defeat) of other objectively justified beliefs, in which case they would not be objectively justified. A person's objectively justified beliefs need not be closed under deductive consequence, however, for the simple reason that a person's beliefs need not be closed under deductive consequence.

The set of propositions warranted for a person, on the other hand, must be both deductively consistent and closed under deductive consequence. The set of warranted propositions is deductively consistent for the same reason the set of objectively justified propositions is deductively consistent. Turning to deductive consequence, suppose $P_1, \ldots, P_n$ are warranted for S

and Q is a deductive consequence of them. Then an argument supporting Q can be constructed by combining arguments for $P_1, \ldots, P_n$ and adding onto the end an argument deducing Q from $P_1, \ldots, P_n$. The last part of the argument consists only of deductive non-defeasible steps of reasoning. If Q is not warranted, it must be possible to reason from S's justified beliefs to some defeater for the argument supporting Q. There can be no defeaters for the final steps, which are nondefeasible, so such a defeater would have to be a defeater for an earlier step. But the earlier steps all occur in the arguments supporting $P_1, \ldots, P_n$, so one of those arguments would have to be defeated, which contradicts the assumption that $P_1, \ldots, P_n$ are warranted. Thus there can be no such defeater, and hence Q is warranted.

Having defended an acceptance rule, I intend to use it to argue that under various circumstances we are warranted in holding various beliefs. In order to do that, we must have a clearer idea of how possibly conflicting prima facie reasons interact to determine what beliefs are warranted. I turn now to constructing an account of this.

3.2 Warrant and Ultimately Undefeated Arguments

Justification proceeds in terms of arguments, the latter being constructed out of reasons. We begin with some propositions we are somehow initially justified in believing, and then we extend and modify this "epistemic basis" through the use of arguments. The traditional view would be that a proposition is warranted for a person iff there is an argument supporting it that proceeds from his epistemic basis. But this traditional view cannot possibly be right. As Gilbert Harman [1980] has pointed out, the traditional view makes it inexplicable how reasoning can lead not only to the adoption of new beliefs but also to the rejection of previously held beliefs. The traditional view of reasoning is simplistic. This is primarily because it overlooks the complexities required to accommodate prima facie reasons. A proposition may fail to be warranted for a person despite its being supported by an argument proceeding from his epistemic basis, because there may be other arguments supporting defeaters for the first argument. These may be defeaters either for individual links of the argument, or for defeasibly justified members of the epistemic basis that are presupposed by the argument. This process is complicated. One argument may be defeated by a second, but then the second argument may be defeated by a third thus reinstating the first, and so on. A proposition is warranted only if it emerges ultimately unscathed from this process.

To make this precise, let us say that an argument σ *defeats* an argument η iff σ supports a defeater for some step of η. Let us also say that a *level 0 argument* is any argument proceeding from a person's epistemic basis.[11] Some level 0 arguments may provide us with defeaters for other level 0 arguments, so let us say that a *level 1 argument* is a level 0 argument not defeated by any level 0 argument. As there are fewer level 1 arguments than level 0 arguments, fewer propositions will be supported by level 1 arguments than level 0 arguments. In particular, fewer defeaters for level 0 arguments will be supported by level 1 arguments than by level 0 arguments. Thus having moved to level 1 arguments, we may have removed some defeaters and thereby reinstated some level 0 arguments. Let us say then that a *level 2 argument* is a level 0 argument not defeated by any level 1 argument. However, level 2 arguments may support some defeaters that were not supported by level 1 arguments, thus defeating some level 0 arguments that were not defeated by level 1 arguments. Thus we take a *level 3 argument* to be any level 0 argument not defeated by level 2 arguments; and so on. In general, a *level k+1 argument* is any level 0 argument not defeated by level k arguments.

A given level 0 argument may be defeated and reinstated many times by this alternating process. Only if we eventually reach a point where it stays undefeated can we say that it warrants its conclusion. Let us say that an argument is *ultimately undefeated* iff there is some m such that the argument is a level n argument for every $n > m$. Epistemological warrant can then be characterized in terms of arguments that are ultimately undefeated:

---

[11] This definition will be modified slightly in Chapter 8.

(3.1) P is warranted for S iff P is supported by some ultimately
         undefeated argument proceeding from S's epistemic basis.

To illustrate, consider the three arguments diagramed in Figure
1. Here β defeats α, and γ defeats β. It is assumed that nothing
defeats γ. Thus γ is ultimately undefeated. Neither α nor β is a
level 1 argument, because both are defeated by level 0 arguments.
As γ is a level n argument for every n, β is defeated at every
level greater than 0, so β is not a level n argument for $n > 0$.
As a result α is reinstated at level 2, and is a level n argument for
every $n > 1$. Hence α is ultimately undefeated, and V is warranted.

       This analysis can be illustrated by considering the principle
(2.2) of collective defeat. It becomes a theorem on this analysis.
The alternating process of defeat and reinstatement renders some
arguments continuously undefeated after a certain point and
renders others continuously defeated after a certain point. But
there will be some arguments whose status remains perpetually in
limbo. This is the explanation of collective defeat. For example,

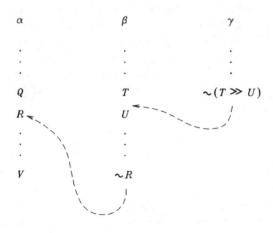

Figure 1. Interacting arguments.

suppose as in (2.2) that some proposition R is warranted and we have equally good prima facie reasons for each member of a minimal set $\{P_1, \ldots, P_n\}$ of propositions deductively inconsistent with R, and none of these prima facie reasons is defeated in any other way. Then there is a level 0 argument for each P_i, and combining the arguments for R and $P_1, \ldots, P_{i-1}, P_{i+1}, \ldots, P_n$ we obtain a level 0 argument for $\sim P_i$. $\sim P_i$ is a rebutting defeater for a step of the argument to the conclusion P_i. Thus for each i, there is a level 0 argument for a defeater for the level 0 argument supporting P_i, and hence the latter argument is not a level 1 argument. But this means that the defeating arguments are not level 1 arguments either, and so all of the arguments get reinstated at level 2. They all get defeated again at level 3, and so on. Hence none of the arguments is ultimately undefeated and so for each i, neither P_i nor its negation is warranted.

3.3 The Structure of Arguments

Thus far little has been said about the form of an argument. The simplest and most natural suggestion is that arguments are simply finite sequences of propositions connected by the 'reason for' relation. More precisely, the proposal would be that an argument is any finite sequence of propositions each of which is either in the epistemic basis or such that earlier propositions in the sequence constitute a reason for it. Despite its initial appeal, this "linear model of arguments" must be rejected. In formal logic, arguments can have a more complicated structure involving subsidiary arguments. For example, an argument for $\ulcorner P \supset Q \urcorner$ may contain a subsidiary argument taking P as a premise and deriving Q. This is what is called 'conditional proof'. Or an argument for $\ulcorner \sim P \urcorner$ may contain a subsidiary argument deriving $\ulcorner Q \ \& \ \sim Q \urcorner$ from P. This is *reductio ad absurdum*. The use of such subsidiary arguments is not unique to formal logic. The same kinds of reasoning occur in epistemic arguments. For example, in reasoning about what would happen if P were true, we might reason as follows (where Q is a reason for believing R):

If P were true then Q, so R.
Therefore, $P > R$.

The natural reconstruction of this argument is:

> | Suppose it were true that *P*.
> |————————————————
> | Then *Q*.
> | So *R*.
> Therefore, *P* > *R*.

The point of this reconstruction is that we make a supposition, then reason linearly using this supposition (in this case, reasoning from *Q* to *R*), and then discharge the supposition to conclude that a conditional is true. The subsidiary arguments that make use of the supposition have the same structure as other arguments. They proceed in terms of reasons and defeaters, and may contain further subsidiary arguments.

Some general features of epistemic warrant emerge from the use of nonlinear arguments. In order to formulate these general principles, we must distinguish between two concepts of a reason. Thus far I have been taking reasons to be the "atomic links" in arguments. If *P* is a reason for *Q*, that is a basic epistemic fact about the concepts involved and is not derivative from anything else. This means that we cannot derive the existence of one reason from the existence of other reasons. For example, if *P* is a reason for *Q* and *Q* is a reason for *R*, it does not follow that *P* is a reason for *R*. As reasons are the atomic links in arguments, the result of stringing them together need no longer be a reason.

It is customary in epistemological contexts to use 'reason' in this narrow sense, but there is a broader concept of a reason according to which, in order for *P* to be a reason for *Q*, all that is required is that there be an argument leading from *P* to *Q*. If *P* is a reason for *Q* in the narrow sense then it is a reason for *Q* in the broad sense (via a two-line argument), but not conversely. Let us call reasons in the narrow sense 'primitive reasons', and I will use the unqualified term 'reason' for reasons in the broad sense. The justification for this choice of terminology is that it is possible to reason about the non-primitive reason relation, giving proofs that one proposition is a reason for another, but that is not possible for the primitive reason relation. For example, although it is not true that if P is a primitive reason for Q and Q is a primitive reason for R then P is a primitive reason for R, it is true that if P is a reason for Q and Q is a reason for R then P is a reason for R. This difference will be important. For example, in Chapter 5, principles of induction will be derived from simpler epistemic principles. It cannot be proven that the inductive evidence constitutes a primitive reason for the inductive generalization (indeed, I doubt that it does). Such a proof is in principle impossible, because primitive reasons cannot be derived from one another. But it will be possible to prove that the inductive evidence constitutes a (non-primitive) reason for the inductive generalization.[12]

Just as primitive reasons can be defeasible or nondefeasible, so can non-primitive reasons. An argument is defeasible iff it involves prima facie reasons, and a non-primitive reason is defeasible iff it is based upon a defeasible argument. It will be convenient to say that P is a (non-primitive) prima facie reason for Q when there is a defeasible argument leading from P to Q. It should be noted that in such a case, the defeaters for P as a prima facie reason for Q may be more complicated than the defeaters for primitive prima facie reasons. We can still talk

---

[12] One rather artificial feature of the concept of a (possibly non-primitive) reason deserves to be mentioned. In order for P to be a reason for Q, all that is required is that there be an argument leading from P to Q. Such an argument could contain defeaters for itself, in which case it is a level 0 argument but not a level 1 argument. In particular, P could be a reason for Q but be automatically defeated by containing a defeater for itself. For example, ⌜That book looks red to me, but there are red lights shining on it and such illumination often makes things look red when they are not⌝ is a (non-primitive) prima facie reason for me to believe that the book is red, but it is automatically defeated. It may be objected that this is an unreasonable concept of a reason, and that we should at least require that reasons not be automatically defeated. The justification for not doing that is is twofold. First, it would make it more difficult to reason about reasons. Second, we lose nothing by adopting the more general concept of a reason, because the other concept is definable in terms of it: P is a not-automatically-defeated reason for Q iff P is a reason for Q and P is not a reason for a defeater for P as a reason for Q.

about rebutting defeaters and undercutting defeaters. Rebutting defeaters attack the conclusion, whereas undercutting defeaters attack the connection between P and Q. But unlike the case of primitive prima facie reasons, undercutting defeaters need not be reasons for $(P \gg Q)$. For example, if we infer Q from R and R from P, then any reason for $\sim R$ will be an undercutting defeater, but it will not usually be a reason for $\sim(P \gg Q)$.

A number of general principles about non-primitive reasons result from the use of nonlinear arguments. Conditional proof for counterfactuals leads immediately to the following principle:

(3.2)   If Q provides a reason for R then $(P > Q)$ provides a reason for $(P > R)$. If S is a defeater for the argument from Q to R, then $(P > S)$ is a defeater for the argument from $(P > Q)$ to $(P > R)$.[13]

Similarly, because we can use conditional proof for material conditionals:

(3.3)   If Q provides a reason for R then $(P \supset Q)$ provides a reason for $(P \supset R)$. If S is a defeater for the argument from Q to R, then $(P \supset S)$ is a defeater for the argument from $(P \supset Q)$ to $(P \supset R)$.

In the same way:

(3.4)   If Q provides a reason for R then $(P \gg Q)$ provides a reason for $(P \gg R)$. If S is a defeater for the argument from Q to R, then $(P \gg S)$ is a defeater for the argument from $(P \gg Q)$ to $(P \gg R)$.

---

[13] Unlike the case of material conditionals, conditional proofs cannot be nested for counterfactuals. That is, we cannot reason to $\ulcorner P > (Q > R))\urcorner$ by first supposing P, then supposing Q, inferring R, discharging Q to conclude $\ulcorner Q > R\urcorner$, and then discharging P to conclude $\ulcorner P > (Q > R))\urcorner$. It can be shown that any conditional allowing such nesting is equivalent to the material conditional. See Chapter 8.

Used in a different way, conditional proof leads immediately to the following:

(3.5) If $(P\&Q)$ provides a reason for R then Q provides a reason for $(P > R)$, and hence also for $(P \supset R)$.

Another important variety of nonlinear reasoning involves the principle of dilemma. Suppose both P and Q are reasons for R, and we have some argument supporting the disjunction $P \lor Q$. Then we might continue the argument as follows:

$P \lor Q$
| Suppose P.
|—————
| Then R.
|═════
| Suppose Q.
|—————
| Then R.
Therefore, R.

Such reasoning gives us the following principle:

(3.6) If P and Q both provide reasons for R, so does $(P \lor Q)$.[14]

An important kind of nonlinear argument in formal logic is universal generalization, and it has an analogue in epistemic arguments. Primitive reasons do not occur individually—they come in schemas. For example, that book's looking red to me gives me a reason for thinking it is red, but it only does so because anything's looking red to me gives me a reason for thinking it is red. Reason schemas can be formulated using free variables: x's looking red to me is a reason for me to think that x is red. The fact that primitive reasons come in schemas is exploited in a kind

___

[14] Note that (3.6) is derivable from (3.3) together with the fact that $\{(P \lor Q), (P \supset R), (Q \supset R)\}$ is a reason for R.

of universal generalization wherein we reason in somewhat the following fashion (where we have the reason schema: $\ulcorner Fx \urcorner$ is a reason for $\ulcorner Gx \urcorner$):

> Suppose every H is an F.
> Let x be some (unspecified) H.
> Then it is an F, so it is a G.
> Therefore, every H is a G.

Just as arguments are constructed out of reasons, *argument schemas* can be constructed out of reason schemas. Let us say that $\ulcorner Fx \urcorner$ is a (possibly non-primitive) *reason schema* for $\ulcorner Gx \urcorner$ just in case there is an argument schema proceeding from $\ulcorner Fx \urcorner$ to $\ulcorner Gx \urcorner$. Then:

(3.7)   If $\ulcorner Fx \urcorner$ is a reason schema for $\ulcorner Gx \urcorner$, then $\ulcorner (\forall x)(Hx \supset Fx) \urcorner$ is a reason for $\ulcorner (\forall x)(Hx \supset Gx) \urcorner$. If $\ulcorner Jx \urcorner$ is a defeater for the argument schema from $\ulcorner Fx \urcorner$ to $\ulcorner Gx \urcorner$, then $\ulcorner (\exists x)(Hx \,\&\, Jx) \urcorner$ is a defeater for $\ulcorner (\forall x)(Hx \supset Fx) \urcorner$ as a reason for $\ulcorner (\forall x)(Hx \supset Gx) \urcorner$.

Two other elementary principles are immediate consequences of the fact that arguments can be constructed by stringing together reasons:

(3.8)   If P is a reason for Q and Q is a reason for R then P is a reason for R.

(3.9)   If P is a reason for Q and P is a reason for R then P is a reason for $(Q\&R)$.

Note that (3.2)–(3.9) all hold for reason schemas as well as for reasons.

In discussing induction we will have to consider the relative strengths of reasons. Principle (3.9) can be strengthened a bit to yield a simple principle of this sort. I have argued that warrant is closed under conjunction. Thus if P is a good enough reason to warrant Q and also a good enough reason to warrant R, then

P is a good enough reason to warrant (*Q&R*). But attributions of warrant are indexical. At different times we require greater or lesser degrees of warrant before we are willing to describe a proposition as warranted *simpliciter*. Thus this relationship must hold for all degrees of warrant:

(3.10)   If *P* is a reason for *Q* and *P* is a reason for *R* then *P* is an equally good reason for (*Q&R*).

A more important principle concerning the strengths of reasons governs those universally quantified reasons of the form of (3.7):

(3.11)   If ⌜*Fx*⌝ is a prima facie reason for ⌜*Gx*⌝, then ⌜(∀*x*)(*Hx* ⊃ *Fx*)⌝ is a prima facie reason for ⌜(∀*x*)(*Hx* ⊃ *Gx*)⌝ and the strength of this prima facie reason is the same as the strength of the prima facie reason ⌜*Fa*⌝ for ⌜*Ga*⌝.

That this must be correct can be seen by considering the case in which it is demonstrably true that there are just finitely many *H*'s, $h_1, \ldots, h_n$. Then we are warranted in believing that ⌜(∀*x*)(*Hx* ⊃ *Fx*)⌝ is materially equivalent to ⌜Fh_1 & ... & Fh_n⌝ and ⌜(∀*x*)(*Hx* ⊃ *Gx*)⌝ is materially equivalent to ⌜Gh_1 & ... & Gh_n⌝. ⌜Fh_1 & ... & Fh_n⌝ is a conclusive reason for each ⌜Fh_i⌝, and the latter is a prima facie reason for ⌜Gh_i⌝, so ⌜Fh_1 & ... & Fh_n⌝ is a prima facie reason of the same strength for ⌜Gh_i⌝. This holds for each *i*, so by (3.10), ⌜Fh_1 & ... & Fh_n⌝ is also a prima facie reason of that same strength for ⌜Gh_1 & ... & Gh_n⌝, and hence ⌜(∀*x*)(*Hx* ⊃ *Fx*)⌝ is a prima facie reason of that same strength for ⌜(∀*x*)(*Hx* ⊃ *Gx*)⌝. This argument works only if there is a fixed finite number of *H*'s, but it does not seem that that should make any difference to the general principle, so I will assume (3.11).

(3.2)–(3.11) can be used to prove some general theorems about reasons. For example, consider a reason schema wherein a proposition *P* is a reason for ⌜*Fx*⌝ (for any '*x*'). As *P* is a proposition, it does not contain a free occurrence of '*x*'. We will encounter an example of this later in the book. An immediate consequence of (3.7) is then:

(3.12) If P is a reason for $\ulcorner Fx \urcorner$ then P is a reason for $\ulcorner (\forall x) Fx \urcorner$.

A second theorem is:

(3.13) If P is a reason for Q then $\ulcorner \Box_p P \urcorner$ is a reason for $\ulcorner \Box_p Q \urcorner$.

Proof: Suppose P is a reason for Q. Then by (3.2), for each possible world w, $\ulcorner w$ obtains $> P \urcorner$ is a reason for $\ulcorner w$ obtains $> Q \urcorner$. Hence by (3.7),

$(\forall w)[w$ is a physically possible world $\supset (w$ obtains $> P)]$

is a reason for

$(\forall w)[w$ is a physically possible world $\supset (w$ obtains $> Q)]$.

But the latter are equivalent to $\ulcorner \Box_p P \urcorner$ and $\ulcorner \Box_p Q \urcorner$, respectively.

These theorems will play a role in our reconstruction of enumerative induction.

4. Generalized Acceptance Rules

The principle $(A1)$ of statistical syllogism was endorsed in section 2. It will now be argued that $(A1)$ is best viewed as a corollary of a more general acceptance rule. To motivate this, note that there is an acceptance rule that is related to $(A1)$ rather like *modus tollens* is related to *modus ponens*:

$(A2)$ If F is projectible with respect to G and $r > .5$, then $\ulcorner \text{prob}(F/G) \geq r \ \& \ {\sim}Fc \urcorner$ is a prima facie reason for $\ulcorner {\sim}Gc \urcorner$, the strength of the reason depending upon the value of r.

In defense of this rule, consider a couple of examples:

(1)   Quantum mechanics enables us to calculate that it is highly probable that an energetic electron will be deflected if it passes

within a certain distance of a uranium atom. We observe that a particular electron was not deflected, and so conclude that it did not pass within the critical distance. Reasoning in this way with regard to the electrons used in a scattering experiment, we arrive at conclusions about the diameter of a uranium atom.

(2)   We are attempting to ascertain the age of a wooden artifact by carbon 14 dating. Carbon 14 dating relies upon probabilistic laws regarding the radioactive decay of carbon 14. We know that it is extremely probable that if an object is at least 7000 years old then the carbon 14 ratio (the ratio of carbon 14 to carbon 12) will be less than .001. Upon discovering that the carbon 14 ratio is greater than .001, we will conclude that the artifact is less than 7000 years old.

In applications of ($A2$), G will typically be a conjunction of properties. For example, in (1) what we know is that prob($D/E\&S$) is high where D is the property of being deflected, E is the property of being an electron with a certain kinetic energy, and S is the property of passing within a certain distance of a uranium atom. Upon learning that the electron is not deflected, we conclude that $\ulcorner Ec\&Sc \urcorner$ is false, and as we know that $\ulcorner Ec \urcorner$ is true we conclude that $\ulcorner Sc \urcorner$ is false.

It is important to recognize that these two examples genuinely illustrate ($A2$) and cannot be handled directly in terms of ($A1$). In the first example, quantum mechanics provides us with the means for calculating the probability of the electron being deflected as a function of the distance the electron passes from the uranium atom, but quantum mechanics *does not* provide us with a way of making the reverse calculation of the probability of the electron having passed within a certain distance of the uranium atom as a function of the degree of deflection. Thus there is *no way* to apply ($A1$) directly in this example. In the second example, the exponential law of radioactive decay enables us to calculate the probability of a given carbon 14 ratio as a function of the age of the artifact, but it must be emphasized that there is *no way* to calculate a probability that the artifact is of any particular age as a function of the carbon 14 ratio. Once again, there is no way to make the inference using ($A1$). But in both cases, the inferences made seem clearly legitimate. Therefore, these examples must be taken to illustrate a *modus tollens* acceptance rule like ($A2$).

The need for the projectibility constraint in ($A2$) is easily illustrated. F's being projectible with respect to G imposes requirements on both F and G. The requirement on F is illustrated as follows. Obviously, if $\text{prob}(F/G)$ is high and $\ulcorner Fc\urcorner$ is true, this fact gives us no reason for thinking that $\ulcorner Gc\urcorner$ is false. It is irrelevant to whether $\ulcorner Gc\urcorner$ is true or false. This should not change if we find a number of F_i's such that $\text{prob}(F_i/G)$ is high in each case and $\ulcorner F_i c\urcorner$ is true. But it follows from the probability calculus that if the F_i's are suitably independent and there are enough of them, we can make $\text{prob}(F_1 \,\&\, \dots \,\&\, F_n \,/\, G)$ as low as we want, and hence can make $\text{prob}((\sim F_1 \vee \dots \vee \sim F_n) \,/\, G)$ as high as we want. As we know that $\ulcorner \sim F_1 c \vee \dots \vee \sim F_n c\urcorner$ is false, ($A2$) without the projectibility constraint would give us a reason for thinking that $\ulcorner Gc\urcorner$ is false. That is clearly unreasonable. This is blocked by the projectibility constraint in ($A2$), because projectibility is closed under neither negation nor disjunction.

The requirement on G is even more easily illustrated. Suppose F is projectible with respect to G and H, but not with respect to ($G \vee H$). Suppose we know that $\ulcorner Fc\urcorner$ is false, and we have $\text{prob}(F/G) \geq r$ and $\text{prob}(F/H) < r$. We should not be able to infer $\ulcorner \sim Hc\urcorner$. But if $\text{prob}(H/G \vee H)$ is small enough, we may have $\text{prob}(F/G \vee H) \geq r$, and then by ($A2$) without the projectibility constraint, we could infer $\ulcorner \sim(Gc \vee Hc)\urcorner$ and hence $\ulcorner \sim Hc\urcorner$.

It is illuminating to consider how ($A2$) works in lottery cases. Suppose we have a lottery consisting of 100 tickets, each with a chance of .01 of being drawn. Let $\ulcorner T_n x\urcorner$ be $\ulcorner x$ has ticket $n\urcorner$ and let $\ulcorner Wx\urcorner$ be $\ulcorner x$ wins$\urcorner$. A person's winning given that he has any particular ticket is extremely improbable (i.e., $\text{prob}(\sim W/T_n)$ is high), but upon discovering that Jones won the lottery we would not infer that he did not have ticket n. As in the case of ($A1$), this results from the principle of collective defeat. Because $\text{prob}(\sim W/T_n)$ is high and we are warranted in believing $\ulcorner Wj\urcorner$, we have a prima facie reason for believing $\ulcorner \sim T_n j\urcorner$. But for each k, $\text{prob}(\sim W/T_k)$ is equally high, so we have an equally strong prima facie reason for believing $\lceil \sim T_k j \rceil$. From the fact that Jones won, we know that he had at least one ticket. We can conclude that if for every $k \neq n$, $\lceil T_k j \rceil$ is false, then $\lceil T_n j \rceil$ is true. Our reason for believing each $\lceil T_k j \rceil$ to be false is as good as our reason for believing $\lceil \sim T_n j \rceil$, so we have as strong a reason for $\lceil T_n j \rceil$ as for $\lceil \sim T_n j \rceil$. Thus we have a case of collective defeat, and we are not warranted in believing $\lceil \sim T_n j \rceil$.

$(A1)$ and $(A2)$ look very much alike, and it seems reasonable to regard them as consequences of a single stronger principle:

$(A3)$ If F is projectible with respect to G and $r > .5$, then $\lceil \text{prob}(F/G) \geq r \rceil$ is a prima facie reason for the conditional $\lceil Gc \supset Fc \rceil$, the strength of the reason depending upon the value of r.

$(A1)$ can then be replaced by an instance of $(A3)$ and *modus ponens*, and $(A2)$ by an instance of $(A3)$ and *modus tollens*. I take it that $(A3)$ is quite an intuitive acceptance rule. It amounts to a rule telling us that, subject to satisfaction of the projectibility constraint, if we know that most G's are F's then it is prima facie reasonable for us to expect of any particular object that it will be an F if it is a G.

In light of the fact that $(A1)$ and $(A2)$ are derivable from $(A3)$, all three of these acceptance rules should have the same defeaters. In discussing $(A1)$, we observed that it licenses an inference on the basis of a limited set of facts about c. That inference should be defeated if the probability can be altered by taking more facts into account. This was captured by the following principle describing subproperty defeaters:

$(D1)$ If F is projectible with respect to H then $\lceil Hc \:\&\: \text{prob}(F/G\&H) \neq \text{prob}(F/G) \rceil$ is an undercutting defeater for $\lceil Gc \:\&\: \text{prob}(F/G) \geq r \rceil$ as a prima facie reason for $\lceil Fc \rceil$.

It follows that the same thing must be a defeater for $(A2)$ and $(A3)$:

(D2)   If F is projectible with respect to H then $\ulcorner Hc$ & $\mathrm{prob}(F/G\&H) \neq \mathrm{prob}(F/G)\urcorner$ is an undercutting defeater for $\ulcorner \sim Fc$ & $\mathrm{prob}(F/G) \geq r\urcorner$ as a prima facie reason for $\ulcorner \sim Gc\urcorner$.

(D3)   If F is projectible with respect to H then $\ulcorner Hc$ & $\mathrm{prob}(F/G\&H) \neq \mathrm{prob}(F/G)\urcorner$ is an undercutting defeater for $\ulcorner \mathrm{prob}(F/G) \geq r\urcorner$ as a prima facie reason for $\ulcorner Gc \supset Fc\urcorner$.

I will refer to all of these as *subproperty defeaters*.

5. Derived Acceptance Rules

This section will be devoted to the derivation of a generalization of (A3) that will be used later in the analysis of inductive reasoning. The formulation of this generalized acceptance rule requires a new notational convention regarding free variables occurring within 'prob'. 'prob' is a variable-binding operator, binding the variables free within its scope. For example, both 'x' and 'y' are bound in $\ulcorner \mathrm{prob}(Axy/Bxy)\urcorner$. But sometimes we want to leave variables unbound within the scope of 'prob' so that we can quantify into these contexts. This can be indicated by subscripting 'prob' with whatever variables are to be left free. Thus, for example, 'x' is free in $\ulcorner \mathrm{prob}_x(Axy/Bxy)\urcorner$, enabling us to quantify into this context and write things like $\ulcorner (\exists x) \, \mathrm{prob}_x(Axy/Bxy) = r\urcorner$.

According to (A3), if F is projectible with respect to G then $\ulcorner \mathrm{prob}(Fxy/Gxy) \geq r\urcorner$ is a reason for believing $\ulcorner Gbc \supset Fbc\urcorner$. This can be represented by the reason schema:

(5.1) $\ulcorner \mathrm{prob}(Fxy/Gxy) \geq r\urcorner$ is a reason for $\ulcorner Gzw \supset Fzw\urcorner$.

A somewhat different application of (A3) yields:

(5.2) $\ulcorner \mathrm{prob}_x(Fxy/Gxy) \geq r\urcorner$ is a reason for $\ulcorner Gxc \supset Fxc\urcorner$.

It then follows by (3.7) that for any H:

(5.3) $\ulcorner(\forall x)[Hx \supset \text{prob}_x(Fxy/Gxy) \geq r]\urcorner$ is a reason for $\ulcorner(\forall x)[Hx \supset (Gxc \supset Fxc)]\urcorner$.

This can be regarded as a strengthened version of (*A3*). A special case of (5.3) arises when 'x' occurs only vacuously in $\ulcorner Fxc\urcorner$. In that case, it can be written more simply as:

(5.4) $\ulcorner(\forall x)[Hx \supset \text{prob}_x(Fy/Gxy) \geq r]\urcorner$ is a reason for $\ulcorner(\forall x)[Hx \supset (Gxc \supset Fc)]\urcorner$.

By elementary logic, $\ulcorner{\sim}Fc \ \& \ (\forall x)[Hx \supset (Gxc \supset Fc)]\urcorner$ is a reason for $\ulcorner(\forall x)(Hx \supset {\sim}Gxc)\urcorner$, so we can conclude from (5.4):

(*A4*) If F is projectible with respect to G and $r > .5$ then $\ulcorner{\sim}Fc \ \& \ (\forall x)[Hx \supset \text{prob}_x(Fy/Gxy) \geq r]\urcorner$ is a prima facie reason for $\ulcorner(\forall x)(Hx \supset {\sim}Gxc)\urcorner$.

This can be regarded as a strengthened version of (*A2*). (*A4*) will play a fundamental role in the derivation of principles of induction. Notice that by (3.11), the strength of the reason described by (*A4*) is a function of the probability r in precisely the same way the strengths of the reasons described by (*A1*), (*A2*), and (*A3*) are functions of r. For a particular value of r, these three acceptance rules all provide reasons of the same strength.

6. The Material Statistical Syllogism

Many philosophers would be happier if they could ignore nomic probability altogether and reason exclusively in terms of relative frequencies. I argued in Chapter 1 that that is not entirely possible, but it is of interest to see just how much of the work of nomic probabilities can be relegated to relative frequencies. In this connection, it is of considerable interest to see that $(A3)$ entails a frequency-based acceptance rule that can replace many of its applications. This can be regarded as a precise version of the Material Statistical Syllogism mentioned in section 2. Recall the following principle from Chapter 2:

$(PFREQ)$ If r is a nomically rigid designator of a real number and $\diamond_p[(\exists x)Gx \;\&\; \text{freq}[F/G] = r]$ then $\text{prob}\big(F \;/\; G \;\&\; \text{freq}[F/G] = r\big) = r.$

This entails

(6.1)   If r is a nomically rigid designator and $\diamond_p[(\exists x)Gx \;\&\; \text{freq}[F/G] \geq r]$ then $\text{prob}\big(F \;/\; G \;\&\; \text{freq}[F/G] \geq r\big) \geq r.$

I assume that whenever F is projectible with respect to G, it is also projectible with respect to $\ulcorner G \;\&\; \text{freq}[F/G] \geq r\urcorner$. $\ulcorner[Gc \;\&\; \text{freq}[F/G] \geq r]\urcorner$ entails the antecedent of (6.1), so it follows by $(A1)$ that (6.1) conjoined with $\ulcorner[Gc \;\&\; \text{freq}[F/G] \geq r]\urcorner$ gives us a prima facie reason for believing $\ulcorner Fc\urcorner$. But (6.1) is demonstrable, so it need not be mentioned in formulating this reason. Therefore, $\ulcorner Gc \;\&\; \text{freq}[F/G] \geq r\urcorner$ gives us a prima facie reason all by itself for believing $\ulcorner Fc\urcorner$, and hence $\ulcorner \text{freq}[F/G] \geq r\urcorner$ gives us a prima facie reason for believing $\ulcorner Gc \supset Fc\urcorner$. The strength of this reason will be the same as the strength of the reason $(PFREQ)$ gives us for $\ulcorner Fc\urcorner$, and by $(A3)$ that is just a function of the value of r. Hence:

$(A3^*)$ If r is a nomically rigid designator then $\ulcorner \text{freq}[F/G] \geq r\urcorner$ is a prima facie reason for $\ulcorner Gc \supset Fc\urcorner$, the strength of this reason being precisely the same as the strength of the reason $\ulcorner \text{prob}(F/G) \geq r\urcorner$ provides for $\ulcorner Gc \supset Fc\urcorner$ by $(A3)$.

Similarly, we can obtain frequency-based analogues of $(A1)$ and $(A2)$. In many cases, uses of $(A1)$–$(A3)$ can be replaced by uses of their frequency-based analogues, and the strengths of the resulting reasons will not suffer as a consequence of the replacement. It is worth noticing, however, that the justification for ($A3*$) still depends upon nomic probabilities. This is because it derives from principles describing the conceptual role of the concept of nomic probability. It could not be maintained that ($A3*$) is a primitive part of the conceptual role of the concept of a relative frequency, because the latter concept is explicitly defined in terms of the cardinalities of sets and hence its conceptual role does not include anything but that definition.

The applicability of ($A3*$) is complicated. Suppose we know that $\{c_1,\ldots,c_n\}$ is the set of all G's, and freq$[F/G] = r < 1$. The latter entails that not all the c_i's are F, so although ($A3*$) provides us with a prima facie reason for judging each c_i to be F, all of these prima facie reasons are subject to collective defeat, and ($A3*$) becomes inapplicable. On the other hand, we often know less. We may know only that most (and perhaps all) G's are F. In this case there is no collective defeat, and we can infer $\ulcorner Fc_i\urcorner$ for each i. But if we also know that freq$[F/G] < 1$, then again we have collective defeat. The most common situation occurs when we know that $r \le$ freq$[F/G] < 1$, but also know of a few particular G's that are not F's. In this case, there is no contradiction in supposing that all of the other G's are F's, and so in this case there is no collective defeat. The point of all this is that exactly what we know about the situation makes a big difference in what we can infer using ($A3*$).

7. Summary

1. The statistical syllogism can be formulated as the rule ($A1$):

> If B is projectible with respect to A and $r > .5$ then $\ulcorner Ac$ & prob$(B/A) \ge r\urcorner$ is a prima facie reason for $\ulcorner Bc\urcorner$.

The projectibility constraint is central to the functioning of this reason.

2. Applications of ($A1$) are defeated, in accordance with ($D1$), by knowing of circumstances that lower the probability. This

constitutes a subproperty defeater.
3. Epistemic warrant is analyzed in terms of ultimately undefeated arguments. This enables us to investigate the logic of defeasible reasoning.
4. The analysis of warrant entails the principle of collective defeat, which describes the way in which reasons for jointly inconsistent conclusions can defeat one another collectively. This enables (*A1*) to deal with the lottery paradox correctly.
5. (*A1*) and (*A2*) are consequences of (*A3*).
6. The principle (*PFREQ*) entails that the material statistical syllogism follows from any of these acceptance rules.

CHAPTER 4
DIRECT INFERENCE AND
DEFINITE PROBABILITIES

1. Introduction

Our task is to characterize nomic probability in terms of its conceptual role–its role in rational thought. One of the most important ingredients of that role concerns the use of probability in practical decisions. To repeat Bishop Butler's famous aphorism, "probability is the very guide to life". In choosing a course of action we often seek to maximize expectation value,[1] and the expectation value of an act is defined in terms of the probabilities of various possible outcomes conditional on that act. The latter are definite probabilities, so an account of the conceptual role of nomic probability must include an account of its relationship to these definite probabilities. This must include an analysis of the definite probabilities themselves and a theory of direct inference explaining how the definite probabilities can be evaluated on the basis of our knowledge of nomic probabilities. That is the topic of the present chapter.

   This chapter will take some rather surprising twists. First, we will find that despite its central role in the theory of nomic probability, direct inference does not constitute a primitive part of that theory. It will be possible to define definite probabilities in terms of nomic probabilities and to derive the requisite theory of direct inference from those parts of the theory of nomic probability already at our disposal. In the course of establishing this we will discover something even more surprising, and ultimately more important. Rules of direct inference warrant inferences from (indefinite) nomic probabilities to definite probabilities, but they will be derived from rules describing parallel inferences from nomic probabilities to nomic probabilities. The latter inferences

---

[1] The notion of maximizing expectation value is not as simple as people generally suppose. See Pollock [1983b] for a critical discussion of this.

are importantly similar to classical direct inference, so I take them to constitute what I call 'nonclassical direct inference'. The rules of nonclassical direct inference will in turn be derived from the acceptance rules and computational principles that have already been defended. Nonclassical direct inference has been entirely overlooked by probabilists, and yet it is of fundamental importance. Not only does it underlie classical direct inference—it also plays a key role in the explanation of induction and is essential to an understanding of how we are able to make any use of probability in our ordinary reasoning. It is the theory of nonclassical direct inference that is the most important product of this chapter.

The design of this chapter is as follows. I will begin by investigating the kind of mixed physical/epistemic definite probability that is involved in decision-theoretic reasoning. I will use 'PROB' to symbolize this kind of definite probability. It is inference from nomic probabilities to these definite probabilities that I will call 'classical direct inference'. The first step of the investigation will consist of an attempt to describe intuitively reasonable principles of classical direct inference. This first step will proceed without any concern for what justifies those principles. Once we have acquired some idea of how classical direct inference ought to work, we can then consider how to justify it, and it will be at this point that we will be led to nonclassical direct inference.

The probabilities required for decision theoretic reasoning are conditional probabilities. We must know the probabilities of various possible outcomes conditional on the different acts available to us. The standard procedure is to define conditional probabilities as ratios of nonconditional probabilities: $\mathrm{PROB}(P/Q) = \mathrm{PROB}(P\&Q) \div \mathrm{PROB}(Q)$. For the time being I will follow this convention.[2] This allows us to regard the theory of classical direct inference as concerned primarily with the evaluation of nonconditional probabilities. Direct inference to conditional probabilities

---

[2] In section 4 I adopt a slightly more general definition that allows $\mathrm{PROB}(P/Q)$ to be defined even when $\mathrm{PROB}(Q) = 0$, but that difference will not be of much practical significance.

can, for the most part, be reduced to direct inference to nonconditional probabilities.

2. Classical Direct Inference

The theory of classical direct inference is the theory of certain kinds of inferences. It tells us how to infer the values of definite probabilities from what we know about nomic probabilities. It is, in this sense, an epistemological theory. The test of such a theory is whether it licenses the correct inferences. In this section I will try formulate a set of rules for classical direct inference. The methodology here is "empirical". The objective is to construct a theory that works, that is, one giving us what seem intuitively to be the right inferences. Only after constructing such a theory can we consider how its inferences are to be justified.

2.1 Reichenbach's Rules

The basic idea behind standard theories of classical direct inference is due to Hans Reichenbach, who stated it as follows: "If we are asked to find the probability holding for an individual future event, we must first incorporate the case in a suitable reference class. An individual thing or event may be incorporated in many reference classes.... We then proceed by considering *the narrowest* reference class for which suitable statistics can be compiled."[3]   In symbols, in determining the probability that an individual c has a property F, we find the narrowest reference class X regarding which we have reliable statistics and then infer that $PROB(Fc) = prob(Fx/x \in X)$. For instance, this is the way insurance rates are calculated. There is almost universal agreement that direct inference is based upon some such principle as this, but it has proven extremely difficult to turn this basic idea into a precise theory capable of withstanding the test of counterexamples. My own theory will be based upon Reichenbach's

---

[3] Reichenbach [1949], p. 374. Reichenbach did not use the term 'probability' for definite probabilities, preferring to call them 'weights' instead.

insight. I will fashion my theory by looking at various problems of detail that arise for Reichenbach's theory and shaping the theory to meet them.[4]

Reichenbach, and most subsequent philosophers, have discussed direct inference in terms of the reference classes in which we *know* something to be contained, but it seems reasonably clear that that is not to be taken literally. What is really at issue in direct inference is what we are warranted in believing. In direct inference we proceed from premises about what propositions are warranted to conclusions about probabilities. This suggests rewriting the Reichenbachian formula for direct inference as follows: if we want to know the value of $\text{PROB}(Fc)$, we find the narrowest reference class X to which we are warranted in believing c to belong, and for which we have a warranted belief regarding the value of $\text{prob}(Fx/x \in X)$, and we take the latter to be the value of $\text{PROB}(Fc)$.

I suggest that the most reasonable way of making Reichenbach's proposal precise is in terms of prima facie reasons. Being warranted in believing that c belongs to X gives us a prima facie reason for thinking that $\text{PROB}(Fc) = \text{prob}(Fx/x \in X)$. But a direct inference from a narrower reference class takes precedence over a direct inference from a broader reference class. That is, being warranted in believing that $c \in Y$ where $Y \subset X$ and $\text{prob}(Fx/x \in Y)$ $\neq \text{prob}(Fx/x \in X)$ defeats our reason for believing that $\text{PROB}(Fc) = \text{prob}(Fx/x \in X)$. This is the principle of *subset defeat*. Symbolizing $\ulcorner P$ is warranted$\urcorner$ as $\ulcorner W(P) \urcorner$, we have *The Rule of Direct Inference*:

(2.1)　$\ulcorner \text{prob}(F/G) = r$ & $W(Gc) \urcorner$ is a prima facie reason for $\ulcorner \text{PROB}(Fc) = r \urcorner$

and *The Rule of Subset Defeat*:

---

[4] The most sophisticated contemporary work on direct inference has been done by Henry Kyburg [1974], and my theory takes its impetus from Kyburg's theory. I begin by embracing a number of Kyburg's insights, but then move off in a different direction.

(2.2)   ⌜∀(H ⊃ G) & **W**(Hc) & prob(F/H) ≠ r⌝ is an undercutting
         defeater for (2.1).

Defeaters of the form of (2.2) will be called *subset defeaters*.

Note that by (2.1), a subset defeater is automatically a
rebutting defeater. What the rule of subset defeat tells us is that
a subset defeater is also an undercutting defeater. Recall the
distinction between rebutting defeaters and undercutting defeaters.
Rebutting defeaters attack the conclusion of a prima facie
inference, while undercutting defeaters attack the connection
between the prima facie reason and the conclusion. If P is a
prima facie reason for R and Q is an equally good prima facie
reason for ~R, then P and Q simply rebut one another and we
are left with no justified conclusion regarding R. But suppose,
as in Figure 1, that Q is also an undercutting defeater for P. An
undercutting defeater removes P from contention. More accurately, although the arguments from P to R and from Q to ~R are
both level 0 arguments, the argument from P to R is not a level
n argument for any n > 0. Thus the fact that P is a rebutting
defeater for Q becomes irrelevant. As a result, nothing rebuts Q
and we are left with Q being an undefeated prima facie reason
for ~R. This logical scenario is important in direct inference,
because it is precisely what is involved in giving preference to a
direct inference from a narrower reference class. To say that a
direct inference from a narrower reference class takes precedence
is to say that when such an inference conflicts with an inference
from a broader reference class, the inference from the narrower

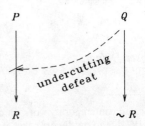

Figure 1. Combined rebutting and undercutting defeat.

reference class constitutes an undercutting defeater for the inference from the broader reference class. Thus we become warranted in ignoring the broader reference class and making the direct inference on the basis of the narrower reference class.

As he lacked the concept of a prima facie reason, Reichenbach himself did not formulate his theory of direct inference as in (2.1) and (2.2), but these seem to be the epistemic principles underlying his theory, so I will persist in calling them 'Reichenbach's rules'. They will be modified in various important ways as we proceed, but they form the basis for my theory of classical direct inference. Before considering the ways in which these rules must be modified, let me contrast them with a competing approach to direct inference. According to Reichenbach's rules, as I construe them, knowing that an object c is a member of a class X for which prob($Fx/x{\in}X$) = r creates a presumption for the conclusion that PROB(Fc) = r. Other information may be relevant, but only in a negative way—as a defeater. Some probabilists think this accords too weak a role to the other kinds of information that may make a direct inference illegitimate. They would require us to rule out such considerations before we can make the direct inference rather than having the considerations relevant only after the fact—as defeaters. For example, Isaac Levi insists that before we can perform a direct inference from the fact that an object c is a member of the reference class X we must know that c is chosen "at random" from the set X. More precisely, Levi requires that c be selected from X by a general method that we know accords the same probability of being selected to each member of X.[5] This theory would make it much harder to perform direct inferences because we would have to know much more before we could get started. However, this kind of theory of direct inference gets things backwards—it would have us start with the set X and then choose c. In actual cases of direct inference we start with the object c and then look for an appropriate reference class. For example, an insurance company is not going to decline to insure someone because his

---

[5] See particularly Levi [1977], [1978], [1980], and [1981].

social security number was not chosen by their random number generator. Their customers are not chosen by lot. Rather, they start with the list of customers and then try to determine what rates to charge them by attempting to determine by direct inference what the probabilities are of various eventualities befalling them.

Levi's theory reflects a common confusion regarding prima facie reasons. The power of prima facie reasons as an epistemological tool is that in order to use them we need not first rule out everything that can go wrong. Instead, prima facie reasons create presumptions for conclusions, and we need only worry about possible difficulties when they actually arise. The failure to appreciate this was largely responsible for both the skeptical problem of perception and the skeptical problem of induction. The skeptical problem of perception arose from asking how perception could possibly provide us with data sufficient to logically ensure that the world around us is as it appears to be. The answer is that it cannot, but it does not have to. Perceptual experience provides us with prima facie reasons for beliefs about the physical world, and we do not have to worry about the untoward eventualities that worried the skeptic unless we have some concrete reasons for thinking they may actually be occurring. Similarly, the skeptical problem of induction arose in part from the observation that our sample of A's always consists of objects sharing properties not possessed by all A's (e.g., being on earth, existing during the twentieth century, observed by human beings, etc.). How can we know that these do not make a difference to whether an A is a B, and if we cannot know this then how can we be justified in concluding on the basis of our sample that all A's are B's? The answer is that we do not have to rule out all such possibilities beforehand. Our sample provides us with a prima facie reason. The sample is "innocent until proven guilty". We do not have to worry about shared properties of the members of the sample unless we acquire a positive reason for thinking they may really make a difference. I take it that these points about perception and induction are fairly obvious and would be granted by most contemporary epistemologists. But it is precisely the same kind of confusion that underlies the idea that in direct inference it is not enough to know that our object is a member of a certain reference class—we must know something more that shows we are not biasing the outcome by choosing that particular reference class. In fact, we do not have to know anything like this beforehand. We need only worry about such possibilities when we acquire a reason for thinking they are actually occurring in a specific case. In other words, they are only relevant as defeaters to our prima facie reason.

2.2 Modifications to Reichenbach's Rules

Reichenbach's rules must be extended and modified in various ways. A minor extension results from considering multi-place relations. Just as $\ulcorner prob(Fx/Gx) = r$ & $\mathbf{W}(Gc)\urcorner$ is a prima facie reason for $\ulcorner \mathsf{PROB}(Fc) = r\urcorner$, so $\ulcorner prob(Rxy/Sxy) = r$ & $\mathbf{W}(Sbc)\urcorner$ is a prima facie reason for $\ulcorner \mathsf{PROB}(Rbc) = r\urcorner$. To explicitly take account of this in our formulation of principles of direct inference tends to make our notation unwieldy, so I will follow the general practice of writing only the one-place case of each principle, letting that stand for the corresponding multiplace principle.

Henry Kyburg [1974] observes that if $\ulcorner Fb \equiv Gc\urcorner$ is warranted, then $\ulcorner Fb\urcorner$ and $\ulcorner Gc\urcorner$ should be assigned the same probability. Certainly, we would adopt the same odds in betting on their truth. But this is not forthcoming from Reichenbach's rules. We can guarantee the desired result by replacing (2.1) with the somewhat stronger:

(2.3) $\ulcorner prob(F/G) = r$ & $\mathbf{W}(Gc)$ & $\mathbf{W}(P \equiv Fc)\urcorner$ is a prima facie reason for $\ulcorner \mathsf{PROB}(P) = r\urcorner$.

Principle (2.3) has an important equivalent reformulation. There is no restriction on what kind of term r may be, so we are free to let it be $\ulcorner prob(F/G)\urcorner$. (2.3) then tells us that $\ulcorner prob(F/G) = prob(F/G)$ & $\mathbf{W}(Gc)$ & $\mathbf{W}(P \equiv Fc)\urcorner$ is a prima facie reason for $\ulcorner \mathsf{PROB}(P) = prob(F/G)\urcorner$. The first conjunct of this conjunction is demonstrable, so the rest of the conjunction constitutes a conclusive reason for the whole conjunction. Hence

(2.4) $\ulcorner W(Gc)$ & $W(P \equiv Fc)\urcorner$ is a prima facie reason for $\ulcorner PROB(P)$
         $= prob(F/G)\urcorner$.

Our information regarding nomic probabilities is often rather
unspecific. Rather than knowing the precise value of prob(F/G),
we may know only that it lies in a certain range, or more
generally that prob(F/G)∈X for some set X. This point has been
emphasized by Kyburg, who adopts complex rules of direct
inference to deal with this case. But in fact, we need no special
rules for this. The preceding rules are adequate for this purpose.
Suppose we are justified in believing $\ulcorner$prob(F/G)∈X & $W(Gc)$ &
$W(P \equiv Fc)\urcorner$. By (2.4), we have a prima facie reason for believing
that PROB(P) = prob(F/G), and hence for believing that PROB(P)∈X.
In other words, either (2.3) or (2.4) implies:

(2.5) $\ulcorner$prob(F/G)∈X & $W(Gc)$ & $W(P \equiv Fc)\urcorner$ is a prima facie
         reason for $\ulcorner PROB(P)∈X\urcorner$.

$\ulcorner$prob(F/G) ≠ $r\urcorner$ is equivalent to $\ulcorner$prob(F/G)∈([0,1] - {r})$\urcorner$,
so it follows from (2.5) that

(2.6) $\ulcorner$prob(F/G) ≠ r & $W(Gc)$ & $W(P \equiv Fc)\urcorner$ is a prima facie
         reason for $\ulcorner PROB(P) ≠ r\urcorner$.

Similarly,

(2.7) $\ulcorner$prob(F/G)∉X & $W(Gc)$ & $W(P \equiv Fc)\urcorner$ is a prima facie
         reason for $\ulcorner PROB(P)∉X\urcorner$.

The Rule of Subset Defeat can be generalized as follows:

(2.8) $\ulcorner$prob(F/H) ≠ prob(F/G) & $W(Hc)$ & $\forall(H \supset G)\urcorner$ is an
         undercutting defeater for each of (2.3)−(2.7).

2.3 Frequencies vs. Nomic Probabilities
Reichenbach and Kyburg base direct inference on relative
frequencies rather than nomic probabilities. It must be

acknowledged that we should base our estimates of definite probability upon actual relative frequencies when we know the value of the latter. For example, suppose we have a die of physical description D and we know that the probability of rolling a 4 with a die of that description is 1/6. Suppose, however, that this die is only rolled three times and we know that two of the rolls were 4's (although we do not know which of the rolls were which). Asked to bet upon whether the second roll was a 4, the correct probability to use would be 2/3 rather than 1/6. This example appears to illustrate that, at least in some cases, it is relative frequency and not nomic probability that provides the appropriate premise for direct inference.

But there are other cases in which it appears that direct inference must proceed from nomic probability. For example, we know quite a bit about how the physical structure of a die affects the probabilities of different faces coming up on a roll. Let D^* be a description of a loaded die that is cubical but with its center of gravity offset by a certain precisely specified amount. We might be able to compute, on the basis of various inductive generalizations, that the probability of rolling a 4 with such a die is 1/8. Suppose that there is just one die of description D^*. This die will be rolled only once and then destroyed. If we are to bet on the probability of a single roll of this die being a 4, it seems clear that the correct probability to use is 1/8. But there may be no reference class containing that roll for which the relative frequency is 1/8. We are supposing that there has never been another die whose center of gravity was offset by exactly the same amount as this one, so we know full well that the frequency of 4's in rolls of dice of description D^* is not 1/8. The frequency is either 0 or 1, depending upon whether the single roll is a 4. Thus this appears to be a case in which direct inference must proceed from nomic probabilities.

The preceding pair of examples suggests that we can make direct inferences from either frequencies or nomic probabilities, but direct inferences from frequencies take precedence when we know their values. If we do not know anything about the frequencies, then we must be content with direct inferences from nomic probabilities. There is a grain of truth in this, but as stated it oversimplifies. The difficulty is that in the second example above, we are not totally ignorant about the relative frequencies. We know that the relative frequency is either 0 or 1; that is, it is in the set {0,1}. If we could make a direct inference from this fact (in accordance with (2.5)) we would have a reason for believing that the probability of a single roll of our die being a 4 is also in the set {0,1}. That conflicts with the direct inference from the nomic probability that led us to conclude that the probability of the single roll being a 4 is 1/8. If direct inference from frequencies always takes precedence, then we should conclude that the probability is *not* 1/8, but that conclusion is obviously unwarranted.

This difficulty can be generalized. Suppose we are attempting to ascertain the value of PROB(Fc). There is always a narrowest reference class to which we know c to belong, namely, {c}. Furthermore, we know that freq[$Fx / x \in \{c\}$] is either 0 or 1. If direct inference from frequencies takes precedence over direct inference from nomic probabilities, this should support a direct inference to the conclusion that PROB(Fc)∈{0,1}. Furthermore, as this is a direct inference from the narrowest possible reference class, it provides a subset defeater for a direct inference to any value intermediate between 0 and 1, and we seem forced to the conclusion that all definite probabilities must be either 0 or 1.

Kyburg avoids this general difficulty by insisting that the frequency information upon which we base direct inference must take the form of reporting the frequency to be in an *interval* rather than an arbitrary set. Given that constraint, all we know is that freq[$Fx / x \in \{c\}$]∈[0,1]. This supports a direct inference to the conclusion that PROB(Fc)∈[0,1], but the latter does not conflict with any other direct inference and so creates no problem. However, this way of avoiding the problem is *ad hoc*. Why should we only be able to make direct inferences from interval evaluations of frequencies? After all, we may know much more. We know that freq[$Fx / x \in \{c\}$]∈{0,1}. Why shouldn't we be able to use that information? I think it must be concluded that without further explanation, this is not an adequate resolution of the difficulty.

It appears that sometimes direct inference proceeds from

frequencies, and other times it proceeds from nomic probabilities. We need a unified account that explains when frequencies are relevant and when nomic probabilities are relevant. It is unlikely that such an account can be based exclusively on relative frequencies. However, a theory taking direct inference to proceed exclusively from nomic probabilities can explain all of this. In particular, it can explain those cases of direct inference that appear to proceed from frequencies. The explanation is in terms of the principle (*PFREQ*) of Chapter 2, according to which if r is a nomically rigid designator for a real number and $\diamond_p[\exists G \,\&\, \text{freq}[F/G] = r]$ then

$$\text{prob}(F \,/\, G \,\&\, \text{freq}[F/G] = r) = r.$$

Suppose we know that $\text{prob}(F/G) = 1/6$ and $\mathbf{W}(Gc)$, and we want to know the value of $\text{PROB}(Fc)$. If we know nothing else, then it is reasonable to infer that $\text{PROB}(Fc) = 1/6$. But suppose we also know that $\text{freq}[F/G] = 2/3$. Then as we have seen, we should infer that $\text{PROB}(Fc) = 2/3$. We can obtain this result as follows. We are warranted in believing that c satisfies both the condition $\ulcorner Gx \urcorner$ and the stronger condition $\ulcorner Gx \,\&\, \text{freq}[F/G] = 2/3 \urcorner$. The nomic probability of being an F is different on these two conditions, and the second condition takes account of more information about c. Hence the probability on the second condition should take precedence. This is in the spirit of subset defeat. By (*PFREQ*), the probability on the second condition is 2/3, so we should infer that $\text{PROB}(Fc) = 2/3$. This illustrates how a theory of direct inference proceeding from nomic probabilities can still give preference to frequencies when their values are known.

Now consider the case in which all we know about the relative frequency is that it is either 0 or 1. If we insist that direct inference always proceeds from nomic probabilities, this case creates no difficulties. We know by (*PFREQ*) that

$$\text{prob}(F \,/\, G \,\&\, \text{freq}[F/G] = r) = r,$$

but it does not follow from this that

(2.9)   prob(F / G & freq[F/G]∈X)∈X.

For example, let $X = \{0,1\}$. From the fact that freq[F/G]∈$\{0,1\}$, all that follows is that there is a certain (perhaps unknown) probability p that the frequency is 0 and the probability $1-p$ that the frequency is 1. Then what follows from (*PFREQ*) by the probability calculus is:

$$\begin{aligned}
&\text{prob}\big(F \ / \ G \ \& \ \text{freq}[F/G]\in\{0,1\}\big)\\
&= \text{prob}\big(F \ / \ G \ \& \ \text{freq}[F/G] = 0\big)\cdot p\\
&\quad + \text{prob}\big(F \ / \ G \ \& \ \text{freq}[F/G] = 1\big)\cdot(1\text{-}p)\\
&= 0\cdot p + 1\cdot(1\text{-}p) = 1\text{-}p.
\end{aligned}$$

But p can be anything from 0 to 1, so all we can conclude is that

$$\text{prob}\big(F \ / \ G \ \& \ \text{freq}[F/G]\in\{0,1\}\big)\in[0,1].$$

Similarly, the most that follows from (*PFREQ*) about the general case is:

(2.10)   prob$\big(F \ / \ G \ \& \ \text{freq}[F/G]\in X\big)\in[glb(X),lub(X)]$.[6]

Thus we get interval estimates of nomic probabilities from discrete estimates of frequencies. In the case of the loaded die that is rolled only once, we know that the frequency is either 0 or 1, but all that we can infer from that by direct inference is that the probability of the roll being a 4 is in the interval [0,1]. That does not conflict with the direct inference to the conclusion that the probability is 1/8, so the latter conclusion is undefeated. What this shows is that by insisting that direct inference proceed from nomic probabilities rather than relative frequencies, we get the effect of Kyburg's restriction to intervals without its *ad hoc* flavor.

For the above reasons, and for additional reasons that will

----

[6] $glb(X)$ is the greatest lower bound of the numbers in X, and $lub(X)$ is the least upper bound.

emerge below, we must regard direct inference as proceeding from nomic probabilities rather than from relative frequencies. This necessitates a general modification to Reichenbach's rules of direct inference. Frequencies are extensional, but nomic probabilities are not. It is possible to have $\text{prob}(A/B) \neq \text{prob}(A/C)$ even though $\forall(B \equiv C)$. For example, if $\text{freq}[A/B] = r$, then

$$\forall[B \equiv (B \ \& \ \text{freq}[A/B] = r)]$$

is true, and

$$\text{prob}(A \ / \ B \ \& \ \text{freq}[A/B] = r) = r$$

but $\text{prob}(A/B)$ need not be r. The general reason for this lack of extensionality is that nomic probabilities pertain to the behavior of B's in other physically possible worlds and not just in the actual world. To accommodate this, subset defeaters must be reformulated. For example, a direct inference from $\ulcorner\text{prob}(A \ / \ B \ \& \ \text{freq}[A/B] = r) = r\urcorner$ should take precedence over a direct inference from $\ulcorner\text{prob}(A/B) = s\urcorner$, but the reference class for the former is not narrower than that for the latter—they are equal. The sense in which the former takes account of more information is that $\ulcorner B \ \& \ \text{freq}[A/B] = r\urcorner$ entails B, but the converse entailment does not hold. Thus it is logical entailment rather than set inclusion that is relevant to direct inference from nomic probabilities. The Rule of Subset Defeat should be revised as follows:

(2.11) $\ulcorner\text{prob}(F/G) \neq \text{prob}(F/H) \ \& \ \mathbf{W}(Hc) \ \& \ \Box\forall(H \supset G)\urcorner$ is an undercutting defeater for (2.3)–(2.7).

It is noteworthy that neither (2.8) nor (2.11) would constitute a correct account of subset defeat if direct inference proceeded from frequencies rather than nomic probabilities. To illustrate, suppose we have a counter c that we know to be contained in a shipment of variously shaped counters in which 50% of the square counters are red. This is frequency information: $\text{freq}[R/S\&C] = .5$ (where C is the property of being a counter in the shipment).

If direct inference proceeded from frequencies, then according to either (2.8) or (2.11), knowing that the counter is contained in a certain box in the shipment within which the proportion of square counters that are red is *not* 50% should defeat the direct inference. But intuitively it would not. The counter is bound to be contained in *some* collection of counters in which the proportion of square ones that are red is different from 50%. On the other hand, if we knew that within the box containing the counter the proportion of square counters that are red was some particular amount other than 50%—for example, 30%—that would defeat the direct inference and make it reasonable to infer instead that the probability is .3. But if all we know is that the proportion in the box is different from the proportion in the entire shipment, we will still infer that PROB(Rc/Sc) = .5. This is readily explicable on the assumption that direct inference proceeds from nomic probabilities rather than frequencies. The original direct inference would proceed from the probability

$$\text{prob}\big(R \ / \ S\&C \ \& \ \text{freq}[R/S\&C] = .5\big) = .5.$$

When we learn that within the box the proportion of square counters that are red is not 50%, what we learn is that freq[$R/S\&B$] $\neq$ .5. But we have no reason to believe that

$$\text{prob}\big(R \ / \ S\&B \ \& \ \text{freq}[R/S\&B] \neq .5\big) \neq .5$$

so this does not provide a subset defeater. On the other hand, suppose we learn that freq[$R/S\&B$] = .3. As freq[$R/S\&C$] = .5, it follows that

$$\text{freq}\big[R \ / \ S\&B \ \& \ \text{freq}[R/S\&C] = .5\big] = .3.^{7}$$

We know that

---

^{7} In general, if P is any true proposition, freq[A/B] = freq[$A/B\&P$].

$$\text{prob}\big(R \ / \ S\&B \ \& \ \text{freq}[R/S\&C] \ = \ .5 \ \& \ \text{freq}[R \ / \ S\&B \ \&$$
$$\text{freq}[R/S\&C] \ = \ .5] \ = \ .3\big) \ = \ .3$$

and so we can infer that

$$\text{prob}\big(R \ / \ S\&C \ \& \ \text{freq}[R/S\&C] \ = \ .5\big) \ \neq$$
$$\text{prob}\big(R \ / \ S\&B \ \& \ \text{freq}[R/S\&C] \ = \ .5 \ \& \ \text{freq}[R \ / \ S\&B \ \&$$
$$\text{freq}[R/S\&C] \ = \ .5] \ = \ .3\big).$$

Thus in this case we *do* have a subset defeater. Consequently, by insisting that direct inference proceeds from nomic probabilities, we can explain what would otherwise be very perplexing behavior on the part of subset defeaters.

2.4 Projectibility

Even acknowledging that direct inference must proceed from nomic probability, the rules of direct inference (2.3) and subset defeat (2.11) do not constitute an adequate theory of direct inference. Given only those principles, direct inferences would almost invariably be defeated. The difficulty is that it is almost always possible to construct prima facie reasons in accordance with (2.3) that conflict with and hence either rebut or undercut any given prima facie reason of that form. For example, consider a coin of some description D such that the probability of a toss landing heads is .5: $\text{prob}(H/T) = .5$ (where T is the property of being a toss of a coin of description D). Suppose we know that c is a toss of this coin, and we do not know anything else that is relevant to whether c lands heads. It is then reasonable to infer that $\text{PROB}(Hc) = .5$. $\ulcorner\text{prob}(H/T) = .5 \ \& \ \mathbf{W}(Tc)\urcorner$ provides a prima facie reason for this direct inference. But now let F be any predicate for which we are justified in believing that $\text{prob}(F/T\&{\sim}H) \neq 1$ and $\mathbf{W}(Fc)$. It follows from the classical probability calculus that

(2.12) If $\text{prob}(F/T\&{\sim}H) \neq 1$ then $\text{prob}\big(H \ / \ (F{\vee}H)\&T\big)$
$\neq \text{prob}(H/T)$,

and this provides a subset defeater for the direct inference to the conclusion that $\mathsf{PROB}(Hc) = \text{prob}(H/T)$.

Most philosophers have overlooked this problem for direct inference, but Henry Kyburg is a notable exception. Kyburg [1974] attempted to resolve the difficulty by imposing restrictions on reference properties that would preclude their being arbitrary disjunctions. His restrictions turned out to be too weak to accomplish this,[8] but nevertheless, that seems to be the right way to go. This is reminiscent of the behavior of disjunctions in connection with acceptance rules, and it suggests that, once again, what is required is a restriction requiring that the consequent property in direct inference be projectible with respect to the reference property.[9]

The projectibility constraint has the effect of imposing constraints on both the consequent property and the reference property. The constraint on the reference property eliminates the

---

[8] Kyburg acknowledges this in his [1982].

[9] When I first encountered this problem, I looked for some formal constraint that could be used to preclude reference properties like $\ulcorner(F \vee H)\&T\urcorner$, but a simple argument shows that no formal constraint can do the job. Consider four properties, A, B, C, and D, that are appropriate for use in direct inference in the absence of defeaters, and suppose we are justified in believing that c entails B, B entails A, $\mathbf{W}(Cc)$, and $\text{prob}(D/A) \neq \text{prob}(D/B)$. In such a case, we want to be able to infer that $\mathsf{PROB}(Dc) = \text{prob}(D/B)$. But we also have that c entails $\ulcorner A \& (C \vee {\sim} B)\urcorner$, and it will generally be the case that $\text{prob}(D \,/\, A \& (C {\sim} B)) \neq \text{prob}(D/B)$. B and $\ulcorner A \& (C \vee {\sim} B)\urcorner$ are logically independent of one another, so neither provides us with a subset defeater for the other. Thus what we have is prima facie reasons for two conflicting direct inferences, and they rebut one another. Symmetry considerations show that no formal constraint can solve this problem by precluding $\ulcorner A \& (C {\sim} B)\urcorner$ from being an allowed reference property. Any such formal constraint would also have to preclude the formally analogous property $\ulcorner A \& (C \vee {\sim}[A \& (C \vee {\sim} B)])\urcorner$ from being an allowed reference property. But the latter is equivalent to B, and we want to be allowed to use B in direct inference. The connections between B and $\ulcorner A \& (C {\sim} B)\urcorner$ are just like the connections between the properties 'grue' and 'green'. That is, each is definable from the other in the same way. Thus there can be no *formal* reason for ruling out one and allowing the other. Any satisfactory constraint must be a substantive constraint having to do with the content of the property and not just its logical form.

problem case involving (2.12). To illustrate the effect of the projectibility constraint on the consequent property, consider a case in which, contrary to the constraint, we appear to make a direct inference regarding a disjunctive consequent property. For example, suppose again that we have a shipment of variously colored and shaped counters, and we know that 25% of them are either red squares or blue triangles. If we know only that c is one of the counters in the shipment, we will judge that $\text{PROB}((Rc\&Sc)\vee(Bc\&Tc)) = .25$. This appears to be a direct inference with regard to the disjunctive (and hence non-projectible) property $((R\&S)\vee(B\&T))$. It is explicable, however, in terms of direct inferences concerning projectible properties. I contend that what we are doing here is making a direct inference to the conclusions that $\text{PROB}(Rc\&Sc) = \text{prob}(R\&S/C)$ and that $\text{PROB}(Bc\&Tc) = \text{prob}(B\&T/C)$ (where c is the property of being a counter in the shipment), and then computing:

$$\begin{aligned}
&\text{PROB}((Rc\&Sc)\vee(Bc\&Tc)) \\
&= \text{PROB}(Rc\&Sc)+\text{PROB}(Bc\&Tc) \\
&= \text{prob}(R\&S/C)+\text{prob}(B\&T/C) \\
&= \text{prob}((R\&S)\vee(B\&T) / C) \\
&= .25.
\end{aligned}$$

That this is the correct description of the inference can be seen as follows. Suppose we know c to be in a particular crate in the shipment, and we know that the proportion of red squares in the crate is .2, while the proportion of red squares in the entire shipment is only .1. We would take that to defeat the disjunctive inference. But for all we know, the proportion of counters in the entire shipment that are either red squares or blue triangles is the same as the proportion in that crate, so we do not have a subset defeater for the direct inference regarding the disjunction. What we do have is a subset defeater for the direct inference to the conclusion that $\text{PROB}(Rc\&Sc) = \text{prob}(R\&S/C)$. This inference is blocked, which in turn blocks the entire calculation and that prevents our concluding that $\text{PROB}((Rc\&Sc)\vee(Bc\&Tc)) = .25$. This is readily explicable on the supposition that the putative direct inference regarding the disjunction is parasitic on the direct inferences regarding the projectible disjuncts and is not really a direct inference in the same sense.[10]

It seems initially puzzling that there should be a projectibility constraint in direct inference. What does projectibility have to do with direct inference? It will turn out that the projectibility constraint in direct inference is derivative from the projectibility constraint in the acceptance rule $(A3)$.

With the requirement that the consequent property in direct inference be projectible with respect to the reference property, the basic rules of classical direct inference and subset defeat become:

(CDI) If F is projectible with respect to G then $\lceil \text{prob}(F/G) = r$ & $\mathbf{W}(Gc)$ & $\mathbf{W}(P \equiv Fc) \rceil$ is a prima facie reason for $\lceil \text{PROB}(P) = r \rceil$.

(CSD) If F is projectible with respect to H then $\lceil \text{prob}(F/H) \neq \text{prob}(F/G)$ & $\mathbf{W}(Hc)$ & $\square \forall (H \supset G) \rceil$ is an undercutting defeater for (CDI).

I believe that these two principles are now correct as formulated. They constitute the basis of a theory of classical direct inference.

---

[10] The astute reader will notice, however, that there is an alternative possible explanation. In the situation described, we have a reason for concluding that $\text{PROB}(Rc\&Sc) \neq \text{prob}(R\&S/C)$. This in turn implies that either $\text{PROB}(Bc\&Tc) \neq \text{prob}(B\&T/C)$ or $\text{PROB}((Rc\&Sc)\vee(Bc\&Tc)) \neq \text{prob}((R\&S)\vee(B\&T) / C)$. If consequent properties need not be projectible, we have no reason to favor one of the latter two direct inferences over the other, so they are both defeated. This is a case of collective defeat. However, this cannot be the correct construal of the way in which the inference that $\text{PROB}((Rc\&Sc)\vee(Bc\&Tc)) = \text{prob}((R\&S)\vee(B\&T) / C)$ is defeated. This is because, although the disjunctive inference should be defeated, the inference to the conclusion that $\text{PROB}(Rc\&Tc) = \text{prob}(R\&T/C)$ should not be defeated, and the latter would be defeated on the collective defeat construal. The only way to get all of the right inferences defeated and none of the wrong inferences defeated is to suppose that the disjunctive inference is parasitic on direct inferences regarding the disjuncts and cannot stand by itself.

3. Nonclassical Direct Inference

It is my contention that the nature of classical direct inference has been fundamentally misunderstood, and I will now attempt to rectify that. The basic rule of classical direct inference (*CDI*) is that if *F* is projectible with respect to *G* and we know $\ulcorner prob(F/G) = r$ & $W(Gc)\urcorner$ but know nothing else about *c* that is relevant, this constitutes a reason to believe that $PROB(Fc) = r$. Typically, we will know *c* to have other projectible properties *H* but not know anything about the value of $prob(F/G\&H)$ and so be unable to use the latter in direct inference. But if the direct inference from $\ulcorner prob(F/G) = r\urcorner$ to $\ulcorner PROB(Fc) = r\urcorner$ is to be reasonable, there must be a presumption to the effect that $prob(F/G\&H) = r$. If there were no such presumption then we would have to regard it as virtually certain that $prob(F/G\&H) \neq r$ (after all, there are infinitely many possible values that $prob(F/G\&H)$ could have), and so virtually certain that there is a true subset defeater for the direct inference. This would make the direct inference to $\ulcorner PROB(Fc) = r\urcorner$ unreasonable.

In the preceding example, (*G&H*) is a subproperty of *G*. Recall the definition from Chapter 2:

(3.1)   *B* is a *subproperty* of *A* iff $\diamond_p \exists B$ & $\square_p \forall (B \supset A)$.

I will symbolize $\ulcorner B$ is a subproperty of $A\urcorner$ as $\ulcorner B \leqslant A\urcorner$. What I allege is that the legitimacy of classical direct inference presupposes that subproperties *H* of a property *G* will normally be such that $prob(F/H) = prob(F/G)$.[11] Perhaps the best way to argue for this presumption is to consider the relationship between definite and indefinite probabilities. It has frequently been observed that an analysis of sorts can be given for indefinite probabilities in terms of definite probabilities. To say that $prob(F/G) = r$ is tantamount to saying that if *b* is an object we

---

[11] There is an obvious connection with the principle of agreement. That connection will be exploited in section 6 to provide foundations for direct inference.

know to have the property G, but we do not know anything else relevant about b, then PROB(Fb) = r. Similarly, to say that prob($F/G\&H$) = r is tantamount to saying that if b is an object we know to have the property H, but we do not know anything else relevant about b, then PROB(Fb) = r. If H is a subproperty of G then the latter is precisely the conclusion obtained from the Rule of Direct Inference given the assumption that prob(F/G) = r. Thus classical direct inference leads directly to the following principle regarding nomic probabilities:

(*DI*)   If F is projectible with respect to G then $\ulcorner H \leqslant G \&$ prob(F/G) = $r\urcorner$ is a prima facie reason for $\ulcorner$prob(F/H) = $r\urcorner$.[12]

Principle (*DI*) licenses inferences from (indefinite) nomic probabilities to nomic probabilities. Such inferences constitute *nonclassical direct inference.* (*DI*) amounts to a kind of principle of insufficient reason. It tells us that if we have no reason to think otherwise, it is reasonable for us to anticipate that conjoining other properties to G will not affect the probability of F.[13] Once it has been pointed out, it seems obvious that this is what is presupposed by classical direct inference. In determining the probability that an object will have the property F, we make use of those nomic probabilities whose values are known to us, and

---

[12] Note that there is no requirement in (*DI*) that F be projectible with respect to H. In Pollock [1983] I erroneously imposed such a requirement. That requirement is incompatible with the reduction of definite probabilities to nomic probabilities given below in section 4, and it is not required by the argument that will be given in section 6 to derive the principles of nonclassical direct inference from the rest of the theory of nomic probability.

[13] The connection between this principle and the Laplacian principle discussed in Chapter 1 is rather tenuous. Both can reasonably be called 'principles of insufficient reason', but they do different jobs and the present principle is not subject to the difficulties encountered by the Laplacian principle. It avoids those difficulties mainly by providing us with only prima facie reasons, whereas the Laplacian principle purports to provide conclusive reasons.

we assume that those whose values are unknown to us would not upset the inference if they were known.

Thus far I have argued on intuitive grounds that classical direct inference presupposes the legitimacy of nonclassical direct inference, but a stronger connection can be established. The principles of classical direct inference *logically entail* an artificially restricted version of principle (*DI*). As we have seen, it follows from (C*DI*) that $\ulcorner$W(Gc)$\urcorner$ is a prima facie reason for $\ulcorner$prob(F/G) = PROB(Fc)$\urcorner$ and $\ulcorner$W(Hc)$\urcorner$ is a prima facie reason for $\ulcorner$prob(F/H) = PROB(Fc)$\urcorner$. Thus $\ulcorner$W(Gc) & W(Hc)$\urcorner$ gives us a prima facie reason for concluding that prob(F/G) = prob(F/H). $\ulcorner$W(Gc) & W(Hc)$\urcorner$ is a deductive consequence of $\ulcorner$W(Hc) & H is a deductive consequence of $G$$\urcorner$, so the latter also gives us a prima facie reason for concluding that prob(F/G) = prob(F/H):

(3.2)   If F is projectible with respect to both G and H then $\ulcorner$W(Hc) & H is a deductive consequence of G & prob(F/G) = $r$$\urcorner$ is a prima facie reason for $\ulcorner$prob(F/G) = $r$$\urcorner$.

(3.2) is a bit weaker than (*DI*), but this argument suffices to show that classical direct inference logically entails the legitimacy of at least some inferences of the general kind I have called 'nonclassical direct inferences'.

I will adopt principle (*DI*) as the basic principle of nonclassical direct inference. It should be emphasized that I am not supposing that the truth of this principle has been established. My only claim is that (*DI*) is presupposed by the inferences we actually make. We have yet to address the question of how those inferences and (*DI*) are to be justified.

A common reaction to principle (*DI*) is that it is absurd—perhaps trivially inconsistent. This arises from the observation that in a large number of cases, (*DI*) will provide us with prima facie reasons for conflicting inferences or even prima facie reasons for inferences to logically impossible conclusions. For example, since in a standard deck of cards a spade is necessarily black and the probability of a black card being a club is .5, (*DI*) gives us a prima facie reason to conclude that the probability of a spade being a club is .5, which is absurd. But this betrays an insensitivity to the functioning of prima facie reasons. A prima facie reason for an absurd conclusion is automatically defeated by the considerations leading us to regard the conclusion as absurd. Similarly, prima facie reasons for conflicting inferences defeat one another. If P is a prima facie reason for Q and R is a prima facie reason for $\sim Q$, then P and R rebut one another and both prima facie inferences are defeated. No inconsistency results. That this sort of case occurs with some frequency in nonclassical direct inference should not be surprising, because it also occurs with some frequency in classical direct inference. In classical direct inference we are very often in the position of knowing that an object has two logically independent properties G and H, where $\text{prob}(F/G) \neq \text{prob}(F/H)$. When that happens, classical direct inferences from these two probabilities conflict with one another, and so each prima facie reason is a defeater for the other, with the result that we are left without an undefeated direct inference to make.

As in the case of classical direct inference, we can obtain a number of variants of (DI), the most useful of which are:

(3.3)   If F is projectible with respect to G then $\lceil H \preccurlyeq G \rceil$ is a prima facie reason for $\lceil \text{prob}(F/H) = \text{prob}(F/G) \rceil$.

(3.4)   If F is projectible with respect to G then $\lceil H \preccurlyeq G \ \& \ \text{prob}(F/G) \in X \rceil$ is a prima facie reason for $\lceil \text{prob}(F/H) \in X \rceil$.

A counterlegal generalization of (DI) will be required in the the theory of induction:

(3.5)   If F is projectible with respect to G then $\lceil \Box \forall (H \supset G) \ \& \ (\diamond_p \exists G \ \rangle \ \diamond_p \exists H) \ \& \ \text{prob}(F/G) = r \rceil$ is a prima facie reason for $\lceil \text{prob}(F/H) = r \rceil$.

Proof: Principle (3.3) of Chapter 3 tells us that if P is a reason for Q, then $(R > P)$ is a reason for $(R > Q)$. Thus it follows from (DI) that

(i) $\diamond_p \exists G > [H \preccurlyeq G \ \& \ \text{prob}(F/G) = r]$

is a prima facie reason for

(ii) $\diamond_p \exists G > \text{prob}(F/H) = r$.

(i) is entailed by:

(iii) $\Box \forall (H \supset G) \ \& \ (\diamond_p \exists G > \diamond_p \exists H) \ \& \ \text{prob}(F/G) = r$.

$\ulcorner \Box \forall (H \supset G) \urcorner$ entails $\ulcorner \diamond_p \exists H > \diamond_p \exists G \urcorner$, so (iii) entails:

(iv) $(\diamond_p \exists G > \diamond_p \exists H) \ \& \ (\diamond_p \exists H > \diamond_p \exists G) \ \& \ \text{prob}(F/G) = r$.

By (6.20) of Chapter 2, the conjunction of (ii) and (iv) entails

(v) $\text{prob}(F/H) = r$.

Therefore, (iii) is a prima facie reason for (v).

The intuitive connection between definite and nomic probability utilized in the defense of (DI) can also be used to derive subset defeaters for (DI) from subset defeaters for classical direct inference. Suppose: (1) F is projectible with respect to G and J; (2) H entails J and J entails G; and (3) we know that $\text{prob}(F/G) = r$ and $\text{prob}(F/J) \neq \text{prob}(F/G)$. The inference to $\ulcorner \text{prob}(F/G) = \text{prob}(F/H) \urcorner$ is warranted iff, if we suppose c to be an object such that the only relevant properties we know c to have are G, J, and H, then the classical direct inference to $\ulcorner \text{PROB}(Fc) \urcorner$ is warranted. But we have a subset defeater blocking the latter direct inference, so it follows that the nonclassical direct inference to $\ulcorner \text{prob}(F/G) = \text{prob}(F/H) \urcorner$ should also be blocked. Therefore:

(SD)   If F is projectible with respect to J then $\ulcorner H \preccurlyeq J \preccurlyeq G \ \& \ \text{prob}(F/J)$
         $\neq \text{prob}(F/G) \urcorner$ is an undercutting defeater for (DI).

Defeaters of the form of (SD) will be called *subset defeaters* because of their connection with subset defeaters in classical direct inference. It follows from the derivation of the variants of (DI) that the defeaters described in (SD) are also defeaters for those variants:

(3.6)   If F is projectible with respect to J then $\ulcorner H \preccurlyeq J \preccurlyeq G$ & $\text{prob}(F/J)$ $\neq \text{prob}(F/G)\urcorner$ is an undercutting defeater for each of (3.3), (3.4), and (3.5).

Note that subset defeaters could be stated in terms of '$\Rightarrow$' rather than '$\preccurlyeq$'. Given that $H \preccurlyeq G$, $\ulcorner(H \Rightarrow J)$ & $(J \Rightarrow G)\urcorner$ holds iff $\ulcorner H \preccurlyeq J \preccurlyeq G\urcorner$ holds.

4. The Reduction of Classical Direct Inference to Nonclassical Direct Inference

There are two kinds of direct inference—classical and nonclassical. Direct inference has traditionally been identified with classical direct inference, but I believe that nonclassical direct inference is more fundamental. The details of classical direct inference are all reflected in nonclassical direct inference. If definite probabilities could be identified with certain (indefinite) nomic probabilities, the theory of classical direct inference could be derived from the theory of nonclassical direct inference. The purpose of this section is to show how this can be accomplished.

PROB(Fc) is a mixed physical/epistemic probability reflecting both the relevant nomic probabilities and our warranted beliefs about the object c. My proposal is that PROB(Fc) can be identified with the indefinite nomic probability of *an* object being F given that it has all the properties we are warranted in believing c to possess. This leads to precisely the classical direct inferences we would expect to be able to make. To illustrate, suppose that the only thing we know about c is that it has the projectible properties G, H, and K, and we want to ascertain the value of PROB(Fc). My proposal is that this should be identified with $\text{prob}(F/G\&H\&K)$. To verify this, suppose we know the values of $\text{prob}(F/G)$, $\text{prob}(F/H)$, and $\text{prob}(F/K)$. The moves we would make in classical direct inference are precisely the same as the moves we would make in attempting to ascertain the value of $\text{prob}(F/G\&H\&K)$. For example, if we know that $\text{prob}(F/G) = \text{prob}(F/H) = \text{prob}(F/K) = r$, then we will infer both that

PROB(Fc) = r and that prob($F/G\&H\&K$) = r. If we know that prob(F/G) $\neq$ r but prob(F/H) = prob(F/K) = r, and we do not know anything else relevant, then we will refrain from concluding that PROB(Fc) = r and we will also refrain from concluding that prob($F/G\&H\&K$) = r. If we know the above and also know that H entails G, then we will again infer both that PROB(Fc) = r and that prob($F/G\&H\&K$) = r. And so on.

My proposal is that PROB(Fc) is the nomic probability of F conditional on the conjunction of all the properties we are warranted in believing c to possess. One property we are warranted in believing c to possess is that of *being c*, that is, $\ulcorner x = c \urcorner$. More generally, if P is any warranted proposition then $\ulcorner x = c \&$ $P \urcorner$ is a property we are warranted in believing c to possess.[14] Consequently, if we let **K** be the conjunction of all warranted propositions, $\ulcorner x = c \& \mathbf{K} \urcorner$ is a property we are warranted in believing c to possess, and it entails every other property we are warranted in believing c to possess.[15] Therefore, my proposed analysis of PROB(Fc) can be expressed as:

(4.1)   PROB(Fc) = prob($Fx/x = c \& \mathbf{K}$).

(4.1) constitutes a precise formulation of the elusive "total evidence" requirement. Philosophers have resisted formulating it in this simple manner because they have supposed probabilities like prob($Fx / x = c \& \mathbf{K}$) to be illegitimate on the grounds either that they must always be 1 or 0 or else that there is no way we could know their values. The objection that they must be either 1 or 0 is dispelled by taking direct inference to pertain to nomic probabilities rather than relative frequencies. The nomic probability need not be either 1 or 0, because although the reference class

---

[14] This might seem like a peculiar property, but recall that properties were defined to be sets of ordered pairs $\langle w,x \rangle$ of possible worlds and objects. The property $\ulcorner x = c \& P \urcorner$ is that set containing just the pairs $\langle w,x \rangle$ such that x is c and P is true at w.

[15] **K** may be only a state of affairs rather than a proposition, because it may be an infinite conjunction, and the set of all propositions is not closed under infinite conjunction.

consists of a single object, the probability averages the behavior of c over many different possible worlds. The objection that there is no way we could know the value of $\text{prob}(Fx \,/\, x = c\ \&\ \mathbf{K})$ is more interesting. Our basic way of knowing the values of nomic probabilities is by statistical induction. It is only possible to ascertain the value of $\text{prob}(F/G)$ inductively if the extension of G is large so that we can compile statistics about the proportion of G's that are F. On the supposition that the only way to ascertain the value of a nomic probability is by induction, it follows that there is no way to ascertain the value of $\text{prob}(Fx \,/\, x = c\ \&\ \mathbf{K})$, but that supposition is mistaken. Non-classical direct inference provides another way of ascertaining the values of nomic probabilities. Induction and direct inference jointly provide the logical or epistemological machinery for dealing with nomic probabilities. By induction, we learn the values of certain nomic probabilities, and then by direct inference we infer the values of others. Without direct inference we would be unable to evaluate many probabilities that everyone agrees should be respectable. For example, there are no redheaded mathematicians who were born in Kintyre, North Dakota (population 7), but barring evidence to the contrary we would regard being redheaded, a mathematician, and having been born in Kintyre as irrelevant to the likelihood of getting lung cancer, and would take the probability of a redheaded mathematician born in Kintyre, North Dakota getting lung cancer to be the same as that for a resident of North Dakota in general. It is direct inference that allows this. But if direct inference legitimates such evaluations, it also allows us to evaluate probabilities like $\text{prob}(Fx \,/\, x = c\ \&\ \mathbf{K})$.

There is a simple and intriguing way of constructing definite probabilities that accord with (4.1). We can think of states of affairs as the limiting case of properties—zero-place properties. We might similarly think of definite probabilities as the corresponding limiting case of nomic probabilities. Applying the probability calculus mechanically to such probabilities yields:

(4.2)   If   (a) $\Box(Q \equiv Sa_1 \ldots a_n)$, and
        (b) $\Box[Q \supset (P \equiv Ra_1 \ldots a_n)]$
    then $\text{prob}(P/Q) =$

$$\text{prob}(Rx_1 \ldots x_n / Sx_1 \ldots x_n \ \& \ x_1 = a_1 \ \& \ \ldots \ \& \ x_n = a_n).$$

Proof: By the probability calculus, $\text{prob}(P/Q) = \text{prob}(Ra_1 \ldots a_n / Sa_1 \ldots a_n)$. By the principle (*IND*) of Chapter 2,

$$\text{prob}(Ra_1 \ldots a_n / Sa_1 \ldots a_n) =$$
$$\text{prob}(Rx_1 \ldots x_n / Sx_1 \ldots x_n \ \& \ x_1 = a_1 \ \& \ \ldots \ \& \ x_n = a_n).$$

Consequently:

$$\text{prob}(P/Q) = \text{prob}(Rx_1 \ldots x_n / Sx_1 \ldots x_n \ \& \ x_1 = a_1 \ \& \ \ldots \ \& \ x_n = a_n).$$

Such a blind application of the probability calculus is suspect, because it is open to the charge that these "definite" indefinite probabilities make no sense. But we can make them make sense by taking (4.2) as a definition. The following theorem of the probability calculus can be proven in the same way as (4.2):

(4.3) If (a) $\Box(Q \equiv Sa_1 \ldots a_n)$,
  (b) $\Box(Q \equiv Bb_1 \ldots b_m)$,
  (c) $\Box[Q \supset (P \equiv Ra_1 \ldots a_n)]$, and
  (d) $\Box[Q \supset (P \equiv Ab_1 \ldots b_m)]$,
then
$$\text{prob}(Rx_1 \ldots x_n / Sx_1 \ldots x_n \ \& \ x_1 = a_1 \ \& \ \ldots \ \& \ x_n = a_n) =$$
$$\text{prob}(Ay_1 \ldots y_m / By_1 \ldots y_m \ \& \ y_1 = b_1 \ \& \ \ldots \ \& \ y_m = b_m).$$

In other words, different ways of reducing P and Q to relational states of affairs will all yield the same probability. Thus we can simply define:

(4.4) $\text{prob}(P/Q) = r$ iff for some n, there are n-place properties R and S and objects $a_1, \ldots, a_n$ such that
  (a) $\Box(Q \equiv Sa_1 \ldots a_n)$,
  (b) $\Box[Q \supset (P \equiv Ra_1 \ldots a_n)]$, and
  (c) $\text{prob}(Rx_1 \ldots x_n / Sx_1 \ldots x_n \ \& \ x_1 = a_1 \ \& \ \ldots \ \& \ x_n = a_n) = r$.

So defined, these definite probabilities satisfy normal axioms for conditional definite probability:

(4.5) $0 \leq \text{prob}(P/Q) \leq 1$.

(4.6)   If $(Q \Rightarrow P)$ then $\text{prob}(P/Q) = 1$.

(4.7)   If $(Q \Leftrightarrow R)$ then $\text{prob}(P/Q) = \text{prob}(P/R)$.

(4.8)   If $\diamond R$ and $[R \Rightarrow \sim(P\&Q)]$ then $\text{prob}(P\lor Q/R) = \text{prob}(P/R)$
        $+ \text{prob}(Q/R)$.

(4.9)   If $[(P\&R) \Rightarrow Q]$ then $\text{prob}(P/R) \leq \text{prob}(Q/R)$.

(4.10) If $\diamond_p R$ then $\text{prob}(P\&Q/R) = \text{prob}(P/R) \cdot \text{prob}(Q/P\&R)$.

　　　　$\text{prob}(P/Q)$ is an *objective* conditional probability. It reflects
the state of the world, not the state of our knowledge. The mixed
physical/epistemic definite probabilities at which we arrive by
classical direct inference are not those defined by (4.4). Instead,
it follows from (4.1) that:

(4.11) $\text{PROB}(Fc) = \text{prob}(Fc/\mathbf{K})$

where $\mathbf{K}$ is the conjunction of all warranted propositions. I
propose that we generalize this and define:

(4.12) $\text{PROB}(P) = \text{prob}(P/\mathbf{K})$

(4.13) $\text{PROB}(P/Q) = \text{prob}(P/Q\&\mathbf{K})$.

These are our physical/epistemic definite probabilities.
　　　　It follows from (4.5)–(4.10) that physical/epistemic probabilities satisfy the following axioms for the probability calculus:

(4.14) $0 \leq \text{PROB}(P/Q) \leq 1$.

(4.15) If $(Q \Rightarrow P)$ then $\text{PROB}(P/Q) = 1$.

(4.16) If $(Q \Leftrightarrow R)$ then $\text{PROB}(P/Q) = \text{PROB}(P/R)$.

(4.17) If $\Diamond R$ and $[R \Rightarrow \sim(P\&Q)]$ then $\mathrm{PROB}(P \vee Q/R) = \mathrm{PROB}(P/R)$
        $+ \mathrm{PROB}(Q/R)$.

(4.18) If $[(P\&R) \Rightarrow Q]$ then $\mathrm{PROB}(P/R) \leq \mathrm{PROB}(Q/R)$.

(4.19) If $\Diamond_p(K\&R)$ then $\mathrm{PROB}(P\&Q/R)$
        $= \mathrm{PROB}(P/R) \cdot \mathrm{PROB}(Q/P\&R)$.

In addition we have:

(4.20) If $\mathbf{W}(P \supset Q)$ then $\mathrm{PROB}(P/Q) = 1$.

(4.21) If $\mathbf{W}(P \equiv Q)$ and $\mathbf{W}(R \equiv S)$ then $\mathrm{PROB}(P/R)$
        $= \mathrm{PROB}(Q/S)$.

(4.22) If $\mathbf{W}[(P\&R) \supset Q]$ then $\mathrm{PROB}(P/R) \leq \mathrm{PROB}(Q/R)$.

Thus the reduction of physical/epistemic probabilities to nomic probabilities endows them with a reasonable structure.[16]

The initial defense of nonclassical direct inference appealed to classical direct inference, but now we can turn the tables.

---

[16] A major point of contention in recent discussions of direct inference concerns the *principle of epistemic conditionalization*. According to this principle, if the set of warranted propositions changes simply by the addition of a state of affairs Q, and PROB_Q is the physical/epistemic probability function resulting from this change, then for any state of affairs P, $\mathrm{PROB}_Q(P) = \mathrm{PROB}(P/Q)$. Most philosophers seem to think that this principle ought to be true, but it fails on Kyburg's theory of direct inference. This has spawned extensive discussion in the literature regarding whether epistemic conditionalization *ought* to hold. (See particularly Harper [1983], Levi [1977], [1978], and [1981]; Kyburg [1977] and [1980]; and Seidenfeld [1978].) It is noteworthy, then, that it holds trivially on the present account:

Let $\mathbf{K}$ be the conjunction of warranted propositions. Suppose our epistemic state alters so that $\mathbf{K}_Q$ becomes the conjunction of propositions we are warranted in believing, where Q is a proposition such that $\Box[\mathbf{K}_Q \equiv (\mathbf{K}\&Q)]$. Let $\mathrm{PROB}_Q(P) = \mathrm{PROB}(P/\mathbf{K}_Q)$. Then $\mathrm{PROB}_Q(P) = \mathrm{PROB}(P/Q)$.

Proof: $\mathrm{PROB}_Q(P) = \mathrm{prob}(P/\mathbf{K}_Q) = \mathrm{prob}(P/\mathbf{K}\&Q) = \mathrm{PROB}(P/Q)$.

Given the reduction of definite probabilities to nomic probabilities, it becomes possible to derive classical direct inference from nonclassical direct inference. The basic rule of classical direct inference is principle (CDI):

> If F is projectible with respect to G then $\ulcorner \text{prob}(F/G) = r$ & $W(Gc)$ & $W(P \equiv Fc) \urcorner$ is a prima facie reason for $\ulcorner \text{PROB}(P) = r \urcorner$.

This can be deduced from the principle (DI) of nonclassical direct inference. Suppose F is projectible with respect to G, and we are warranted in believing $\ulcorner \text{prob}(F/G) = r$ & $W(Gc)$ & $W(P \equiv Fc) \urcorner$. $\ulcorner W(P \equiv Fc) \urcorner$ entails $\ulcorner \Box[K \supset (P \equiv Fc)] \urcorner$. Thus by (4.4) and (4.12), $\text{PROB}(P) = \text{prob}(P/K) = \text{prob}(Fx \,/\, x = c$ & $K)$. $\ulcorner W(Gc) \urcorner$ entails $\ulcorner \Box(K \supset Gc) \urcorner$ and hence entails $\ulcorner \Box(\forall x)[(x = c$ & $K) \supset Gx] \urcorner$. By the definition of K, we are warranted in believing $\ulcorner (\exists x)(x = c$ & $K) \urcorner$. Therefore, we are warranted in believing $\ulcorner \Diamond_p(\exists x)(x = c$ & $K) \urcorner$, and hence we are warranted in believing that $\ulcorner x = c$ & $K \urcorner$ is a subproperty of G. Thus by (DI), we have a prima facie reason for believing that $\text{prob}(Fx \,/\, x = c$ & $K) = r$ and hence for believing that $\text{PROB}(P) = r$.

Similar reasoning enables us to deduce the principle (CSD) of subset defeat for classical direct inference from the principle (SD) of subset defeat for nonclassical direct inference. According to (CSD):

> If F is projectible with respect to H then $\ulcorner \text{prob}(F/H) \neq \text{prob}(F/G)$ & $W(Hc)$ & $\Box\forall(H \supset G) \urcorner$ is an undercutting defeater for (CDI).

Suppose F is projectible with respect to H and we are warranted in believing $\ulcorner \text{prob}(F/H) \neq \text{prob}(F/G)$ & $W(Hc)$ & $\Box\forall(H \supset G) \urcorner$. The classical direct inference to $\ulcorner \text{PROB}(P) = r \urcorner$ is derived from the nonclassical direct inference to $\ulcorner \text{prob}(Fx \,/\, x = c$ & $K) = r \urcorner$, so a defeater for the latter also defeats the former. Because $\ulcorner W(Hc) \urcorner$ entails $\ulcorner \Box(\forall x)[(x = c$ & $K) \supset Hx] \urcorner$, it follows from (SD) that we have a subset defeater for the nonclassical direct inference, and hence the corresponding classical direct inference is defeated.

Thus classical direct inference can be regarded as nonclassical direct inference to the nomic probabilities with which the physical/epistemic definite probabilities are identified, and defeaters for classical direct inference are defeaters for the corresponding nonclassical direct inferences.

I remarked at the beginning of the chapter that decision theory requires conditional probabilities rather than nonconditional probabilities. Thus far I have discussed only classical direct inference to nonconditional physical/epistemic probabilities, but rules for conditional classical direct inference can also be derived from the rules for nonclassical direct inference:

(4.23) If F is projectible with respect to $(G\&H)$ then $\ulcorner W(Gc \supset Hc)$ & $W(P \equiv Fc)$ & $W(Q \equiv Gc)$ & $\text{prob}(F/G\&H) = r\urcorner$ is a prima facie reason for $\ulcorner \text{PROB}(P/Q) = r\urcorner$.

Proof: As above, $\ulcorner W(Gc \supset Hc)$ & $W(P \equiv Fc)$ & $W(Q \equiv Gc)\urcorner$ entails that $\text{PROB}(P/Q) = \text{prob}(Fx \,/\, Gx \,\&\, x = c \,\&\, \mathbf{K})$ and that $\square(\forall x)[(Gx \,\&\, x = c \,\&\, \mathbf{K}) \supset (Gx\&Hx)]$, so nonclassical direct inference gives us a prima facie reason for believing that $\text{prob}(Fx \,/\, Gx \,\&\, x = c \,\&\, \mathbf{K}) = r$ and hence that $\text{PROB}(P/Q) = r$.

Similar reasoning establishes the following principle of subset defeat for conditional classical direct inference:

(4.24) If F is projectible with respect to $(G\&J)$ then $\ulcorner \text{prob}(F/G\&J) \neq \text{prob}(F/G\&H)$ & $W(Gc \supset Jc)$ & $\square\forall[(G\&J) \supset H]\urcorner$ is an undercutting defeater for (4.23).

The upshot of all this is that classical direct inference can be explained in terms of nonclassical direct inference, and definite probabilities can be defined in terms of nomic probabilities. This constitutes a very important simplification to the overall theory of nomic probability.

The mixed physical/epistemic probabilities symbolized by 'PROB' are not the only definite probabilities of interest. Although of less central importance, propensities have played a role in probabilistic thought and they may play a significant role in science. Having seen how to define 'PROB' in terms of 'prob' and how to derive the theory of classical direct inference from the theory of nonclassical direct inference it is a simple matter to construct similar definitions for various kinds of propensities and to derive principles of direct inference for them as well. However, to avoid disrupting the continuity of the discussion, I will postpone the details of this until Chapter 10.

5. Foundations for Direct Inference

The reduction of classical direct inference to nonclassical direct inference brings an important unity to the theory of nomic probability, but it must still be considered how nonclassical direct inference is to be justified. Thus far the only reason for endorsing the principles of nonclassical direct inference has been that they seem to be required to license classical direct inferences that are intuitively correct. The principles of direct inference are descriptive of inferences we actually make, but that is not yet to justify them. What will now be shown is that the principles of direct inference, both classical and nonclassical, are derivable from the acceptance rules and computational principles that have already been endorsed.

The theory of direct inference is based upon the following two principles:

(DI)   If F is projectible with respect to G then $\ulcorner H \preccurlyeq G$ & $\mathrm{prob}(F/G) = r\urcorner$ is a prima facie reason for $\ulcorner\mathrm{prob}(F/H) = r\urcorner$;

(SD)   If F is projectible with respect to J then $\ulcorner H \preccurlyeq J \preccurlyeq G$ & $\mathrm{prob}(F/J) \neq \mathrm{prob}(F/G)\urcorner$ is an undercutting defeater for (DI);

Thus the objective is to derive these principles. For the purpose of deriving (DI), it suffices to derive principle (3.3):

If F is projectible with respect to G then $\ulcorner H \preccurlyeq G \urcorner$ is a prima facie reason for $\ulcorner \text{prob}(F/H) = \text{prob}(F/G) \urcorner$.

This is equivalent to (DI). This, in turn, follows from a seemingly weaker version of (DI) in the same way it followed from (DI):

(DI^*) If F is projectible with respect to G and r is a nomically rigid designator of a real number then $\ulcorner H \preccurlyeq G \ \& \ \text{prob}(F/G) = r \urcorner$ is a prima facie reason for $\ulcorner \text{prob}(F/H) = r \urcorner$;

(3.3) follows from (DI^*) because $\ulcorner \text{prob}(F/G) \urcorner$ is a nomically rigid designator. This shows that (DI^*) is not really weaker than (DI). So the strategy is to derive (DI^*).

The key element in providing foundations for direct inference is the principle of agreement, $(AGREE)$ of Chapter 2. First recall the definition of 'strict subproperty' (henceforth abbreviated '$\prec$'):

(5.1) $H \prec G$ iff (1) $H \preccurlyeq G$ and (2) if $\langle w, x_1, \ldots, x_n \rangle \in H$ then w is a physically possible world.

Set-theoretically, strict subproperties are the restrictions of subproperties to the set of all physically possible worlds. Using this notation, I list the principle of agreement for easy reference:

$(AGREE)$ If F and G are properties and there are infinitely many physically possible G's and $\text{prob}(F/G) = p$ (where p is a nomically rigid designator) then for every $\delta > 0$, $\text{prob}\big(\text{prob}(F/X) \approx_\delta p \ / \ X \prec G\big) = 1$.

Recall that $(A1)$ and its defeater $(D1)$ are as follows:

$(A1)$ If A is projectible with respect to B and $r > .5$ then $\ulcorner Bc \ \& \ \text{prob}(A/B) \geq r \urcorner$ is a prima facie reason for the conditional $\ulcorner Ac \urcorner$, the strength of the reason depending upon the value of r.

(D1)   If A is projectible with respect to c then $\ulcorner Cc$ and prob$(A/B\&C)$ $\neq$ prob$(A/B)\urcorner$ constitutes an undercutting defeater for $(A1)$.

The derivation of (DI^*) and (SD) is now simple. Suppose p is a nomically rigid designator. We have the following instance of $(A1)$:

(5.2)   If $\ulcorner$prob$(F/X) \approx_\delta p\urcorner$ is projectible with respect to $\ulcorner X \ll G\urcorner$ then $\ulcorner H \ll G \ \& \ $prob$\big($prob$(F/X) \approx_\delta p \ / \ X \ll G\big) = 1\urcorner$ is a prima facie reason for $\ulcorner$prob$(F/H) \approx_\delta p\urcorner$.

If we assume that the property $\ulcorner$prob$(F/X) \approx_\delta p\urcorner$ is projectible with respect to $\ulcorner X \ll G\urcorner$ whenever F is projectible with respect to G, then it follows that:

(5.3)   If F is projectible with respect to G then
          $\ulcorner H \ll G \ \& \ $prob$\big($prob$(F/X) \approx_\delta p \ / \ X \ll G\big) = 1\urcorner$
          is a prima facie reason for $\ulcorner$prob$(F/H) \approx_\delta p\urcorner$.

Let us symbolize $\ulcorner$There are infinitely many physically possible G's$\urcorner$ as $\ulcorner \exists_\infty G\urcorner$. By $(AGREE)$, for each $\delta > 0$, $\ulcorner \exists_\infty G \ \& \ $prob$(F/G) = p\urcorner$ entails $\ulcorner$prob$\big($prob$(F/X) \approx_\delta p \ / \ X \ll G\big) = 1\urcorner$, so it follows that:

(5.4)   If F is projectible with respect to G then for each $\delta > 0$, $\ulcorner \exists_\infty G \ \& \ $prob$(F/G) = p \ \& \ H \ll G\urcorner$ is a prima facie reason for $\ulcorner$prob$(F/H) \approx_\delta p\urcorner$.

Consider the requirement in (5.4) that we be warranted in believing $\ulcorner \exists_\infty G\urcorner$. To require that there are infinitely many physically possible G's is to require very little more than that it is physically possible for there to be G's. It is extremely difficult to to construct properties G that are not counterlegal but such that there are only finitely many physically possible G's. In this connection, recall that a physically possible G is not just a physically possible object that is G in some world. Rather, a physically possible G is a pair $\langle w,x \rangle$ where w is a physically possible world and x is G at w. Consequently, for there to be infinitely many physically possible G's it suffices for there to be infinitely many physically possible worlds at which there are G's. Any reasonable noncounterlegal property will satisfy this condition. Even if G is a property like that of *being Bertrand Russell*, which can only be possessed by a single object, there are infinitely many physically possible G's because there are infinitely many physically possible worlds at which Bertrand Russell has the property of being Bertrand Russell. It appears that the only way there can be physically possible G's but only finitely many of them is for G to be a very contrived property. For example, picking some particular possible world α, we might consider the property of *being Bertrand Russell and such that α is the actual world*. This peculiar property can only be possessed by Bertrand Russell, and it can only be possessed by him at the world α. But any normal noncounterlegal property G will be such that there are infinitely many physically possible G's. In particular, any noncounterlegal "qualitative" property will satisfy this condition. Specifically, I will assume that any noncounterlegal projectible property satisfies this condition necessarily. It follows that if we are warranted in believing $\ulcorner \Diamond_p \exists G \urcorner$ then we are warranted in believing $\ulcorner \exists_\infty G \urcorner$. But $\ulcorner H \ll G \urcorner$ entails $\ulcorner \Diamond_p \exists G \urcorner$, so (5.4) can be simplified as follows:

(5.5)   If F is projectible with respect to G then for each $\delta > 0$, $\ulcorner \mathrm{prob}(F/G) = p \ \& \ H \ll G \urcorner$ is a prima facie reason for $\ulcorner \mathrm{prob}(F/H) \approx_\delta p \urcorner$.

According to principle (3.7) of Chapter 3, given any reason schema of the form $\ulcorner Ax$ is a prima facie reason for $Bx \urcorner$ it follows that $\ulcorner (\forall x)Ax \urcorner$ is a prima facie reason for $\ulcorner (\forall x)Bx \urcorner$. Applying this to (5.5), it follows that:

(5.6)   If F is projectible with respect to G then
$$(\forall \delta)[\delta > 0 \supset [\mathrm{prob}(F/G) = p \ \& \ H \ll G]]$$
is a prima facie reason for
$$(\forall \delta)[\delta > 0 \supset \mathrm{prob}(F/H) \approx_\delta p].$$

⌜$(\forall \delta)[\delta > 0 \supset [\text{prob}(F/G) = p \ \& \ H \ll G]]$⌝ is equivalent to ⌜$\text{prob}(F/G) = p \ \& \ H \ll G$⌝, and ⌜$(\forall \delta)[\delta > 0 \supset \text{prob}(F/H) \approx_\delta p]$⌝ is equivalent to ⌜$\text{prob}(F/H) = r$⌝. Thus it follows from (5.6) that:

(5.7)   If F is projectible with respect to G then ⌜$\text{prob}(F/G) = p \ \& \ H \ll G$⌝ is a prima facie reason for ⌜$\text{prob}(F/H) = p$⌝.

Principle (5.7) differs from the principle (DI) of nonclassical direct inference only in that it is about strict subproperties rather than subproperties in general. That can be rectified as follows. Where G and H are n-place properties, define:

(5.8) H^0 is the *truncation* of H iff, for every w, $x_1,\ldots,x_n$, $\langle w,x_1,\ldots,x_n\rangle \in H^0$ iff $\langle w,x_1,\ldots,x_n\rangle \in H$ and w is a physically possible world.

Note that if H^0 is the truncation of H then $\mathbf{H}^0 = \mathbf{H}$.[17] Consequently, if H is not counterlegal, $\text{prob}(F/H) = \text{prob}(F/H^0)$. With this observation, (DI) follows:

If F is projectible with respect to G then ⌜$\text{prob}(F/G) = p \ \& \ H \leqq G$⌝ is a prima facie reason for ⌜$\text{prob}(F/H) = p$⌝.

Proof: Suppose F is projectible with respect to G. Let H^0 be the truncation of H. ⌜$\text{prob}(F/G) = p \ \& \ H \leqq G$⌝ entails ⌜$\text{prob}(F/G) = p \ \& \ H^0 \ll G$⌝, which by (5.7) is a prima facie reason for ⌜$\text{prob}(F/H^0) = p$⌝, which entails ⌜$\text{prob}(F/H) = p$⌝.

Note that this derivation explains the projectibility constraint in (DI). It arises directly out of the projectibility constraint in $(A3)$. Furthermore, it explains a previously unexplained feature of the projectibility constraint. The most natural constraint would require F to be projectible with respect to both G and H. It is easy enough to find examples illustrating the apparent need for F to be projectible with respect to G, but not for F to be project-

---

[17] In fact, the definition of a physically possible object has the consequence that $H^0 = H$.

ible with respect to H. It becomes important that F not be required to be projectible with respect to H when we reduce classical direct inference to nonclassical direct inference. In classical direct inference we are making nonclassical direct inferences to probabilities of the form $\ulcorner\mathrm{prob}(Fx \,/\, x = c\ \&\ \mathbf{K})\urcorner$, but properties of the form $\ulcorner x = c\ \&\ \mathbf{K}\urcorner$ will not usually be projectible. The explanation for why (DI) does not require F to be projectible with respect to H is now evident. (DI) is obtained from the instance (5.2) of (AI), and that is the source of the projectibility constraint. F is required to be projectible with respect to G in order to make the property A projectible with respect to B as is required by (AI). But in (5.2), H is playing the role of the individual c in (AI), and thus there is no reason why it should have to be projectible.

Principle (SD) (The Rule of Subset Defeat) follows with the help of (DI). According to the present account, the direct inference from $\ulcorner\mathrm{prob}(F/G) = p\urcorner$ to $\ulcorner\mathrm{prob}(F/H) = p\urcorner$ proceeds via instance (5.2) of (AI), and hence it will be defeated by any defeater for that instance of (AI). By (DI) and our assumptions about projectibility, the following is such a defeater:

(5.9)   If F is projectible with respect to J then $\ulcorner H^0 \ll J^0$ and $\mathrm{prob}\big(\mathrm{prob}(F/X) \approx_\delta p \,/\, X \ll G^0$ and $X \ll J^0\big) < 1\urcorner$ is an undercutting defeater for (5.9).

To establish (SD), suppose we are warranted in believing $\ulcorner H \leqslant J \leqslant G\ \&\ \mathrm{prob}(F/J) \neq \mathrm{prob}(F/G)\urcorner$. As $J \leqslant G$, it follows that $J^0 \ll G^0$ and hence any strict subproperty of J^0 is a strict subproperty of G^0. Therefore,

$$\mathrm{prob}\big(\mathrm{prob}(F/X) \approx_\delta p \,/\, X \ll G^0 \text{ and } X \ll J^0\big)$$
$$= \mathrm{prob}\big(\mathrm{prob}(F/X) \approx_\delta p \,/\, X \ll J^0\big).$$

By hypothesis, we are warranted in believing that $\mathrm{prob}(F/J) \neq \mathrm{prob}(F/G)$. Let $s = \mathrm{prob}(F/J)$. Then we are warranted in believing that $\mathrm{prob}(F/J^0) = s$. It follows from $(AGREE)$ that we are warranted in believing that for every $\delta > 0$,

$$\text{prob}\big(\text{prob}(F/X) \approx_\delta s \:/\: X \blacktriangleleft J^0\big) = 1$$

and therefore that

$$\text{prob}\big(\text{prob}(F/X) \approx_\delta s \:/\: X \blacktriangleleft G^0 \text{ and } X \blacktriangleleft J^0\big) = 1.$$

Choosing $\delta < \frac{1}{2}|r\text{-}s|$, it follows that we are warranted in believing that

$$\text{prob}\big(\text{prob}(F/X) \approx_\delta p \:/\: X \blacktriangleleft G^0 \text{ and } X \blacktriangleleft J^0\big) < 1.$$

Furthermore, because $H \underset{\sim}{\leq} G$ it follows that $H^0 \blacktriangleleft G^0$. Thus we have a defeater of the form of (5.9). Consequently, the Rule of Subset Defeat follows from this account of direct inference.

The main result of this section is a very important simplification in the overall theory of nomic probability. The fundamental assumptions of the theory have been reduced to two classes of principles. The most uniquely philosophical principles are $(A3)$ and $(D3)$. The second class of principles consists of the computational principles governing the logical and mathematical structure of nomic probability. The latter principles are complicated, but not surprising from a philosophical point of view. They amount to no more than an elaboration of the classical probability calculus. Thus nomic probability is found, so far, to have a rather simple philosophical foundation. One important kind of probabilistic reasoning remains to be examined, and that is induction. But it will turn out that induction requires no new principles. Thus the entire theory will ultimately be based upon this simple foundation.

6. Nonclassical Direct Inference for Frequencies

In keeping with our project of comparing inferences from frequencies with inferences from nomic probabilities, consider to what extent we could construct a similar theory of direct inference based upon relative frequencies.   It is possible to construct a somewhat analogous theory of nonclassical direct inference for frequencies, but there will be important differences. The difficulty is that the above theory is based upon (*AGREE*), and (*AGREE*) does not hold for finite sets (which are the only sets for which relative frequencies exist). To a certain extent we can replace the use of (*AGREE*) in the above arguments by the use of the finitary principle (9.2) of Chapter 2, which is what first led us to (*AGREE*). More precisely, if $\#B = n$, then it is possible to compute the value of

(6.1) $\text{freq}\big[\text{freq}[A/X] \approx_\delta p \, / \, X \subseteq B\big].$

According to principle (9.3), this value goes to 1 as n goes to infinity. But for any particular n, the value is less than 1, and if δ is chosen to be sufficiently small, the value of (6.1) will not even be close to 1.   On the basis of (6.1) and the frequency-based acceptance rule ($A3^*$), we can construct a frequency-based theory of nonclassical direct inference allowing the defeasible inference that if B is a large finite set and $C \subseteq B$, then for large enough δ, $\text{freq}[A/C] \approx_\delta \text{freq}[A/B]$.   This is weaker than the theory of nonclassical direct inference for nomic probability in two respects. First, it only provides a reason for thinking that $\text{freq}[A/C]$ is approximately equal to $\text{freq}[A/B]$, and unless B is quite large the approximation may not be very good.   Second, the probability involved in (*AGREE*) is 1, but the probability of (6.1) is always less than 1, so the reason provided by nonclassical direct inference for frequencies will be weaker than the reason provided by nonclassical direct inference for nomic probabilities (and this is true no matter how large a δ is chosen). Thus, although a theory of nonclassical direct inference for relative frequencies can be constructed, it will not be nearly as useful as the theory of nonclassical direct inference for nomic probabilities. In addition, it is not possible to define definite probabilities in terms of frequencies in the same way they were defined in terms of nomic probabilities. This is because $\text{freq}[Fx \, / \, x = c \, \& \, \mathbf{K}]$ must be either 0 or 1. Thus it appears that direct inference is an arena in which there is no possibility of replacing all appeal to nomic probabilities by appeal to relative frequencies.

7. Summary

1. Reichenbach's theory of classical direct inference can be formulated in terms of prima facie reasons and subset defeaters.
2. Direct inference from relative frequencies is explicable in terms of direct inference from nomic probabilities derived from (*PFREQ*).
3. Classical direct inference both presupposes and entails special cases of nonclassical direct inference.
4. Definite probabilities can be identified with nomic probabilities in such a way that the principles of classical direct inference become derivable from the principles of nonclassical direct inference.
5. The principles of nonclassical direct inference (and so indirectly the principles of classical direct inference) are derivable from the acceptance rule (*A3*) and (*AGREE*).

CHAPTER 5
INDUCTION

1. Introduction

The objective of this book is to provide an analysis of nomic probability in terms of its conceptual role. This requires an account both of (1) what inferences can be drawn from nomic probabilities, and (2) how nomic probabilities can be evaluated on the basis of nonprobabilistic information. We have an account of (1). That consists of the acceptance rules and the theory of direct inference. The theory of nonclassical direct inference also provides a partial account of (2). Some nomic probabilities can be evaluated in terms of others by using nonclassical direct inference. But in order to get this process started in the first place, there must be some other way of evaluating nomic probabilities that does not require any prior contingent knowledge of the values of such probabilities. It seems intuitively clear that that is accomplished by some kind of statistical induction. In statistical induction, we observe a sample of B's, determine the relative frequency of A's in that sample, and then estimate prob(A/B) to be approximately equal to that relative frequency. A close kin to statistical induction is enumerative induction, wherein it is observed that all of the B's in the sample are A's, and it is concluded that any A would be a B, that is, $A \Rightarrow B$.

There are two possibilities regarding statistical and enumerative induction. They could be derivable from more basic epistemic principles, or they might be irreducible constituents of the conceptual role of nomic probability and nomic generalizations. These two possibilities reflect what have come to be regarded as two different problems of induction. The traditional problem of induction was that of justifying induction. But most contemporary philosophers have forsaken that for Goodman's "new riddle of induction", which I am construing here as the problem of giving an accurate account of correct principles of induction. This change in orientation reflects the view that principles of induction are basic epistemic principles, partly constitutive of rationality, and not reducible to or justifiable on the basis of anything more fundamental. I endorsed the latter view in my [1974], but now I am convinced that it is false. In this chapter I will address both problems of induction, showing that precise principles of induction can be derived from (and hence justified on the basis of) the various principles regarding nomic probability that have already been endorsed. This will simultaneously provide a solution to the new riddle of induction, and a solution to the traditional problem of induction.

On the face of it, statistical induction seems more complicated than enumerative induction and has usually been taken to be a generalization of enumerative induction. This suggests that we should solve the problem of enumerative induction before addressing statistical induction. But I will argue that statistical induction is actually the more fundamental of the two. Enumerative induction presupposes statistical induction, but goes beyond it.

Principles of statistical induction are principles telling us how to estimate probabilities on the basis of observed relative frequencies in finite samples. The problem of constructing such principles is often called 'the problem of inverse inference'. All theories of inverse inference are similar in certain respects. In particular, they all make use of some form of Bernoulli's theorem. In its standard formulation, Bernoulli's theorem tells us that if we have n objects $b_1,\ldots,b_n$ and for each i, $\mathrm{PROB}(Ab_i) = p$, and any of these objects being A is statistically independent of which others of them are A, then the probability is high that the relative frequency of A's among $b_1,\ldots,b_n$ is approximately p. Furthermore, the probability increases and the degree of approximation improves as n is made larger. These probabilities can be computed quite simply by noting that on the stated assumption of independence, it follows from the probability calculus that

$$\mathrm{PROB}(Ab_1 \ \& \ \ldots \ \& \ Ab_r \ \& \ {\sim}Ab_{r+1} \ \& \ \ldots \ \& \ {\sim}Ab_n)$$

$$= \mathrm{PROB}(Ab_1) \cdot \ldots \cdot \mathrm{PROB}(Ab_r)$$
$$\cdot \mathrm{PROB}({\sim}Ab_{r+1}) \cdot \ldots \cdot \mathrm{PROB}({\sim}Ab_n)$$

$$= p^r(1-p)^{n-r}.$$

There are $n!/r!(n-r)!$ distinct ways of assigning A-hood among $b_1,\dots,b_n$ such that freq$[A/\{b_1,\dots,b_n\}] = r/n$, so it follows by the probability calculus that

$$\text{PROB}\Big(\text{freq}[A/\{b_1,\dots,b_n\}] = \frac{r}{n}\Big) = \frac{n!p^r(1-p)^{n-r}}{r!(n-r)!}.$$

The right side of this equation is the formula for the binomial distribution. An interval $[p-\delta,p+\delta]$ around p will contain just finitely many fractions r/n with denominator n, so we can calculate the probability that the relative frequency has any one of those values, and then the probability of the relative frequency being in the interval is the sum of those probabilities.

Thus far everything is uncontroversial. The problem is what to do with the probabilities resulting from Bernoulli's theorem. Most theories of inverse inference, including most of the theories that are part of contemporary statistical theory, can be regarded as variants of a single intuitive argument that goes as follows. Suppose prob$(A/B) = p$, and all we know about $b_1,\dots,b_n$ is that they are B's. By classical direct inference it can be inferred that for each i, PROB$(A_i) = p$. If the b_i's seem intuitively unrelated then it seems reasonable to suppose they are statistically independent and so Bernoulli's theorem can be used to conclude that it is extremely probable that the observed relative frequency r/n lies in a small interval $[p-\delta,p+\delta]$ around p. This entails conversely that p is within δ of r/n; that is, p is in the interval $[(r/n)-\delta,(r/n)+\delta]$. This becomes the estimate of p.

This general line of reasoning seems plausible until we try to fill in the details. Then it begins to fall apart. There are basically two problems. The first is the assumption of statistical independence that is required for the calculation involved in Bernoulli's theorem. That is a probabilistic assumption. Made precise, it is the assumption that for each Boolean conjunction B of the conjuncts $\ulcorner Ab_j\urcorner$ or their negations (for $j \neq i$), PROB(Ab_i/B) = PROB(Ab_i). To know this we must already have probabilistic information. In practice, it is supposed that if the b_i's "have nothing to do with one another" then they are independent in this sense, but it is hard to see how that can be justified noncircularly.

We might try to solve this problem by adopting a fundamental postulate allowing us to assume independence unless we have some reason for thinking otherwise. In a sense, I think that is the right way to go, but it is terribly *ad hoc*. It will turn out below that such a postulate can be replaced by a derived principle following from the parts of the theory of nomic probability that have already been established.

A much deeper problem for the intuitive argument concerns what to do with the conclusion that it is very probable that the observed frequency is within δ of p. It is tempting to use the acceptance rule *(A1)* and reason:

If $\mathrm{prob}(A/B) = p$ then
$$\mathrm{PROB}\Big(\mathrm{freq}[A/\{b_1,\ldots,b_n\}]\in[p-\delta,p+\delta]\Big) \approx 1$$

so

if $\mathrm{prob}(A/B) = p$ then $\mathrm{freq}[A/\{b_1,\ldots,b_n\}]\in[p-\delta,[p+\delta]$.

The latter entails

If $\mathrm{freq}[A/\{b_1,\ldots,b_n\}] = \frac{r}{n}$ then $\mathrm{prob}(A/B)\in[\frac{r}{n}-\delta,\frac{r}{n}+\delta]$.

An immediate difficulty for this reasoning is that it is an incorrect use of *(A1)*. *(A1)* concerns indefinite probabilities, whereas Bernoulli's theorem provides definite probabilities. But let us waive that difficulty for the moment, because there is a much more profound difficulty. This is that the probabilities obtained in this way have the structure of the lottery paradox. Given any point q in the interval $[p-\delta,p+\delta]$, we can find a small interval λ around it such that if we let I_q be the union of two intervals $[0,q-\lambda]\cup[q+\lambda,1]$, the probability of $\mathrm{freq}[A/\{b_1,\ldots,b_n\}]$ being in I_q is as great as the probability of its being in $[p-\delta,p+\delta]$. This is diagramed in Figure 1. The probability of the frequency falling in any interval is represented by the area under the curve corresponding to that interval. The curve is reflected about the x axis so that the probability for the interval $[p-\delta,p+\delta]$ can be represented above the axis and the probability for the interval I_q

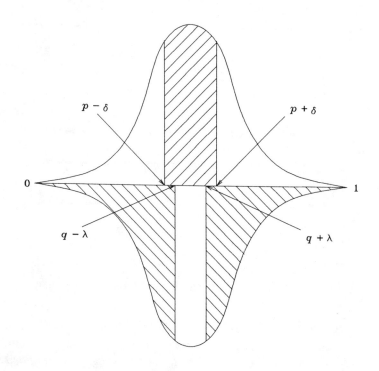

Figure 1. Intervals on the tails with the
same probability as an interval at the maximum.

represented below the axis. Next notice that we can construct a
finite set $q_1,\ldots,q_k$ of points in $[p-\delta,p+\delta]$ such that the "gaps"
in the I_{q_i} collectively cover $[p-\delta,p+\delta]$. This is diagramed in
Figure 2. For each $i \leq k$, we have as good a reason for believing
that freq$[A/\{b_1,\ldots,b_n\}]$ is in I_{q_i} as we do for thinking it is in
$[p-\delta,p+\delta]$, but these conclusions are jointly inconsistent. The
result is a case of collective defeat, and thus we are unwarranted
in concluding that the relative frequency is in the interval
$[p-\delta,p+\delta]$.

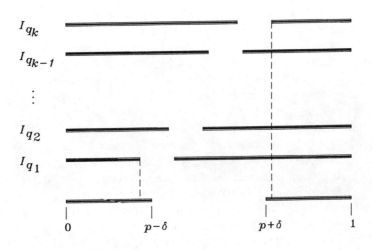

Figure 2. A collection of small gaps covers a large gap.

The intuitive response to the "lottery objection" consists of noting that $[p-\delta, p+\delta]$ is an interval, whereas the I_q are not. Somehow, it seems right to make an inference regarding intervals when it is not right to make the analogous inference regarding "gappy" sets. That is the line taken in orthodox statistical inference when confidence intervals are employed.[1] But it is very hard to see why this should be the case, and some heavy-duty argument is needed here to justify the whole procedure.

In sum, when we try to make the intuitive argument precise,

---

[1] The orthodox theory of confidence intervals is subject to the same difficulty. Of course, J. Neyman and E. S. Pearson, principal architects of the orthodox theory, always insisted that it was not intended as a theory of hypothesis testing. But as Alan Birnbaum [1962] observes, "Neyman himself has been foremost in insisting that standard statistical methods ... cannot appropriately be interpreted as methods of inference.... Nevertheless, the latter interpretation is taken as basic, implicitly if not explicitly, in a major part of the exposition and application of the Neyman-Pearson theory."

it becomes apparent that it contains major lacunae. This does not constitute an utter condemnation of the intuitive argument. Because it is so intuitive, it would be surprising if it were not at least approximately right. Existing statistical theories tend to be *ad hoc* jury-rigged affairs without adequate foundations, but it seems there must be some sound intuitions that statisticians are trying to capture with these theories.[2] The problem is to turn the intuitive argument into a rigorous and defensible argument. In what follows, the intuitive argument will undergo three kinds of repairs, creating what I call *the statistical induction argument*. First, it will be reformulated in terms of indefinite probabilities, thus enabling us to make legitimate use of our acceptance rules. Second, it will be shown that the gap concerning statistical independence can be filled by nonclassical direct inference. Third, the final step of the argument will be scrapped and replaced by a more complex argument not subject to the lottery paradox. This more complex argument will employ a principle akin to the *Likelihood Principle* of classical statistical inference, and it is to the justification of that principle that I turn next.

2. A Generalized Acceptance Rule

The principle required for the last part of the statistical induction argument is a generalization of the acceptance rules already adopted. Recall that $(A2)$ is the following acceptance rule:

$(A2)$　　If F is projectible with respect to G and $r > .5$, then $\lceil \text{prob}(F/G) \geq r \ \& \ \sim Fc \rceil$ is a prima facie reason for $\lceil \sim Gc \rceil$, the strength of the reason depending upon the value of r.

---

[2] In thus chastising existing statistical theory, I take it that I am not saying anything controversial. It is generally acknowledged that the foundations of statistical theory are a mess. For discussions of some of this, see Alan Birnbaum [1969], D. R. Cox [1958], A. P. Dempster [1964] and [1966], Henry Kyburg [1974] (51ff), John Pratt [1961], and Teddy Seidenfeld [1979].

To prepare the way for generalizing $(A2)$, consider how $(A2)$ works in some simpler cases that will be similar in various respects to the case of statistical induction. First, recall how $(A2)$ works in a simple lottery case. Suppose we have a lottery consisting of 100 tickets, each with a chance of .01 of being drawn. Let $\ulcorner T_n x\urcorner$ be $\ulcorner x$ has ticket $n\urcorner$ and let $\ulcorner Wx\urcorner$ be $\ulcorner x$ wins$\urcorner$. $\text{prob}(\sim Wx/T_n x)$ is high, but upon discovering that Jones won the lottery we would not infer that he did not have ticket n. The reason we would not do so is that this is a case of collective defeat. Because $\text{prob}(\sim Wx/T_n x)$ is high and we are warranted in believing $\ulcorner Wj\urcorner$, we have a prima facie reason for believing $\ulcorner \sim T_n j\urcorner$. But for each k, $\text{prob}(\sim Wx/T_k x)$ is equally high, so we have an equally strong prima facie reason for believing each $\ulcorner \sim T_k j\urcorner$. From the fact that Jones won, we know that he had at least one ticket. Thus we can construct the counterargument diagramed in Figure 3 for the conclusion that $\ulcorner T_n j\urcorner$ is true. Our reason for believing each $\ulcorner \sim T_k j\urcorner$ is as good as our reason for believing $\ulcorner \sim T_n j\urcorner$, so we have as strong a reason for $\ulcorner T_n j\urcorner$ as for $\ulcorner \sim T_n j\urcorner$. Hence

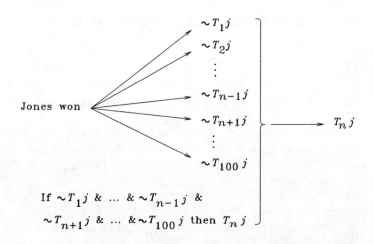

Figure 3. The counterargument for $T_n j$.

our prima facie reason for the latter is defeated and we are not warranted in believing $\ulcorner \sim T_n j \urcorner$.

Next, consider a biased lottery consisting of ten tickets. The probability of ticket 1 being drawn is .000001, and the probability of any other ticket being drawn is .111111. It is useful to diagram these probabilities as in Figure 4. If Jones wins it is reasonable for us to infer that he did not have ticket 1, because the probability of his having any other ticket is more than 100,000 times greater. This inference is supported by $(A2)$ as follows. As before, for each n, prob($\sim Wx/T_n x$) is fairly high. Combining this with the fact that Jones wins gives us a prima facie reason for believing $\ulcorner \sim T_n j \urcorner$, for each n. But these reasons are no longer of equal strength. Because Jones would be much less likely to win if he had ticket 1 than if he had any other ticket, we have a much stronger reason for believing that he does not have ticket 1. As

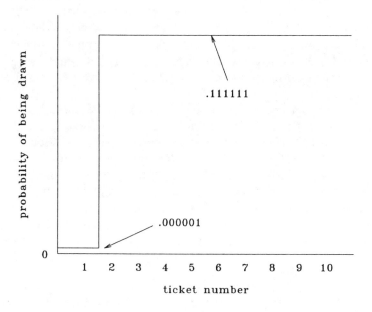

Figure 4. Lottery 2.

before, for $n \neq 1$, we have the counterargument diagramed in Figure 3 for $\ulcorner T_n j \urcorner$, and that provides as good a reason for believing $\ulcorner T_n j \urcorner$ as we have for believing $\ulcorner \sim T_n j \urcorner$. Thus the prima facie reason for $\ulcorner \sim T_n j \urcorner$ is defeated. But we do not have as good a reason for believing $\ulcorner T_1 j \urcorner$ as we do for believing $\ulcorner \sim T_1 j \urcorner$. An argument is only as good as its weakest link, and the counterargument for $\ulcorner T_1 j \urcorner$ employs the prima facie reasons for $\ulcorner \sim T_n j \urcorner$. These reasons are based upon lower probabilities (of value .888889) and hence are not as strong as the prima facie reason for $\ulcorner \sim T_1 j \urcorner$ (based upon a probability of value .999999). Thus, although we have a reason for believing $\ulcorner T_1 j \urcorner$, we have a much better reason for believing $\ulcorner \sim T_1 j \urcorner$, and so on sum we are warranted in believing the latter.

This reasoning is reminiscent of orthodox statistical significance testing. The *likelihood* of T_n on W is $\text{prob}(Wx/T_n x)$.[3] In the present lottery (call it 'lottery 2') the likelihood of Jones having ticket 1 given that he wins is .000001. A statistician would tell us that we can reject the hypothesis that Jones has ticket 1 at the extraordinarily low significance level of .000001. The significance level is just the likelihood, and is supposed to be a measure of how good our reason is for believing $\ulcorner \sim T_1 j \urcorner$. Statisticians normally operate with significance levels around .1 or .01.

Although our conclusion that $\ulcorner T_1 j \urcorner$ should be rejected agrees with that of the orthodox statistician, it is not difficult to see that significance levels are unreasonable measures of how strong our warrant is for $\ulcorner \sim T_1 j \urcorner$. Contrast lottery 2 with lottery 3, which consists of 10,000 tickets. In lottery 3, the probability of ticket 1 being drawn is still .000001, but the probability of any other ticket being drawn is .000011. This is diagramed in Figure 5. If Jones wins lottery 3, it may still be reasonable to infer that he did not have ticket 1, but now the warrant for this conclusion is not nearly

---

[3] Both the concept of likelihood and the theory of significance tests is due to R. A. Fisher [1921]. The standard concept of likelihood concerns definite probabilities rather than indefinite probabilities, but in most cases these two concepts will agree because direct inference will give us a reason for thinking that the definite and indefinite probabilities have the same value.

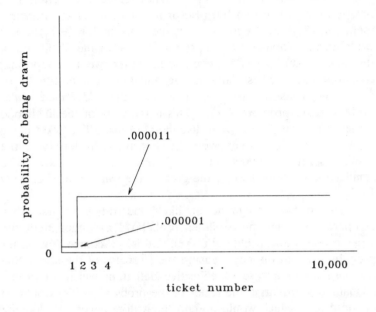

Figure 5. Lottery 3.

so strong. This is because although we have the same prima facie reason for $\ulcorner {\sim} T_1 j \urcorner$, the reasons involved in the counterargument for $\ulcorner T_1 j \urcorner$ are now much better, being based upon a probability of .999989. They may still not be strong enough to defeat the reason for $\ulcorner {\sim} T_1 j \urcorner$, but they are strong enough to significantly weaken the warrant. Thus the warrant for $\ulcorner {\sim} T_1 j \urcorner$ is much weaker in lottery 3 than it is in lottery 2, but the likelihood is unchanged and so the statistician would tell us that we can reject $\ulcorner T_1 j \urcorner$ at the same level of significance. Consequently, the significance level cannot be regarded as an appropriate measure of our degree of warrant.[4]

---

[4] It may be responded, correctly I think, that no statistician would use a

Significance level is simply a measure of the strength of the prima facie reason for $\ulcorner\sim T_1 j\urcorner$. The degree of warrant is a function not just of that but also of how strong our counterargument for $\ulcorner T_1 j\urcorner$ is. A measure of the warrant for $\ulcorner\sim T_1 j\urcorner$ must compare the likelihood of $\ulcorner T_1 x\urcorner$ on $\ulcorner Wx\urcorner$ with the likelihood of the other $\ulcorner T_i x\text{'s}\urcorner$ on $\ulcorner Wx\urcorner$, because the latter (we are supposing) constitute the weakest links in the counterargument for $\ulcorner T_1 j\urcorner$. Such a measure is provided by the *likelihood ratio* prob($Wx/T_1 x$)÷prob($Wx/T_i x$).[5] This is the ratio of the likelihood of ticket 1 being drawn to the likelihood of any other ticket being drawn (given that Jones won the lottery). In lottery 2 the likelihood ratio is .000009, but in lottery 3 it is only .09. The smaller the likelihood ratio, the greater the warrant for believing $\ulcorner\sim T_1 j\urcorner$.

Notice two things about likelihood ratios. First, they compare the small probabilities prob($Wx/T_i x$) rather than the large probabilities prob($\sim Wx/T_i x$). It is, of course, the latter probabilities that directly measure the strengths of the competing reasons, but in a case of collective defeat in which all of the probabilities are large, the ratios of the prob($\sim Wx/T_i x$) would all be close to 1, which would obscure their differences. By looking at the ratios of the prob($Wx/T_i x$) instead we emphasize the differences. The second thing to note is that because we look at the likelihood ratio prob($Wx/T_1 x$)÷prob($Wx/T_i x$) rather than the likelihood ratio prob($Wx/T_i x$)÷prob($Wx/T_1 x$), a smaller likelihood ratio corresponds to a greater degree of warrant. That seems unnatural, but to define the likelihood ratio the other way around would be to flaunt the standard convention and confuse those familiar with that convention.

Likelihood ratios play an important part in statistical theory,

---

significance test in a case like this. But all that seems to show is that though statisticians give lip service to the orthodox theory of significance testing, they do not really accept it at an intuitive level.

[5] This is reminiscent of the Neyman-Pearson theory of hypothesis testing, in light of Neyman's theory that a best test (in the sense of that theory) of a simple hypothesis is always a likelihood ratio test. See paper #6 in Neyman and Pearson [1967]. See also Neyman [1967] and Pearson [1966].

where a number of authors have endorsed what is known as *The Likelihood Principle.* A. W. F. Edwards states this principle as follows:

> Within the framework of a statistical model, *all* the information which the data provide concerning the relative merits of two hypotheses is contained in the likelihood ratio of those hypotheses on the data.[6]

There are important connections between this principle and the acceptance rule $(A2)$. In effect, $(A2)$ licenses conclusions on the basis of likelihoods. Note, however, that the need for the projectibility constraint in $(A2)$, and the fact that $(A2)$ formulates a reason that is defeasible, shows that the likelihood principle of statistics is not defensible without qualifications. As far as I know, statisticians have taken no notice of projectibility, but nonprojectibility creates the same problems for statistical theories as it does for ordinary induction. It is also worth noting that on the present account likelihoods do not play a primitive role but only a derivative role. Given some elementary epistemological principles, the role of likelihoods is determined by $(A2)$. $(A2)$ follows from $(A3)$, and $(A3)$ is the extremely intuitive principle that knowing most B's to be A's gives us a prima facie reason for expecting of a particular object that if it is a B then it is an A.

Care must be taken in the use of likelihood ratios. It is tempting to compare the support of different hypotheses H_1 and H_2 in different experimental situations in terms of their likelihood ratios, taking H_1 to be better supported than H_2 iff the likelihood ratio of H_1 on its evidence is lower than that of H_2 on its evidence. For example, we might use likelihood ratios to compare (1) our justification for believing that one ticket in one biased lottery will not be drawn with (2) our justification for believing that another ticket in a different biased lottery will not be drawn. Statisticians often use likelihood ratios in this way, but

---

[6] Edwards [1972], p. 30. The Likelihood Principle is due to Fisher, and versions of it have been endorsed by a variety of authors, including G. A. Barnard ([1949] and [1966]), Alan Birnbaum [1962], and Ian Hacking [1965].

such comparisons are not warranted by our acceptance rules in their present form. According to those acceptance rules, the strength of a probabilistic reason is a function of the probability, but the rules do not tell us how to sum conflicting probabilistic reasons and measure the resulting "on-balance" warrant. Likelihood ratios measure "on-balance" warrant in terms of ratios of the strengths of the individual reasons. That is unexceptionable within a fixed setting in which the reasons *against* the different hypotheses all have the same strength, because in that case all the likelihood ratios have the same denominator. But in order for likelihood ratios to be used in making meaningful comparisons between hypotheses in different experimental situations (or in experimental situations having more complicated structures), our acceptance rules would have to be augmented with the following principle:

> Given probabilistic reasons of the form (*A2*) with probabilities r for P, s for $\sim P$, t for Q, and u for $\sim Q$, the strength of the on-balance warrant for P is the same as the strength of the on-balance warrant for Q iff $r/s = t/u$.

This is not an implausible principle, and I suspect that it is actually true. But it is not *obviously* true, and it is not required for the defense of statistical induction, so I will refrain from assuming it. Without it, the use of likelihood ratios must be confined to the comparisons of alternatives within a fixed experimental situation in which the arguments against the different alternatives all have the same strength.

Now consider a fourth lottery, this consisting of 1 million tickets, and suppose that the first 999 tickets each have a probability of 1/1000 of being drawn. For k between 1,000 and 999,999, the probability of ticket k being drawn is $2^{-k-999}/1000$, and the probability of ticket 1,000,000 being drawn is $2^{-999000}/1000$ (the latter to make the probabilities sum to 1). In other words, the probability of drawing each ticket with a number greater than 999 is one-half the probability of drawing its predecessor. This is diagramed as in Figure 6. Upon learning that

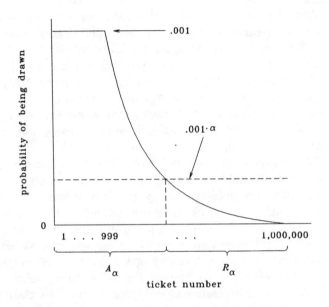

Figure 6. Lottery 4.

Jones won the lottery, we would infer that he did not have a high-numbered ticket. The reasoning is the same as before. For each k, prob($\sim Wx/T_kx$) is high, so we have a prima facie reason for $\ulcorner\sim T_kJ\urcorner$. For each k > 999, the counterargument employs the reasons for $\ulcorner\sim T_1J\urcorner - \ulcorner\sim T_{999}J\urcorner$, and these reasons are weaker than the reason for $\ulcorner\sim T_kJ\urcorner$. To decide whether they are weak enough to warrant $\ulcorner\sim T_kJ\urcorner$, we must compare the strength of the weakest steps of the counterargument with the strength of the prima facie reason for $\ulcorner\sim T_kJ\urcorner$. The weakest step is that supporting $\ulcorner\sim T_1\urcorner$. Thus the comparison proceeds in terms of the likelihood ratio prob(Wx/T_kx)÷prob(Wx/T_1x). The important thing to be emphasized about this example is that although all of the prob($\sim Wx/T_kx$) can be as close to 1 as we like, we can still distinguish between them by comparing the likelihood ratios prob(Wx/T_kx)÷prob($Wx/T_{min}x$)

where T_{min} (in this case, T_1) is that T_i (or one of the T_i's) with minimum likelihood.

Any given degree of warrant determines a likelihood ratio α such that for each ticket k, we are warranted in concluding that ticket k was not drawn iff $prob(Wx/T_k x) \div prob(Wx/T_{min} x) \leq \alpha$. The set of all such k's comprise *the rejection class* R_α. In this example, $prob(Wx/T_{min} x) = .001$, so $R_\alpha = \{k| \ prob(Wx/T_k x) \leq .001 \cdot \alpha\}$. The complement of R_α, A_α, is *the acceptance class*. Because we are warranted in believing that no member of R_α was drawn, we are warranted to degree α in believing that *some* member of A_α was drawn.

The reasoning that will be involved in the statistical induction argument involves something more like an infinite lottery, because there are infinitely many possible values the nomic probability may have. Consider an infinite lottery in which participants purchase the use of a positive integer and a mechanism of some sort selects an integer k as the winner. As before, if $1 \leq k \leq 999$ then the probability of k being selected is $1/1000$, and if $k > 999$ then the probability of k being selected is $2^{-k-999}/1000$. This is diagramed in Figure 7. It seems clear that the mere fact that the lottery is infinite should not affect which $\ulcorner T_k j \urcorner$'s we are warranted in rejecting. As before, upon learning that Jones won the lottery, we would infer that he did not have a high-numbered ticket. But we can no longer reconstruct our reasoning quite so simply as we did in the preceding lottery. That would require the counterarguments for the $\ulcorner T_k j \urcorner$'s to be infinitely long, which is impossible. We must instead take the counterarguments to have the following form:

> Wj
> For each $i \neq k$, $prob(\sim Wx/T_i x) > r$.
> Therefore, for each $i \neq k$, $\sim T_i j$.
> If for each $i \neq k$, $\sim T_i j$, then $T_k j$.
> Therefore, $\ulcorner T_k j \urcorner$ is true.

In order to reason in this way, we must take $\ulcorner Wj$ and for each $i \neq k$, $prob(\sim Wx/T_i x) > r\urcorner$ to be a prima facie reason not only for each $\ulcorner \sim T_i j\urcorner$ separately, but also for the conclusion that *all* of

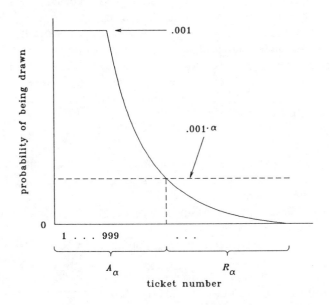

Figure 7. Lottery 5.

the $\ulcorner T_i j \urcorner$s for $i \neq k$ are false. This is justified by the derived acceptance rule $(A4)$ of Chapter 3:

$(A4)$ If $r > .5$ and F is projectible with respect to B then $\ulcorner \sim Fc$ and $(\forall x)[Gx \supset \text{prob}_x(Fy/Bxy) \geq r]\urcorner$ is a prima facie reason for believing $\ulcorner(\forall x)(Gx \supset \sim Bxc)\urcorner$, the strength of the reason depending upon the value of r in the same way the strength of reasons described by $(A1)$, $(A2)$, and $(A3)$ depend upon the value of r.

Recall that $(A4)$ was derived from the basic acceptance rule $(A3)$.

$(A4)$ can be employed in the infinite lottery to conclude that Jones did not have a high-numbered ticket. The reasoning is analogous to the use of $(A2)$ in the finite lottery, but there are some subtle differences.   For each individual high-numbered ticket, $(A2)$ provides an undefeated prima facie reason for thinking that Jones did not have that ticket.   In the infinite lottery, we want the stronger conclusion that Jones did not have *any* of the high-numbered tickets. $(A4)$ provides a prima facie reason for believing this, so what remains to be shown is that this prima facie reason is undefeated.   As in the finite lottery, there is a counterargument for the conclusion that Jones did have a high-numbered ticket:

Wj
For each low-numbered ticket i, prob$(\sim Wx/T_ix) \geq r$.
Therefore (by $(A4)$), for each low-numbered ticket i, Jones did not have i.
If Jones had no low-numbered ticket then he had a high-numbered ticket.
Therefore, Jones had a high-numbered ticket.

In order to conclude that Jones did not have a high-numbered ticket, it must be established that this counterargument does not defeat the prima facie reason provided by $(A4)$ for believing that Jones did not have a high-numbered ticket. But this follows from the fact that the strength of the reasons described by $(A4)$ depends upon the value of the probability in the same way as the strengths of the reasons described by $(A2)$,[7] and hence we can evaluate a use of $(A4)$ in terms of likelihood ratios in the precisely the same way we evaluate a use of $(A2)$. This means that the strengths of the reasons and counterarguments in the infinite lottery are the same as the strengths of the reasons and counterarguments in the preceding finite lottery (because the probabilities are the same), and hence a conclusion is warranted for the infinite lottery iff the analogous conclusion is warranted for the finite lottery.

The form of the argument used in connection with the infinite lottery will be the same as the form that will be used in

---

[7] This was defended in Chapter 3.

the statistical induction argument, with one minor twist. Suppose
F is projectible with respect to B, and we know that $\ulcorner Fc \urcorner$ is false.
Suppose further that X is a set for which we know that $(\exists x)[x \in X$
$\& \ Bxc]$. Then for each object b in X, a reason for believing that
none of the other members of X stand in the relation B to c
gives us a reason for believing $\ulcorner Bbc \urcorner$. Suppose further that for
every b in X, we know that if $\ulcorner Bbc \urcorner$ is true then $\text{prob}(Fy/Bby) \geq$
r. Notice that we do not assume that we know the values of
$\text{prob}(Fy/Bby)$ — only that we know what each value is if $\ulcorner Bbc \urcorner$ is
true. The reason for this weaker assumption will become
apparent when we turn to the statistical induction argument. It
will emerge below that weakening the assumption in this way
makes no difference to the argument.

As in the previous examples, we can diagram the *low* probabilities $\text{prob}(\sim Fy/Bby)$ as in Figure 8, where for expository
purposes I have supposed that they form a bell curve. If b is any
member of X, $(A2)$ provides a prima facie reason for believing
that if $\ulcorner Bbc \urcorner$ is true, and hence the probability is as represented,
then $\ulcorner Bbc \urcorner$ is false. Hence $(A2)$ provides a reason for believing
simply that $\ulcorner Bbc \urcorner$ is false. (This shows that the weaker assumption about the probabilities does not affect the argument.) As in
the case of the lottery, by using $(A4)$ we can also construct a
counterargument for the conclusion that $\ulcorner Bbc \urcorner$ is true. Suppose
we have no other potential defeaters for the prima facie reasons
resulting from these uses of $(A2)$ and $(A4)$. Let γ be the least
upper bound of $\{\text{prob}(\sim Fy/Bby) \mid b \in X\}$. This corresponds to the
weakest reason (the b for which the probability $\text{prob}(Fy/Bby)$ is
smallest). Then we are warranted in believing that $\ulcorner Bbc \urcorner$ is false
iff the likelihood ratio $\text{prob}(\sim Fy/Bby) \div \gamma$ is low enough. For any
particular degree of warrant, there is a *maximum acceptable
likelihood ratio* α such that for any b in X, one is warranted to
that degree in believing $\ulcorner \sim Bbc \urcorner$ iff the likelihood ratio
$\text{prob}(\sim Fy/Bby) \div \gamma$ is less than or equal to α.[8] We can thus employ
α as a measure of the corresponding degree of warrant. Let the

---

[8] For the reasons discussed earlier, the correlation between degrees of
warrant and likelihood ratios might vary from one setting to another.

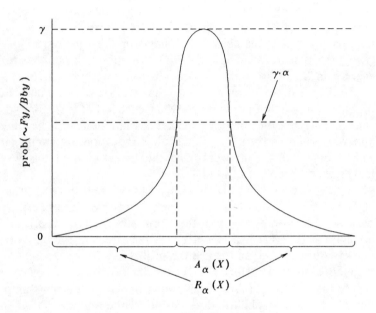

Figure 8. The structure of (*A5*).

α-*rejection class* of *X* be the set of all the *b*'s such that we are warranted in rejecting ⌜*Bbc*⌝ on the basis of its having a low likelihood ratio:

(2.1) $R_\alpha(X) = \{b \mid b \in X \ \& \ \text{prob}(\sim Fy/Bby) \div \gamma \leq \alpha\}$.

If *b*∈*X*, we are warranted to degree α in believing that ⌜*Bbc*⌝ is false iff *b*∈$R_\alpha(X)$. Note that $R_\alpha(X)$ consists of the tails of the bell curve.

The next thing to be established is that we are also warranted in believing that *no* member of $R_\alpha(X)$ stands in the relation *B* to *c*, that is, $(\forall x)[x \in R_\alpha X \supset \sim Bxc]$. Just as in the infinite lottery, (*A4*) provides a prima facie reason for believing this, but we also have a counterargument to the conclusion that $(\exists x)[x{\in}R_\alpha(X) \ \& \ Bxc]$. The counterargument goes as follows. Let $A_\alpha(X)$ be the complement of $R_\alpha(X)$:

(2.2) $A_\alpha(X) = X - R_\alpha(X).$

The counterargument to the conclusion that $(\exists x)[x{\in}R_\alpha(X) \ \& \ Bxc]$ is then:

> $\sim Fc$
> For each b in $A_\alpha(X)$, prob$(Fy/Bby) \geq 1-\gamma$.
> Therefore, $(\forall x)[x{\in}A_\alpha(X) \supset \sim Bxc]$.
> If $(\forall x)[x{\in}A_\alpha(X) \supset \sim Bxc]$ then $(\exists x)[x{\in}R_\alpha(X) \ \& \ Bxc]$.
> Therefore, $(\exists x)[x{\in}R_\alpha(X) \ \& \ Bxc]$.

To determine whether this counterargument defeats the prima facie reason for $\ulcorner(\forall x)[x{\in}R_\alpha X \supset \sim Bxc]\urcorner$, we must compare their strengths.   The strength of the reason provided by $(A4)$ is determined by the probability involved, and that is the greatest lower bound β of the probabilities prob(Fy/Bby) for b in $R_\alpha(X)$. It follows that $(1-\beta)/\gamma$ is the least upper bound of the likelihood ratios of $\ulcorner Bbc\urcorner$ for b in $R_\alpha(X)$. By hypothesis, those likelihood ratios are all less than or equal to α, so $(1-\beta)/\gamma$ is also less than or equal to α. We saw above that we can assess a use of $(A4)$ in terms of likelihood ratios just as we can assess a use of $(A2)$ in terms of likelihood ratios. A likelihood ratio less than or equal to α guarantees that we are warranted to degree α in reasoning in accordance with $(A2)$, so the fact that $(1-\beta)/\gamma \leq \alpha$ guarantees that we are warranted to degree α in reasoning in accordance with $(A4)$ and concluding that $(\forall x)[x{\in}R_\alpha X \supset \sim Bxc]$. As we are also warranted in believing that for some b in X, $\ulcorner Bbc\urcorner$ is true, it follows that we are warranted in concluding that $(\exists x)[x{\in}A_\alpha(X) \ \& \ Bxc]$. For this reason I will call $A_\alpha(X)$ the *α-acceptance class* of X.

The upshot of this is that if X satisfies our assumptions, then it can be divided into two subclasses—the acceptance class and the rejection class—and we are warranted in believing that some member of the acceptance class stands in the relation B to c and no member of the rejection class does. Precisely:

(A5)   If F is projectible with respect to B and X is a set such that
  - (a) we are warranted in believing that $(\exists x)[x \in X \;\&\; Bxc]$; and
  - (b) we are warranted in believing $\ulcorner \sim Fc$ and for each b in X, if $\ulcorner Bbc \urcorner$ is true then $\text{prob}(Fy/Bby) \geq r \urcorner$; and
  - (c) the only defeaters for the instances of (A2) and (A4) that result from (b) are those "internal" ones that result from those instances conflicting with one another due to (a);

then we are warranted to degree α in believing that $(\exists x)[x \in A_\alpha(X) \;\&\; Bxc]$ and $(\forall x)[x \in R_\alpha(X) \supset \;\sim Bxc]$.[9]

It should be emphasized that (A5) is not being proposed here as a primitive acceptance rule. Rather, it has been derived from the much simpler and more obvious (A3). (A5) can be regarded as a heavily qualified and sanitized version of the Likelihood Principle.

3. The Statistical Induction Argument

We now have the ingredients for the statistical induction argument. Suppose we have observed a sample $X = \{b_1, \ldots, b_n\}$ of

---

[9] Gillies [1973], p. 169, gives a counterexample to a related principle that suggests examples at first appearing to be counterexamples to (A5). Consider a lottery in which ticket 1 has a probability of 10^{-2} of being drawn, and any other ticket has a probability of 10^{-6} of being drawn. If we know that Jones won the lottery, it appears that (A5) commits us to the conclusion that Jones had ticket 1. But as Gillies points out, if we reason this way we will be wrong 99% of the time, so surely that conclusion is unreasonable. What this amounts to is the observation that the probability of Jones having ticket 1 is .01, and that provides a reason by (A1) for denying that he had ticket 1. This is a stronger reason for denying that Jones had ticket 1 than (A4) provides for affirming it. This is not a counterexample to (A5) because clause (c) is not satisfied. This use of (A1) provides an "outside" source of defeaters.

B's and noted that only $b_1,\ldots,b_r$ are A's (where A and $\sim A$ are projectible with respect to B).[10] Then the relative frequency freq$[A/X]$ of A's in X is r/n. From this we want to infer that prob(A/B) is approximately r/n. The reasoning proceeds in two stages, the first stage employing the theory of nonclassical direct inference, and the second stage employing the derived acceptance rule ($A5$).

Stage I

Suppose prob$(A/B) = p$. Let us abbreviate $\ulcorner x_1,\ldots,x_n$ are distinct & Bx_1 & $\ldots$ & Bx_n & prob$(Ax/Bx) = p\urcorner$ as θ_p. We are given that $b_1,\ldots,b_n$ satisfy θ_p, and hence θ_p is not counterlegal. When $r \leq n$, it follows from principle (6.13) of Chapter 2 (the generalization of the multiplicative axiom) that:

(3.1) $\mathrm{prob}(Ax_1 \ \& \ \ldots \ \& \ Ax_r \ \& \ \sim Ax_{r+1} \ \& \ \ldots \ \& \ \sim Ax_n \ / \ \theta_p) =$
  $\mathrm{prob}(Ax_1 \ / \ Ax_2 \ \& \ \ldots \ \& \ Ax_r \ \& \ \sim Ax_{r+1} \ \& \ \ldots \ \& \ \sim Ax_n \ \& \ \theta_p)$
  $\cdot \ \ldots \ \cdot \ \mathrm{prob}(Ax_r \ / \ \sim Ax_{r+1} \ \& \ \ldots \ \& \ \sim Ax_n \ \& \ \theta_p) \ \cdot$
  $\mathrm{prob}(\sim Ax_{r+1} \ / \ \sim Ax_{r+2} \ \& \ \ldots \ \& \ \sim Ax_n \ \& \ \theta_p) \ \cdot \ \ldots$
  $\cdot \ \mathrm{prob}(\sim Ax_n \ / \ \theta_p).$

Notice that this turns on the assumption that θ_p is not counterlegal, and hence on the assumption that prob$(A/B) = p$. Making θ_p explicit:

(3.2) $\mathrm{prob}(Ax_i \ / \ Ax_{i+1} \ \& \ \ldots \ \& \ Ax_r \ \& \ \sim Ax_{r+1} \ \& \ \ldots \ \& \ \sim Ax_n \ \& \ \theta_p)$
  $= \mathrm{prob}\big(Ax_i \ / \ x_1,\ldots,x_n$ are distinct & Bx_1 & $\ldots$ & Bx_n &
  Ax_{i+1} & $\ldots$ & Ax_r & $\sim Ax_{r+1}$ & $\ldots$ & $\sim Ax_n$ &
  $\mathrm{prob}(A/B) = p\big).$

Projectibility is closed under conjunction, so $\ulcorner Ax_i \urcorner$ is projectible with respect to $\ulcorner Bx_1$ & $\ldots$ & Bx_n & $x_1,\ldots,x_n$ are distinct & Ax_{i+1} & $\ldots$ & Ax_r & $\sim Ax_{r+1}$ & $\ldots$ & $\sim Ax_n \urcorner$. On the assumption that prob$(A/B) = p$, it follows that $\square_p[\mathrm{prob}(A/B) = p]$.

---

[10] The requirement that $\sim A$ be projectible with respect to B is surprising. It will be discussed in section 5.

Consequently, if $\ulcorner Ax\urcorner$ is projectible with respect to $\ulcorner Fx\urcorner$, it is also projectible so with respect to $\ulcorner Fx$ & $\text{prob}(Ax/Bx) = p\urcorner$. Therefore, $\ulcorner Ax_i\urcorner$ is projectible with respect to the reference property of (3.2). Thus nonclassical direct inference provides a reason for believing that

$$\text{prob}(Ax_i \ / \ Ax_{i+1} \ \& \ ... \ \& \ Ax_r \ \& \ \sim Ax_{r+1} \ \& \ ... \ \& \ \sim Ax_n \ \& \ \theta_p)$$
$$= \text{prob}\Big(Ax_i \ / \ Bx_i \ \& \ \text{prob}(A/B) = p\Big),$$

which by the principle (*PPROB*) of Chapter 2 equals p.[11] Similarly, nonclassical direct inference provides a reason for believing that if $r < i \leq n$ then

(3.3) $\text{prob}(\sim Ax_i \ / \ \sim Ax_{i+1} \ \& \ ... \ \& \ \sim Ax_n \ \& \ \theta_p) = 1-p.$

Then from (3.1) we have:

(3.4) $\text{prob}(Ax_1 \ \& \ ... \ \& \ Ax_r \ \& \ \sim Ax_{r+1} \ \& \ ... \ \& \ \sim Ax_n \ / \ \theta_p)$
$= p^r(1-p)^{n-r}.$

$\ulcorner \text{freq}[A \ / \ \{x_1,...,x_n\}] = r/n\urcorner$ is equivalent to a disjunction of $n!/r!(n-r)!$ pairwise incompatible disjuncts of the form $\ulcorner Ax_1 \ \& \ ... \ \& \ Ax_r \ \& \ \sim Ax_{r+1} \ \& \ ... \ \& \ \sim Ax_n\urcorner$, so by the probability calculus:

(3.5) $\text{prob}\Big(\text{freq}[A \ / \ \{x_1,...,x_n\}] = \tfrac{r}{n} \ / \ \theta_p\Big) = \dfrac{n!p^r(1-p)^{n-r}}{r!(n-r)!}.$

This, of course, is the formula for the binomial distribution. All of this is predicated on the supposition that $\text{prob}(A/B) = p$. Discharging that supposition, we conclude that *if* $\text{prob}(A/B) = p$ then (3.5) holds.

This completes Stage I of the statistical induction argument. This stage reconstructs the first half of the intuitive argument

---

[11] (*PPROB*) tells us that $\text{prob}(F \ / \ G \ \& \ \text{prob}(F/G) = p) = p$. Note that if $\text{prob}(A/B) \neq p$ then the direct inference here involves counterlegal properties and hence proceeds via (3.5) of Chapter 4 rather than via (*DI*).

described in section 1. Note that it differs from that argument in that it proceeds in terms of indefinite probabilities rather than definite probabilities, and it avoids unwarranted assumptions about independence by using nonclassical direct inference instead. In effect, nonclassical direct inference provides a reason for expecting independence unless we have explicit evidence to the contrary.

Stage II

The second half of the intuitive argument ran afoul of the lottery paradox and seems to me to be irreparable. I propose to replace it with an argument using (*A5*).

(*A5*) contains a built-in projectibility constraint, so let us first verify that that is not a problem. Suppose we have a shipment of cartons containing red and blue counters. Each carton is supposed to contain half red counters and half blue counters. But upon inspecting a number of the cartons and finding that none of them contain half red counters, we would inductively infer that none of the cartons in the shipment contain half red counters. What this illustrates is that if A is a projectible property then $\lceil freq[A/X] \neq r/n \rceil$ is a projectible property of X. Thus the following conditional probability, derived from (3.5), satisfies the projectibility constraint of (*A5*):

$$(3.6) \quad prob\left(freq[A \; / \; \{x_1,\ldots,x_n\}] \neq \tfrac{r}{n} \; / \; \theta_p\right)$$

$$= 1 - \frac{n!p^r(1-p)^{n-r}}{r!(n-r)!}$$

Let $b(n,r,p) = n!p^r(1-p)^{n-r}/r!(n-r)!$. For sizable n, $b(n,r,p)$ is almost always quite small. For example, $b(50,20,.5) = .04$. Thus by (*A2*) and (3.6), for each choice of p we have a prima facie reason for believing that if $prob(A/B) = p$ then $\langle b_1,\ldots,b_n \rangle$ does not satisfy θ_p, that is, for believing that if $prob(A/B) = p$ then

$$\sim\!\!\left(b_1,\ldots,b_n \text{ are distinct } \& \; Bb_1 \; \& \; \ldots \; \& \; Bb_n \; \& \; prob(A/B) = p\right).$$

As we know that ⌈$b_1,\ldots,b_n$ are distinct & Bb_1 & ... & Bb_n⌉ is true, this provides a prima facie reason for believing that if prob(A/B) = p then prob(A/B) ≠ p, and hence for believing simply that prob(A/B) ≠ p. But, of course, for some p, prob(A/B) = p. Thus we have a situation of the sort described by ($A5$). These probabilities are diagramed in Figure 9, using the model of Figure 8. For each α, we can divide the interval [0,1] into a rejection class $R_\alpha([0,1])$ and an acceptance class $A_\alpha([0,1])$. The minimum value γ of $b(n,r,p)$ always occurs when $p = r/n$. Letting $f = r/n$, the likelihood ratio for p is

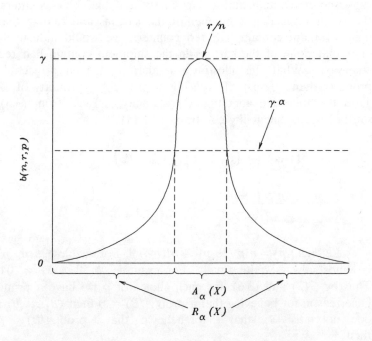

Figure 9. The statistical induction argument.

$$\left(\frac{p}{f}\right)^{nf} \cdot \left(\frac{1-p}{1-f}\right)^{n(1-f)}$$

Let us abbreviate this ratio as $\ulcorner L(n,f,p)\urcorner$. By definition,

(3.7) $R_\alpha([0,1]) = \{p \mid p \in [0,1] \ \& \ L(n,f,p) \le \alpha\}.$

By $(A5)$, we are warranted to degree α in believing that every p in $R_\alpha([0,1])$, $\mathrm{prob}(A/B) \ne p$, and hence we are warranted to degree α in believing that $\mathrm{prob}(A/B) \in A_\alpha([0,1])$. To make the observed frequency and sample size explicit, let us rewrite $A_\alpha([0,1])$ as $A_\alpha(f,n)$. We have:

(3.8) $A_\alpha(f,n) = \{p \mid \left(\frac{p}{f}\right)^{nf} \cdot \left(\frac{1-p}{1-f}\right)^{n(1-f)} > \alpha\}.$

$A_\alpha(f,n)$ is an interval around f that becomes quite narrow as n increases. I will call $A_\alpha(f,n)$ the *acceptance interval at level* α, as we are warranted to degree α in believing that the value of $\mathrm{prob}(A/B)$ lies in this interval. Some typical values of the acceptance interval are listed in Table 1.

The plots in Figure 10 are dramatic illustrations of how rapidly the acceptance intervals become small. In interpreting these plots, recall that reference to the acceptance level reflects the fact that attributions of warrant are indexical. Sometimes an acceptance level of .1 may be reasonable, at other times an acceptance level of .01 may be required, and so forth.

This completes the statistical induction argument. This argument makes precise the way in which observation of the relative frequency of A's in the sample warrants the conclusion that $\mathrm{prob}(A/B)$ is approximately the same as that relative frequency. This conclusion has been derived from epistemic principles not specifically aimed at induction, thus justifying statistical induction on the basis of more general epistemic principles.

The reader familiar with the literature on statistical inference may get a feeling of déjà vu when reading the statistical induction argument, because there are repeated similarities to standard statistical arguments. As I indicated above, it would be

Table 1. Values of $A_\alpha(f,n)$.

$A_\alpha(.5,n)$

α	10	10^2	10^3	10^4	10^5	n 10^6
.1	[.196,.804]	[.393,.607]	[.466,.534]	[.489,.511]	[.496,.504]	[.498,.502]
.01	[.112,.888]	[.351,.649]	[.452,.548]	[.484,.516]	[.495,.505]	[.498,.502]
.001	[.068,.932]	[.320,.680]	[.441,.559]	[.481,.519]	[.494,.506]	[.498,.502]

$A_\alpha(.9,n)$

α	10	10^2	10^3	10^4	10^5	n 10^6
.1	[.596,.996]	[.823,.953]	[.878,.919]	[.893,.907]	[.897,.903]	[.899,.901]
.01	[.446,1.00]	[.785,.967]	[.868,.927]	[.890,.909]	[.897,.903]	[.899,.901]
.001	[.338,1.00]	[.754,.976]	[.861,.932]	[.888,.911]	[.897,.903]	[.899,.901]

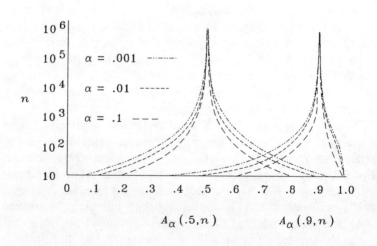

Figure 10. Acceptance intervals.

very surprising if that were not so. All theories of inverse inference are based upon variants of the intuitive argument discussed in section 1, and it would be preposterous to suppose there are no sound intuitions underlying current statistical theory even if the details of current statistical theory will not withstand scrutiny. But it should also be emphasized that if you try to list the exact points of similarity between the statistical induction argument and orthodox statistical reasoning concerning confidence intervals or significance testing, there are as many important differences as there are similarities. These include the use of indefinite probabilities in place of definite probabilities, the use of nonclassical direct inference in place of assumptions of independence, the *derivation* of the principle governing the use of likelihood ratios rather than the simple postulation of the likelihood principle, and the projectibility constraint in that derived likelihood principle. The latter deserves particular emphasis. A few writers have suggested general theories of statistical inference that are based on likelihood ratios and agree closely with the present account when applied to the special case of statistical induction. However, such theories have been based merely on statistical intuition, without an underlying rationale of the sort given here, and they are all subject to counterexamples having to do with projectibility.[12]

4. Enumerative Induction

The explanation of statistical induction is an essential ingredient in the analysis of nomic probability. But by and large, philosophers since Hume have been more interested in enumerative induction. Enumerative induction is that kind of induction wherein upon observing that all members of a sample of B's are A's, it is inferred any B would be an A, that is, $B \Rightarrow A$. Initially,

---

[12] See particularly Ian Hacking [1965], D. A. Gillies [1973], and A. W. F. Edwards [1972]. See also R. A. Fisher [1921] and [1922], and G. A. Barnard [1949], [1966], and [1967].

it is apt to seem that enumerative induction should be a special case of statistical induction. But there is an important difference between them that makes it impossible to regard enumerative induction as *just* a special case of statistical induction. This is that the conclusion of statistical induction is always that a probability falls in a certain interval—never that it has a single precise value. Thus all that can be inferred by statistical induction is that prob(A/B) is high, that is, most (and *perhaps* all) B's are A's.[13]

What will now be shown is that although enumerative induction is not just a special case of statistical induction, principles of enumerative induction can nevertheless be derived from the account of statistical induction. For this purpose, use will be made of the following four principles that were defended in Chapters 2 and 3:

(4.1) If P is a prima facie reason for Q, then $\ulcorner \Box_p P \urcorner$ is a prima facie reason for $\ulcorner \Box_p Q \urcorner$. ((3.13) of Chapter 3.)

(4.2) If P is a prima facie reason for $\ulcorner Fx \urcorner$ (i.e., this is a reason schema), then P is a prima facie reason for $\ulcorner (\forall x)Fx \urcorner$. ((3.12) of Chapter 3.)

(4.3) For each number r, $\ulcorner \text{prob}(A/B) = r \urcorner$ is equivalent to $\ulcorner \Box_p(\text{prob}(A/B) = r) \urcorner$.[14]

(4.4) $\ulcorner (\exists x)Bx \,\&\, \Box_p(\forall x)(Bx \supset Ax) \urcorner$ entails $\ulcorner B \Rightarrow A \urcorner$. (From (3.17) of Chapter 2.)

Let X be our sample of n B's, and suppose freq$[A/X] = 1$. Enumerative induction is supposed to provide a prima facie reason for $\ulcorner B \Rightarrow A \urcorner$. This prima facie inference can be justified by extending the statistical induction argument, adding a third

[13] And, of course, contrary to what many philosophers often suppose, $\ulcorner \text{prob}(A/B) = 1 \urcorner$ does not entail $\ulcorner B \Rightarrow A \urcorner$ anyway.

[14] (4.3) is trivial. If $\ulcorner \text{prob}(A/B) = r \urcorner$ is true, then it *is* a law, and hence is entailed by a law; that is, it is physically necessary. Conversely, if it is physically necessary then it is true.

stage to it to form the *enumerative induction argument*. Given
(4.1)–(4.4), we can reason as follows. By definition, $A_\alpha(1,n) =$
$\{p \mid L(n,f,p) > \alpha\}$. By (3.5), $L(n,1,p) = p^n \div 1^n = p^n$. Thus $A_\alpha(1,n)$
$= [\alpha^{1/n},1]$. Hence by the statistical induction argument, observation of the sample provides prima facie warrant at the α level for
believing that $\text{prob}(A/B) > \alpha^{1/n}$. By $(A3)$ this constitutes a prima
facie reason for $\ulcorner Bc \supset Ac \urcorner$, and then by (4.2) it provides a prima
facie reason for $\ulcorner (\forall x)(Bx \supset Ax) \urcorner$. Then by (4.1), $\ulcorner \square_p[\text{prob}(A/B)$
$> \alpha^{1/n}] \urcorner$ is a prima facie reason for $\ulcorner \square_p(\forall x)(Bx \supset Ax) \urcorner$. Hence, it
follows from (4.3) that $\ulcorner \text{prob}(A/B) > \alpha^{1/n} \urcorner$ is a prima facie reason
for $\ulcorner \square_p(\forall x)(Bx \supset Ax) \urcorner$. Thus observation that $\text{freq}[A/X] = 1$
provides a prima facie reason for $\ulcorner \square_p(\forall x)(Bx \supset Ax) \urcorner$. Observation
that the sample contains at least one B (i.e., that it is non-empty)
provides $\ulcorner (\exists x)Bx \urcorner$, and by (4.4) these jointly entail $\ulcorner B \Rightarrow A \urcorner$, which
is the desired conclusion. Consequently, observation that the
sample consists of n B's all of which are A's provides prima facie
warrant for $\ulcorner B \Rightarrow A \urcorner$. This completes the enumerative induction
argument.

The fact that the enumerative induction argument proceeds
by adding a third stage to the statistical induction argument
explains both the similarities and the differences between
statistical induction and enumerative induction.   Enumerative
induction automatically inherits many of the characteristics of
statistical induction. For example, the reasons involved are only
prima facie, any defeaters for statistical induction are also
defeaters for enumerative induction, and the applicability of
enumerative induction is restricted to projectible properties. It is
by adding the third stage to the statistical induction argument that
enumerative induction is able to arrive at exceptionless generalizations rather than merely interval estimates of probabilities.

5. Material Induction

As I have described them, statistical induction pertains to nomic
probabilities and enumerative induction confirms nomic generalizations. But it seems clear that induction can also lead to con-

clusions about relative frequencies and material generalizations in circumstances in which the analogous conclusions concerning nomic probabilities and nomic generalizations are unwarranted. For example, while walking in the city park I may note that all the benches I see are painted green, and conclude that all the benches in the park are painted green. I know very well that this is not physically necessary, but nevertheless, my generalization appears to be warranted. This is an example of induction confirming a material generalization. Similarly, pollsters may survey 1000 residents of Philadelphia, finding that 3/5 of them intend to vote for Senator Schlopp, and conclude that approximately 3/5 of the residents of Philadelphia will vote for Senator Schlopp. This is a case of statistical induction to a relative frequency rather than a nomic probability. I will call induction to a material generalization or relative frequency *material induction*. An adequate theory of induction must accommodate material induction.

Material induction results from a variant of the statistical and enumerative induction arguments. The statistical induction argument makes essential use of the principle (*PPROB*) according to which:

$$\text{prob}\Big(A \ / \ B \ \& \ \text{prob}(A/B) = p\Big) = p.$$

Stage I of the statistical induction argument proceeds by using this principle and nonclassical direct inference to conclude that:

(5.1) $\text{prob}\Big(Ax_i \ / \ Ax_{i+1} \ \& \ \dots \ \& \ Ax_r \ \& \ {\sim}Ax_{r+1} \ \& \ \dots \ \& \ {\sim}Ax_n$
$\& \ x_1, \dots, x_n \text{ are distinct} \ \& \ Bx_1 \ \& \ \dots \ \& \ Bx_n \ \&$
$\text{prob}(A/B) = p\Big)$

$= \text{prob}\Big(Ax_i \ / \ Bx_i \ \& \ \text{prob}(A/B) = p\Big)$

$= p.$

We then compute:

(5.2) $\mathrm{prob}\Big(\mathrm{freq}\{x_1,\ldots,x_n\} = \frac{r}{n} \ / \ x_1,\ldots,x_n$ are distinct

$\& \ Bx_1 \ \& \ \ldots \ \& \ Bx_n \ \& \ \mathrm{prob}(A/B) = p\Big)$

$$= \frac{n!p^r(1-p)^{n-r}}{r!(n-r)!}.$$

Material induction to relative frequencies replaces the use of
(*PPROB*) with the use of the principle (*PFREQ*) of Chapter 2,
according to which:

$$\mathrm{prob}\Big(A \ / \ B \ \& \ \mathrm{freq}[A/B] = p\Big) = p.$$

Stage I of the *material statistical induction argument* proceeds by
using this principle together with nonclassical direct inference to
conclude that:

(5.3) $\mathrm{prob}\Big(Ax_i \ / \ Ax_{i+1} \ \& \ \ldots \ \& \ Ax_r \ \& \ {\sim}Ax_{r+1} \ \& \ \ldots \ \& \ {\sim}Ax_n$

$\& \ x_1,\ldots,x_n$ are distinct $\& \ Bx_1 \ \& \ \ldots \ \& \ Bx_n \ \&$

$\mathrm{freq}[A/B] = p\Big)$

$= \mathrm{prob}\Big(Ax_i \ / \ Bx_i \ \& \ \mathrm{freq}[A/B] = p\Big)$

$= p.$

The rest of the argument is just like the statistical induction
argument, but now the conclusion locates freq[*A/B*] in the interval
$A_\alpha(f,n)$ rather than locating prob(*A/B*) in that interval. Note that
for any particular acceptance level α, the material statistical
induction argument locates the relative frequency in precisely the
same interval as that in which the statistical induction argument
would locate the nomic probability (were the latter not defeated
by our already knowing that that is not the value of the nomic

probability).[15] Thus material induction to relative frequencies is easily accommodated.

The enumerative induction argument proceeds by adding a step to the end of the statistical induction argument. We can construct an analogous *material enumerative induction argument* by adding a step to the end of the material statistical induction argument. Suppose we have a sample of n B's and we observe that they are all A's. The material statistical induction argument warrants us in believing that freq$[A/B] > \alpha^{1/n}$. The frequency-based acceptance rule ($A3^*$) provides a reason for $\ulcorner Bc \supset Ac \urcorner$. As in the enumerative induction argument, this is a reason schema and 'c' does not occur in the reason, so by (3.12) of Chapter 3 we have a reason for the material generalization $\ulcorner(\forall x)(Bx \supset Ax)\urcorner$. Thus we have enumerative induction to a material generalization. Note further that because the reason provided by ($A3^*$) has the same strength as the analogous reason provided by ($A3$), it follows that the material enumerative induction argument provides just as good a reason for $\ulcorner(\forall x)(Bx \supset Ax)\urcorner$ as the enumerative induction argument does for $\ulcorner B \Rightarrow A \urcorner$.

Although the preceding argument shows that we can make sense of induction to relative frequencies and material generalizations, it does not show that a theory of induction could be based entirely on relative frequencies and without appeal to nomic probability. This is because essential use is made of nonclassical direct inference in (5.3), and we saw in Chapter 4 that nonclassical direct inference to frequencies is inherently weaker than nonclassical direct inference to nomic probabilities. On the other hand, if we know the cardinality of B to be some number N, we can avoid the use of direct inference altogether, computing directly:

$$(5.4) \quad \mathrm{freq}\left[\mathrm{freq}[A/X] = \tfrac{r}{n} \;/\; \#X = n \;\&\; X \subseteq B \;\&\; \mathrm{freq}[A/B] = p\right]$$

---

[15] Actually, the conclusion can be strengthened a bit. A relative frequency can never be an irrational number, so all irrationals can be deleted from the acceptance interval (in which case it is no longer an interval).

$$= \frac{n!(pN)!(N-n)!}{r!N!(pN-r)!}.$$

We can then reconstruct the material induction argument on the basis of (5.4) instead of (5.3). In fact, it is mandatory to do so because, with a bit of manipulation, (5.4) provides us with a subset defeater for the application of (*A3*) to (5.3). The mathematics gets messy because, even ignoring the irrational points, the values of (5.4) do not lie on a smooth curve. Appeal to (5.4) makes it possible to reconstruct the statistical induction argument and obtain the conclusion that freq[*A/B*] $\approx$ freq[*A/X*] (where *X* is our sample and '$\approx$' symbolizes 'is approximately equal to'), but the characterization of the acceptance interval will be inelegant.

The material enumerative induction argument can be reconstructed using (5.4), and this time the results are somewhat more satisfactory. Just as before, if $X \subseteq A$ then we acquire a prima facie reason for $\ulcorner(\forall x)(Bx \supset Ax)\urcorner$. The reason will, however, be weaker than the reason generated by (5.3). This is an essential feature of material induction because, as indicated above, if we know the cardinality of *B* then we are precluded from using (5.3). In general, holding freq[*X/B*] constant, the smaller we know *B* to be, the weaker the reason for $\ulcorner(\forall x)(Bx \supset Ax)\urcorner$ becomes.

6. Concluding Remarks

This completes the reconstruction of statistical and enumerative induction. Just what has been accomplished? I announced at the beginning of the chapter that I was addressing both the "original" problem of justifying induction and the "new" problem of giving a precise statement of the principles of induction, so let us consider our results in light of those two problems.

Talk of "justifying" induction is not to be taken seriously, any more than is talk of justifying belief in the material world or belief in other minds. Induction does not need justifying. It cannot reasonably be doubted that induction is a rational procedure. What goes under the rubric 'justifying induction' is really a matter of explaining induction, that is, explaining why the principles of induction are reasonable principles. The answer might have been that they are reasonable because they are basic epistemic principles, not reducible to anything else, and partly constitutive of our conceptual framework. That was the view I defended in Pollock [1974], but I take it that I have now shown that view to be false. I have "justified" induction by showing that natural principles of induction can be derived from other more basic epistemic principles that can plausibly be regarded as fundamental.

Turning to the so-called new riddle of induction, I am inclined to regard this as a more serious problem for statistical induction than for enumerative induction. In the case of enumerative induction, we all have a pretty good intuitive grasp of how to go about confirming generalizations. Accordingly, I am inclined to regard the problem of giving an accurate statement of the principles of enumerative induction as being primarily (although not wholly) of theoretical interest. But when we turn to statistical induction, we find that there are serious practical problems regarding its employment. For instance, our untutored intuitions are of little help for deciding in how narrow an interval around the observed relative frequency we are justified in locating the actual probability. In Chapter 11 we will see that there are other practical problems as well. These concern the defeaters for induction. For example, when is an apparent bias in the sample sufficient to destroy the warrant? These are matters regarding which we do not have a good intuitive grasp. Consequently, the statistical induction argument is of more than merely theoretical interest. It can provide practical guidance in using statistical induction.

To the best of my knowledge, this chapter provides the first precise account of how we can acquire knowledge of indefinite probabilities (and hence, by direct inference, knowledge of definite probabilities). According to the present account, statistical induction looks much like orthodox statistical inference proceeding in terms of significance testing, confidence intervals, fiducial probabilities, and so on. It is of interest to compare statistical induction with orthodox statistical inference. Beside the fact that the foundations of all such statistical inference are in doubt, there is a major difference between statistical induction and statistical inference. All standard forms of statistical inference presuppose some knowledge of distributions regarding the phenomena under consideration. For example, in attempting to ascertain the probability of a coin landing heads when flipped, a statistician will typically begin by assuming that flips are statistically independent of one another. But you cannot know that without already knowing something about probabilities. Thus orthodox statistical inference only provides knowledge of probabilities if we already have some prior knowledge of probabilities. In contrast, the statistical induction argument starts from scratch and provides knowledge of indefinite probabilities without assuming any prior knowledge of probabilities.

The statistical and enumerative induction arguments purport to reconstruct our intuitive reasoning. If they are to be successful, they must endow induction with the various logical properties we expect it to have. Chapter 11 is devoted to a discussion of a number of those properties and to the resolution of some well-known puzzles regarding induction. The most notable of the latter is the paradox of the ravens.

7. Summary

1. There is an intuitive argument that underlies all theories of inverse inference, but its details do not withstand scrutiny. Specifically, it requires an undefended assumption of statistical independence in order to calculate the probabilities used, and it employs (A1) in a manner that is subject to collective defeat.

2. In biased lotteries, the tickets can be divided into a rejection class and an acceptance class by comparing likelihood ratios, and it can be concluded that some member of the acceptance class is drawn. This reasoning is made precise by (A5).

3. Stage I of the statistical induction argument shows that nonclassical direct inference can replace the illegitimate assumption of independence that is employed in the intuitive

argument. Stage II uses (*A5*) in place of the incorrect use of (*A1*).

4. Enumerative induction is justified by adding a third stage to the statistical induction argument.

5. Material induction to material generalizations or relative frequencies can be justified by modifying the statistical and enumerative induction arguments in various ways so that they appeal to frequencies rather than nomic probabilities.

CHAPTER 6
RECAPITULATION

The purpose of this book is to clarify probability concepts and analyze the structure of probabilistic reasoning. The intent is to give an account that is precise enough to actually be useful in philosophy, decision theory, and statistics. An ultimate objective will be to implement the theory of probabilistic reasoning in a computer program that models human probabilistic reasoning. The result will be an AI system that is capable of doing sophisticated scientific reasoning.[1] However, that takes us beyond the scope of the present book. The purpose of this chapter is to give a brief restatement of the main points of the theory of nomic probability and provide an assessment of its accomplishments.

1. The Analysis of Nomic Probability

The theory of nomic probability has a parsimonious basis. This consists of two sets of principles. First, there are the epistemic principles (*A3*) and (*D3*):

(*A3*)   If F is projectible with respect to G and $r > .5$, then $\ulcorner \text{prob}(F/G) \geq r \urcorner$ is a prima facie reason for the conditional $\ulcorner Gc \supset Fc \urcorner$, the strength of the reason depending upon the value of r.

(*D3*)   If F is projectible with respect to H then $\ulcorner Hc\ \&\ \text{prob}(F/G\&H) < \text{prob}(F/G) \urcorner$ is an undercutting defeater for $\ulcorner \text{prob}(F/G) \geq r \urcorner$ as a prima facie reason for $\ulcorner Gc \supset Fc \urcorner$.

Second, there are some computational principles that generate a calculus of nomic probabilities. These principles jointly constitute

---

[1] This is part of the OSCAR project at the University of Arizona. For more detail, see Pollock [1989a] and [1989b].

the conceptual role of the concept of nomic probability and are the basic principles from which the entire theory of nomic probability follows.

The epistemic principles presuppose a prior epistemological framework governing the interaction of prima facie reasons and defeaters. Certain aspects of that framework play an important role in the theory of nomic probability. For example, the principle of collective defeat is used recurrently throughout the book. The details of the epistemological framework are complicated, but they are not specific to the theory of probability. They are part of general epistemology.

The computational principles are formulated in terms of what some will regard as an extravagant ontology of sets of possible objects and possible worlds. It is important to realize that this ontology need not be taken seriously. The purpose of the computational principles is to generate some purely formal computational principles regarding nomic probability. These constitute the calculus of nomic probabilities. This calculus includes a version of the classical probability calculus:

$$0 \leq \text{prob}(F/G) \leq 1.$$

If $(G \Rightarrow F)$ then $\text{prob}(F/G) = 1$.

If $\diamond_p \exists H$ and $[H \Rightarrow {\sim}(F\&G)]$ then $\text{prob}(F \vee G/H)$
$= \text{prob}(F/H) + \text{prob}(G/H)$.

If $\diamond_p \exists H$ then $\text{prob}(F\&G/H) = \text{prob}(F/H) \cdot \text{prob}(G/F\&H)$.

If $\diamond \exists G$ then $\text{prob}({\sim}F/G) = 1 - \text{prob}(F/G)$.

If $[(F\&G) \Rightarrow H]$ then $\text{prob}(F/G) \leq \text{prob}(H/G)$.

If $[H \Rightarrow (F \equiv G)]$ then $\text{prob}(F/H) = \text{prob}(G/H)$.

If $(G \Leftrightarrow H)$ then $\text{prob}(F/G) = \text{prob}(F/H)$.

It also includes the following four "nonclassical" principles:

(*PPROB*) If r is a rigid designator of a real number and $\diamond \big[\exists G$
         & $\text{prob}(F/G) = r \big]$ then $\text{prob}\big(F \ / \ G \ \& \ \text{prob}(F/G) = r \big)$
         $= r$.

(*PFREQ*) If r is a nomically rigid designator of a real number
         and $\diamond \big[\exists G \ \& \ \text{freq}[F/G] = r \big]$ then $\text{prob}\big(F \ / \ G \ \&$
         $\text{freq}[F/G] = r \big) = r$.

(*IND*) $\text{prob}(Axy \ / \ Rxy \ \& \ y{=}b) = \text{prob}(Axb/Rxb)$.

(*AGREE*) If F and G are properties and there are infinitely
         many physically possible G's and $\text{prob}(F/G) = p$ then for
         every $\delta > 0$, $\text{prob}\big(\text{prob}(F/X) \approx_\delta p \ / \ X$ is a strict
         subproperty of $G \big) = 1$.

One can take an "as if" attitude toward the ontology of possible worlds and possible objects (van Fraassen's [1981] "suspension of disbelief") and use it merely as a heuristic motivation for adopting these principles as axioms. The rest of the theory will be unchanged.

    The theory of nomic probability is basically a theory of probabilistic reasoning. My claim is that the epistemic and computational principles described constitute the conceptual role of the concept of nomic probability and as such they also constitute an analysis of that concept. They do not constitute a definition of the concept, but I have urged that that is an outworn demand that should no longer be regarded as philosophically obligatory.

2. Main Consequences of the Theory

Starting from its parsimonious epistemic and computational basis, the theory of nomic probability has powerful consequences. These include the theory of direct inference and the theory of induction.

2.1 Direct Inference

The theory of direct inference has two parts. Classical direct inference is inference from indefinite probabilities to definite probabilities, and this has been the traditional focus of theories of direct inference. But I have argued that nonclassical direct inference, which is inference from indefinite probabilities to indefinite probabilities, is more fundamental. Nonclassical direct inference proceeds in terms of the following two principles:

(DI)   If F is projectible with respect to G then $\ulcorner H \preccurlyeq G$ & $\text{prob}(F/G) = r\urcorner$ is a prima facie reason for $\ulcorner \text{prob}(F/H) = r\urcorner$.

(SD)   If F is projectible with respect to J then $\ulcorner H \preccurlyeq J \preccurlyeq G$ & $\text{prob}(F/J) \neq \text{prob}(F/G)\urcorner$ is an undercutting defeater for (DI).

Nonclassical direct inference is a new subject not previously discussed in probability theory. Nevertheless, it is of fundamental importance to nomic probability. Given the principles of nonclassical direct inference it becomes possible to give a precise definition of definite probabilities in terms of nomic probabilities, and a theory of (classical) direct inference for these definite probabilities can be derived from the theory of nonclassical direct inference. It is shown in Chapter 10 that this procedure can be extended to include several kinds of propensities. My claim will be that this makes propensities philosophically respectable in a way they have not previously been. It is the theory of nonclassical direct inference that makes all these definitions possible. Of course, the definitions could have been given without the theory of nonclassical direct inference, but this would not have illuminated the concepts defined because it would have remained mysterious how we could ever learn the values of the resulting definite probabilities. It is nonclassical direct inference that makes all these kinds of definite probabilities epistemologically manageable.

Nonclassical direct inference is very powerful and conceptually beautiful, but it would remain philosophically suspect if it had to be posited on its own strength. One of the most important results in the theory of nomic probability is that the principles constituting the theory of nonclassical direct inference are *theorems* of nomic probability. They are all derivable from the basic principles that describe the conceptual role of nomic probability. The main tool in the derivation of the theory of nonclassical direct inference is the Principle of Agreement.

2.2 Induction

If nomic probability is to be useful we must have some way of arriving at numerical values for nomic probabilities. This is the problem of inverse inference. Inverse inference has been a problem for all previous theories of probability. The theory of nomic probability solves the problem of inverse inference with the statistical induction argument. Statistical induction enables us to estimate nomic probabilities on the basis of relative frequencies in finite samples. Of course, there is nothing new about this. Virtually everyone agrees that we should be able to evaluate probabilities by using some kind of statistical induction. The problem has been to give a precise formulation of principles of statistical induction and justify their use. Most theories of probability have been forced to simply posit some (usually inconsistent) form of statistical induction. In the theory of nomic probability, on the other hand, precise principles of statistical induction follow as theorems from the initial principles.

In general philosophy, more interest has attached to enumerative induction than to statistical induction, and the connection between enumerative induction and statistical induction has remained murky. That is partly because at least some philosophers have recognized that even if we could conclude by statistical induction that $\text{prob}(A/B) = 1$, it would not follow that $B \Rightarrow A$. Within the theory of nomic probability it is possible to base enumerative induction on statistical induction in a more indirect fashion. Thus, in a sense, the theory of nomic probability contains a solution to Hume's problem of justifying enumerative induction. Of course, the justification is only relative to the basic principles of nomic probability, but any justification has to stop somewhere. You cannot get something from nothing.

3. Prospects: Other Forms of Induction

Statistical and enumerative induction are just the simplest forms of inductive reasoning. There are other forms of reasoning that are inductive in somewhat the same sense. These include curve fitting, inference to the best explanation, the confirmation of scientific theories, and general patterns of statistical inference. Although these forms of inference involve more than statistical and enumerative induction, it is reasonable to hope that they will succumb to analysis within the theory of nomic probability. Statistical and enumerative induction turn ultimately on probabilistic reasoning and the acceptance rule (A3). It would be a bit surprising if this were not true of other forms of induction as well, although for at least some of these kinds of inductive reasoning the underlying structures are probably more complicated and the arguments required to legitimate them more involved. I have spent some time thinking about these various kinds of induction, and although in no case do I have a theory to propose, I will make some preliminary remarks that may or may not be useful to those who want to carry the matter further.

3.1 Statistical Inference

Statistical inference is a very broad subject that includes many diverse kinds of reasoning, some of it not even remotely connected with induction or the confirmation of hypotheses. But even if we confine our attention to inductive varieties of statistical inference, the theories we find are *ad hoc* jury-rigged affairs without adequate foundations. They have grown up as piecemeal attempts to capture reasoning that seems intuitively correct. It would be surprising if there weren't some sound intuitions that statisticians are trying to capture with these theories. But no one believes that existing statistical theories are really adequate. It is universally acknowledged that such theories do not always lead to reasonable conclusions. It is noteworthy that practicing statisticians do not use the theories in those cases in which they yield unreasonable conclusions. Statistics is not an axiomatized science. There is a good deal of art involved in knowing when to apply what kind of statistical test.

I would urge that the foundations for much statistical practice can be found in the theory of nomic probability. The key to working this out is the statistical induction argument, which can be modified and generalized in various ways to make it applicable to familiar kinds of statistical reasoning involving confidence intervals, significance tests, and so forth. But I say this without having worked out the details. The latter is a large undertaking, and I will leave it to others.

3.2 Curve Fitting

A problem that has interested me for years is that of curve fitting. Presented with a finite set of data points, there are infinitely many different curves that a scientist could draw through those points. It is remarkable then that any two scientists will draw approximately the same curve through the same data points, and they do this independently of any theory about how such curves should be drawn. What is involved here is a form of induction. The scientist is taking the finite set of data points to confirm the general numerical hypothesis represented by the curve. At first glance, this looks like a case of enumerative induction. The scientist has a finite set of instances of the generalization represented by the curve, and takes that as a prima facie reason for the generalization. The difficulty is that, as already noted, there are infinitely many curves that can be drawn through the points. If the generalizations represented by these infinitely many curves are all projectible then the data points provide prima facie reasons for all of the generalizations. But the generalizations are inconsistent with one another, and so we have a case of collective defeat.

One might try to solve this problem by denying that the

competing generalizations are all projectible. But that is not satisfactory. The curve we want to draw is the one that is, in some sense, the simplest. No one knows how to analyze the requisite sense of simplicity, but let us waive that for now. Thus we take the data points to confirm the generalization represented by the simplest curve. But it would be a mistake to suppose that the less simple curves all represent nonprojectible generalizations. If we acquire new data points that, when added to the original data points, rule out the previously simplest curve, we take the expanded set of data points to confirm a different generalization— the one that is represented by the simplest curve passing through the expanded set of points. This new generalization must be projectible. But of course, the new curve also passes through the original points, so as the new generalization is projectible it seems that the original set of points should also have provided a prima facie reason for it, and hence we would have collective defeat after all.

There is a purely formal ploy that would avoid this problem. What actually happens in curve fitting is that there are lots of different curves passing through the data points, and any of these curves could be confirmed by an appropriate set of points, but we take each curve to be confirmed only when we have acquired data ruling out all the simpler curves. This could be explained by supposing that the generalizations represented by the curves are not projectible *simpliciter*, but are only projectible with respect to the additional assumption that no simpler competing generalization is true. More accurately, where f is a function describing a curve, the generalizations have the form $\ulcorner Rxy \Rightarrow y = f(x) \urcorner$. We might suppose that $\ulcorner y = f(x) \urcorner$ is not projectible with respect to $\ulcorner Rxy \urcorner$, but only with respect to $\ulcorner Rxy$ & no simpler competing generalization is true$\urcorner$. Then positive instances of $\ulcorner Rxy \Rightarrow y = f(x) \urcorner$ will only confirm this generalization insofar as they also rule out all competing simpler hypotheses. It seems to me that the solution to the problem of curve fitting might conceivably lie in this direction, but there are major difficulties for working it out. First, of course, is the problem of analyzing the requisite notion of simplicity. But what bothers me even more is the heavy burden this puts on the notion of projectibility. It has the effect of imposing a very rich *a priori* structure on projectibility, and that will require justification. There are enough problems concerning projectibility already. I hesitate to add to them. I will say more about projectibility in section 4.

3.3 Inference to the Best Explanation

Inference to the best explanation is very common. It ranges from the mundane to the highly theoretical. I may infer that it is windy outside because I hear the shutters rattling and see leaves whisking briskly by the window. I may infer that hadrons (protons, neutrons, etc.) are composed of quarks because that is the best explanation for their having the rich taxonomic structure they do and the best explanation for the observed decay products from high-energy interactions. My own feeling is that the confirmation of scientific theories is best described as proceeding via inference to the best explanation, and it is this mode of inference that should replace the now largely defunct hypothetico-deductive method in the philosophy of science. But I will not argue for that here.

Inference to the best explanation often looks superficially like the use of (*A1*). For example, it is tempting to suppose we know it probable that it is windy outside given that the shutters are rattling and leaves whisking briskly by the window. I am not at all sure, however, that we have adequate evidence for regarding this nomic probability as high. I am inclined to think that, if not in this case then at least in many other cases, we can only assert this to be probable in the sense of epistemic probability, and to say that the epistemic probability is high is merely to *report* the reasonableness of the inference—not to explain it or justify it.

An analysis that I find tempting is that in confirming a hypothesis H by seeing that it explains various observations, we are confirming by enumerative induction that all the consequences of H are true. If we could confirm this latter generalization it would entail that H is true, because H is one of its own consequences. It seems to me that there may be something right about this general approach, but major difficulties arise when we try to spell it out in detail. One difficulty concerns how we count consequences. For example, if P is a consequence of H, does

⌜P∨Q⌝ count as a separate consequence? And if R and S are consequences of H, is ⌜R&S⌝ another consequence? A second difficulty arises from noting that the consequences that are taken to confirm a hypothesis are often only probabilistic consequences. But it is unreasonable to expect *all* of the probabilistic consequences of a true hypothesis to be true. It is only reasonable to expect most of them to be true. But then it is equally unreasonable to try to confirm by enumerative induction that all of the consequences of the hypothesis are true. We might try instead to confirm by statistical induction that most of the consequences are true (the problem of counting consequences recurs here). This no longer entails the truth of the hypothesis, but perhaps we can infer the latter by (*A1*). Despite the difficulties I have enumerated, I find this approach appealing, but I have not been able to make it work so I merely throw it out as something for the reader to think about.

4. The Problem of Projectibility

Now we come to the part of the book where the author confesses its shortcomings. To my mind, the principal problem remaining for the theory of nomic probability is the problem of projectibility. Projectibility plays a remarkably pervasive role in the theory. Since Nelson Goodman, everyone knows that principles of induction must be constrained by projectibility. Henry Kyburg was the first to note that a similar constraint is required in direct inference. He did not talk about projectibility as such, but that is what the constraint comes to. And projectibility recurs as a constraint in probabilistic acceptance rules. In fact, it was the occurrence of projectibility constraints in all three of these areas that first led me to look for connections between them. It seemed to me that they must have a common source and I traced that to the acceptance rules, thus arguing that projectibility has first and foremost to do with acceptance rules and has to do with direct inference and induction only derivatively.

To complete the theory of nomic probability we must have an analysis of projectibility. Unfortunately, I have none to offer.

When Goodman first called our attention to projectibility, a spate of proposals followed regarding the analysis of projectibility. Goodman himself proposed to analyze the concept in terms of entrenchment. Most other authors favored analyses in terms of "positionality". The latter analyses ruled out as unprojectible those properties that made reference to particular individuals or particular places and times. All of these theories labored under the misapprehension that most properties are projectible and that it is only a few funny ones that have to be ruled out. But as I have shown in the present book, projectibility is the exception rather than rule. Most properties are unprojectible. In addition, projectibility has a rich logical structure—conjunctions of projectible properties are projectible, but disjunctions are not generally projectible, and so forth. Any adequate theory of projectibility must explain this, but no extant theory does.

In Pollock [1974], I alleged that the problem of projectibility had a simple solution—the projectibility of a concept is part of its conceptual role. As the conceptual role of a concept is constitutive of that concept and not derivable from anything deeper, that is the end of the matter. The trouble with this account is that it overlooks the "logic" of projectibility. The projectibility of a logically simple concept could be an irreducible feature of its conceptual role, but logically complex concepts (disjunctions, conjunctions, etc.) are characterized by the way they are constructed out of simpler concepts, and their conceptual roles are derivative from that. Thus their projectibility or nonprojectibility would have to follow somehow from the way in which they are constructed out of simpler projectible or nonprojectible concepts. The kind of account of projectibility proposed in Pollock [1974] might conceivably be right for logically simple concepts, but the theory of nomic probability makes heavy use of closure conditions for projectibility. These closure conditions pertain to the projectibility of logically complex concepts. An adequate account of projectibility must determine what closure conditions hold and explain why they hold. No account of projectibility that talks only about logically simple concepts can do that. At this point, I have no idea what a correct theory of projectibility is going to look like.

There is some evidence that projectibility may have an even more fundamental role to play than in acceptance rules. It appears to play a general role in prima facie reasons. As I have pointed out elsewhere, all prima facie reasons are subject to "reliability defeaters".[2] For example, something's looking red to me gives me a prima facie reason for thinking that it is red, but if I know that color vision is unreliable under the present circumstances, this constitutes an undercutting defeater. To formulate reliability defeaters precisely, let us symbolize ⌜x (the epistemic agent) is in circumstances of type P⌝ as ⌜Px⌝. Suppose ⌜Px⌝ is a prima facie reason for ⌜Qx⌝. Then as a first approximation it seems that for an agent s, ⌜Rs and prob($Q/P\&R$) is low⌝ should be a reliability defeater for this prima facie reason. But arguments having familiar-looking structures show that we must impose restrictions on R. For example, suppose $R = (S \lor T)$, where prob($S/S \lor T$) is high, prob($Q/P\&S$) is low, prob($Q/P\&T$) is unknown, and Ts & $\sim Ss$. Then it follows that Rs and that prob($Q/P\&R$) is low, but intuitively this should not be a defeater. For example, let ⌜Px⌝ be ⌜x is appeared to redly⌝ and let ⌜Qx⌝ be ⌜there is something red before x⌝. So ⌜Px⌝ is a prima facie reason for ⌜Qx⌝. Pick some highly improbable property T possessed by the epistemic agent, for instance, the property of having been born in the first second of the first minute of the first hour of the first year of the twentieth century, and let S be the considerably more probable property (not actually possessed by the agent) of wearing rose colored glasses. Then prob($Q/P\&(S \lor T)$) is low, and the agent has the property $(S \lor T)$, but this should not constitute a defeater because the agent has the property $(S \lor T)$ only by virtue of having the property T, and by hypothesis we do not know whether prob($Q/P\&T$) is low.

It seems reasonably clear that we need a projectibility constraint here. In order for ⌜Rs and prob($Q/P\&R$) is low⌝ to be a reliability defeater for ⌜Px⌝ as a prima facie reason for ⌜Qx⌝, it is required that Q be projectible with respect to $(P\&R)$. This is a bit puzzling, however. There is no obvious connection

---

[2] In Pollock [1984c].

between this projectibility constraint and the one in our acceptance rules. On the other hand, it also seems unlikely that there should be two separate sources of constraints in epistemology both appealing to the same concept of projectibility. I am unsure what to make of this except to suggest that the problem of projectibility is an even deeper one than we realized before.

In the course of developing the theory of nomic probability, we have learned quite a bit about projectibility, but we have not discovered a general theory of projectibility. This does not disqualify the theory of nomic probability, because by relying upon intuitions we are able to discover what properties are projectible and what properties are unprojectible, but we still need a general theory of projectibility. Perhaps the ultimate result of the theory of nomic probability is to underscore the importance of the problem of projectibility, both for probability and for epistemology in general.

PART II

ADVANCED TOPICS

This part consists of technical material that is not essential for an understanding of the general theory, but is important for the application of the theory to concrete cases.

CHAPTER 7
EXOTIC COMPUTATIONAL PRINCIPLES

1. Introduction

Exotic computational principles are those not derivable from the theory of proportions constructed in Chapter 2. The most important of these principles are (*PFREQ*) and (*AGREE*). The purpose of this chapter is to show that these principles can be derived from a strengthened theory of proportions—what we might call 'the exotic theory of proportions'.

The principles of the theory of proportions constructed in Chapter 2 seem completely unproblematic, and accordingly the derivations of principles of nomic probability can reasonably be regarded as proofs of those principles. That ceases to be the case when we turn to the exotic theory of proportions and the corresponding exotic principles of nomic probability. Although quite intuitive, the exotic axioms for proportions are also very strong and correspondingly riskier. Furthermore, although the exotic axioms are intuitive, intuitions become suspect at this level. The problem is that there are a large number of intuitive candidates for exotic axioms, and although each is intuitive by itself, they are jointly inconsistent. This will be illustrated below. It means that we cannot have unqualified trust in our intuitions. In light of this, (*PFREQ*) and (*AGREE*) seem more certain than the exotic principles of proportions from which they can be derived. As such, those derivations cannot reasonably be regarded as justifications for the probability principles. Instead they are best viewed as explanations for why the probability principles are true given the characterization of nomic probabilities in terms of proportions. The derivations play an explanatory role rather than a justificatory role.

2. The Exotic Theory of Proportions

The exotic principles of proportions concern proportions in relational sets. Recall that we can compare sizes of sets with the relation $\ulcorner X \approx Y \urcorner$, which was defined as $\ulcorner \wp(X/X \cup Y) = \wp(Y/X \cup Y) \urcorner$. Our first exotic principle relates the size of a binary relation (a "two-dimensional set") to the sizes of its one-dimensional segments. If x is in the domain $\mathbf{D}(R)$ of R, let R_x be the R-*projection* of x, i.e., $\{y \mid Rxy\}$. Suppose $\mathbf{D}(R) = \mathbf{D}(S)$, and for each x in their domain, $R_x \approx S_x$. Then their "linear dimensions" are everywhere the same. If we graph R and S, we can think of S as resulting from sliding some of the segments of R up or down. It seems that this should not affect the measure of S. This is the *Principle of Translation Invariance*:

(2.1)   If $\mathbf{D}(R) = \mathbf{D}(S)$ and $(\forall x)[x \in \mathbf{D}(R) \supset R_x \approx S_x]$ then $R \approx S$.

The principle of translation invariance has a number of important consequences. We first prove:

(2.2)   If n is an integer, $S \subseteq T$ and $R \subseteq T$, $\mathbf{D}(R) = \mathbf{D}(S)$, and $(\forall x)[x \in \mathbf{D}(R) \supset \wp(R_x/T_x) = n \cdot \wp(S_x/T_x)]$ then $\wp(R/T) = n \cdot \wp(S/T)$.

Proof: By (5.12) of Chapter 2, we can divide each R_x into n disjoint subsets $R_{x,1}, \ldots, R_{x,n}$ such that for each i, $\wp(R_{x,i}/T_x) = \wp(S_x/T_x)$. Let $R_i = \{\langle x,y \rangle \mid x \in \mathbf{D}(R) \ \& \ y \in R_{x,i}\}$. Then by translation invariance, $\wp(R_i/T) = \wp(S/T)$. By (5.6) of Chapter 2, $\wp(R/T)$ is the sum of the $\wp(R_i/T)$, so $\wp(R/T) = n \cdot \wp(S/T)$.

Analogously:

(2.3)   If n,m are integers, $S \subseteq T$ and $R \subseteq T$, $\mathbf{D}(R) = \mathbf{D}(S)$, and $(\forall x)[x \in \mathbf{D}(R) \supset \wp(R_x/T_x) = n/m \cdot \wp(S_x/T_x)]$ then $\wp(R/T) = n/m \cdot \wp(S/T)$.

Then by looking at sequences of rationals (and using the denseness principle) we obtain:

(2.4) If r is a real number, $S \subseteq T$ and $R \subseteq T$, $\mathbf{D}(R) = \mathbf{D}(S)$, and $(\forall x)[x \in \mathbf{D}(R) \supset \wp(R_x/T_x) = r \cdot \wp(S_x/T_x)]$, then $\wp(R/T) = r \cdot \wp(S/T)$.[1]

Proof: If r is rational, this follows from (2.3). If r is irrational, consider a sequence $\{k_i/m_i\}_{i \in \omega}$ of rationals such that for each i, $k_i/m_i < r < (k_i+1)/m_i$. By the denseness principle (5.12) of Chapter 2 and its corollary (5.13) of Chapter 2, for each $i \in \omega$ we can find an R^i and and R_i such that $\mathbf{D}(R^i) = \mathbf{D}(R_i) = \mathbf{D}(R)$ and for each $x \in \mathbf{D}(R)$, $R_{i,x} \subseteq R_x \subseteq R^i_x$ and $\wp(R_{i,x}/T_x) = (k_i/m_i) \cdot \wp(S_x/T_x)$ and $\wp(R^i_x/Tx) = ((k_i+1)/m_i) \cdot \wp(S_x/T_x)$. By (2.3), $\wp(R_i/T) = (k_i/m_i) \cdot \wp(S/T)$ and $\wp(R^i/T) = ((k_i+1)/m_i) \cdot \wp(S/T)$. By (5.6) of Chapter 2, $\wp(R_i/T) \leq \wp(R/T) \leq \wp(R^i/T)$. Thus for each i, $(k_i/m_i) \cdot \wp(S/T) \leq \wp(R/T) \leq ((k_i+1)/m_i) \cdot \wp(S/T)$. As the sequence of rationals converges to r, it follows that $\wp(R/T) = r \cdot \wp(S/T)$.

Letting $S = T$, *The Constancy Principle* is an immediate consequence of (2.4):

(2.5) If r is a real number, $(\forall x)\Big[x \in \mathbf{D}(T) \supset \wp(R_x/T_x) = r\Big]$, and $\mathbf{D}(R) = \mathbf{D}(T)$ then $\wp(R/T) = r$.

This principle will be very important. Note that it is actually equivalent to translation invariance.

Principle (2.1) secures "vertical" translation invariance. It would be natural to suppose that we should also add an axiom securing "horizontal" translation invariance. One of the most surprising features of the general theory of proportions is that we cannot have both. We can have either, but the conjunction of the two is inconsistent. This can be seen as follows. Just as vertical translation invariance implies the constancy principle, horizontal translation invariance would imply the following "horizontal" constancy principle:

If $(\forall y)\Big[y \in \mathbf{R}(R) \supset \wp(\{x \mid Axy \ \& \ Rxy\} / \{x \mid Rxy\}) = r\Big]$ and $\mathbf{R}(R) = \mathbf{R}(T)$ then $\wp(A/R) = r$.

---

[1] This implies that probability spaces generated by proportion functions are generalizations of what Renyi [1955] calls 'Cavalieri spaces'.

These two constancy principles jointly imply:

$$\text{If } (\forall y)\Big[y{\in}\mathbf{R}(R) \supset \wp(\{x\mid Axy \ \& \ Rxy\} \ / \ \{x\mid Rxy\}) = r\Big]$$
$$\text{and } (\forall x)\Big[x{\in}\mathbf{D}(R) \supset \wp(A_x/R_x) = s\Big] \text{ and } \mathbf{D}(R) = \mathbf{D}(T) \text{ and}$$
$$\mathbf{R}(R) = \mathbf{R}(T) \text{ then } r = s.$$

Now consider the set

$$R = \{\langle x,y\rangle \mid x,y{\in}\omega \ \& \ x < y\}$$

and let $T = \omega{\times}\omega$. For each $x{\in}\omega$, $R_x = \{y\mid y{\in}\omega \ \& \ x < y\}$, so it follows by the extended frequency principle that $\wp(R_x/\omega) = 1$. Similarly, the extended frequency principle implies that for each $y{\in}\omega$, $\wp(\{x\mid x{\in}\omega \ \& \ x < y\} \ / \ \omega) = 0$. Thus the two constancy principles jointly imply that $1 = 0$. This shows that we cannot have both vertical and horizontal translation invariance.[2] Which we adopt is just a convention regarding the orientation of the axes. I have opted for vertical translation invariance.

Let us define:

(2.6) R is a *rectangle of height* s iff R is a binary relation and $s > 0$ and $(\forall x)[x{\in}\mathbf{D}(R) \supset \wp(R_x/\mathbf{R}(R)) = s]$.

Rectangles are "rectangular sets", in the sense that every projection has the same nonzero "height" s relative to the range of the set. A quick corollary of (2.5) is:

(2.7)   If R is a rectangle of height s then $\wp(R \ / \ \mathbf{D}(R){\times}\mathbf{R}(R)) = s$.

---

[2] This is related to a theorem of Hausdorff [1949] (469-472). Suppose we consider finitely additive normalized measures defined on all subsets of a Euclidean sphere of radius 1. Hausdorff proved that no such measure can satisfy the condition that congruent subsets receive the same measure. (My attention was drawn to this theorem by the mention in van Fraassen [1981], 180ff.)

When I first began thinking about relational principles, it seemed to me that the following existential generalization principle should be true:

$$\wp(Ax/Rxy) = \wp(Ax \: / \: (\exists y)Rxy).$$

However, this principle is subject to simple counterexamples. Let $R = \{\langle 1,1 \rangle, \langle 2,1 \rangle, \langle 2,2 \rangle\}$ and $A = \{1\}$. Then $\wp(Ax/Rxy) = 1/3$, but $\wp(Ax \: / \: (\exists y)Rxy) = 1/2$. The reason the principle fails is that there may be more y's related to some x's than to others, and that biases the proportion $\wp(Ax/Rxy)$ but does not bias the proportion $\wp(Ax \: / \: (\exists y)Rxy)$. To say that there are no more y's related to some x than to any other x^* can be captured by requiring that $\wp(R_x/\mathbf{R}(R)) = \wp(R_{x^*}/\mathbf{R}(R))$, that is, by requiring that R is a rectangle. Accordingly, we should have the following two *Principles of Existential Generalization*:

(2.8)   If R is a rectangle then $\wp(Ax/Rxy) = \wp(Ax/(\exists y)Rxy)$.

(2.9)   If the converse of R is a rectangle then $\wp(Ay/Rxy) = \wp(Ay/(\exists x)Rxy)$.

With the help of translation invariance, we can prove (2.8):

Proof: Suppose that $(\forall x)[x \in \mathbf{D}(R) \supset \wp(R_x \: / \: \mathbf{R}(R)) = s]$, where $s > 0$. The *restriction of R to A* is $R{\upharpoonright}A = \{\langle x,y \rangle \mid x \in A \: \& \: \langle x,y \rangle \in R\}$. By definition, $\wp(Ax/Rxy) = \wp(R{\upharpoonright}A/R)$. It is a theorem of the classical probability calculus that if $X \subseteq Y \subseteq Z$ then $\wp(X/Z) = \wp(X/Y) \cdot \wp(Y/Z)$. Suppose $x \in A \cap \mathbf{D}(R)$. Then $(R{\upharpoonright}A)_x = R_x$. Thus we have:

$$s = \wp((R{\upharpoonright}A)_x \: / \: \mathbf{R}(R))$$
$$= \wp((R{\upharpoonright}A)_x \: / \: \mathbf{R}(R{\upharpoonright}A)) \cdot \wp(\mathbf{R}(R{\upharpoonright}A) \: / \: \mathbf{R}(R)).$$

Thus

$$\wp((R{\upharpoonright}A)_x \: / \: \mathbf{R}(R{\upharpoonright}A)) = s \div \wp(\mathbf{R}(R{\upharpoonright}A) \: / \: \mathbf{R}(R))$$

and hence $R{\upharpoonright}A$ is also a rectangle. Thus using (2.7) and the cross product principle (7.2) of Chapter 2 we have:

$$\wp(R{\restriction}A \;/\; \mathbf{D}(R){\times}\mathbf{R}(R))$$

$$= \wp(R{\restriction}A \;/\; \mathbf{D}(R{\restriction}A){\times}\mathbf{R}(R)) \cdot \wp(\mathbf{D}(R{\restriction}A){\times}\mathbf{R}(R) \;/\; \mathbf{D}(R){\times}\mathbf{R}(R))$$

$$= \wp(R{\restriction}A \;/\; \mathbf{D}(R{\restriction}A){\times}\mathbf{R}(R{\restriction}A))$$
$$\cdot \wp(\mathbf{D}(R{\restriction}A){\times}\mathbf{R}(R{\restriction}A) \;/\; \mathbf{D}(R{\restriction}A){\times}\mathbf{R}(R))$$
$$\cdot \wp(\mathbf{D}(R{\restriction}A){\times}\mathbf{R}(R) \;/\; \mathbf{D}(R){\times}\mathbf{R}(R))$$

$$= \wp((R{\restriction}A)_x \;/\; \mathbf{R}(R{\restriction}A)) \cdot \wp(\mathbf{R}(R{\restriction}A) \;/\; \mathbf{R}(R))$$
$$\cdot \wp(\mathbf{D}(R{\restriction}A) \;/\; \mathbf{D}(R))$$

$$= s \cdot \wp(A/\mathbf{D}(R)).$$

But we also have:

$$\wp(R{\restriction}A \;/\; \mathbf{D}(R){\times}\mathbf{R}(R))$$
$$= \wp(R{\restriction}A \;/\; R) \cdot \wp(R \;/\; \mathbf{D}(R){\times}\mathbf{R}(R))$$
$$= \wp(R{\restriction}A \;/\; R) \cdot s.$$

Therefore, $\wp(Ax/Rxy) = \wp(R{\restriction}A \;/\; R) = \wp(A \;/\; \mathbf{D}(R)) = \wp(Ax/(\exists y)Rxy)$.

Interestingly, it is not possible to give an analogous proof of (2.9). The difficulty is that the proof of (2.8) used vertical translation invariance, but as noted above, horizontal translation invariance cannot be endorsed at the same time. Nevertheless, (2.9) seems equally true, so I will adopt it as an additional axiom.

It seems intuitively that if R and S are binary relations, and $\widetilde{R}$ and $\widetilde{S}$ are their converses, then it should be the case that $\wp(R/S) = \wp(\widetilde{R}/\widetilde{S})$. I will say that a proportion is *reversible* if this holds. The principle telling us that all proportions are reversible is the *Reversibility Principle*. A surprising feature of the general theory of proportions is that the reversibility principle is false. For example, consider once more the set

$$R = \{\langle x,y\rangle \mid x,y{\in}\omega \;\&\; x < y\}$$

and let us evaluate the proportion $\wp(R/\omega{\times}\omega)$. In discussing horizontal translation invariance we noted that it follows from the extended frequency principle that for each $x \in \omega$, $\wp(R_x/\omega) = 1$, and for each $y \in \omega$, $\wp(\{x \mid x \in \omega \ \& \ x < y\} \ / \ \omega) = 0$. It follows by the constancy principle that $\wp(R/\omega \times \omega) = 1$ and $\wp(\widetilde{R}/\omega \times \omega) = 0$. Therefore, $\wp(R/\omega \times \omega) \neq \wp(\widetilde{R}/\omega \times \omega)$.

The only possibility for making the reversibility principle true is to restrict the domain of $\wp$. I have been assuming that $\wp(X/Y)$ exists for all sets X and Y, but it might be ruled that some sets, like R above, are "nonmeasurable" and are excluded from the domain of $\wp$. I do not see any good reason for doing this, however. Its effect is simply to reduce the class of measurable sets. It is more useful to have a general proportion function not satisfying the reversibility principle and then investigate the conditions under which proportions are reversible. That is the course followed here.

The question of when proportions are reversible has proven to be very difficult. The two principles of existential generalization give us one rather trivial condition under which reversibility holds:

(2.10) If R is a rectangle and $A \subseteq \mathbf{D}(R)$ then $\wp(x \in A \ / \ \langle x,y \rangle \in R)$
$= \wp(y \in A \ / \ \langle x,y \rangle \in \widetilde{R})$.

But I have been unable to find a more interesting reversibility condition.

A perplexing consequence for the theory of nomic probability is that if proportions are not reversible, neither are nomic probabilities. That is, it is not automatically the case that $\mathrm{prob}(Axy/Rxy) = \mathrm{prob}(Ayx/Ryx)$. This follows from (8.4) of Chapter 2, according to which proportions are special nomic probabilities. We could avoid this by giving a slightly different analysis of nomic probability, ruling that $\mathrm{prob}(Axy/Rxy)$ is undefined if the proportion with which it is identified is not reversible. But the effect would merely be to throw away some probabilities. It would not make any further probabilities reversible. There can be little point to such a modification.

What makes the nonreversibility of nomic probability so

puzzling is that the probabilities we ordinarily think about seem almost invariably to be reversible. The examples of nonreversible probabilities that we have encountered are all recherché mathematical examples. When we turn to logically contingent probabilities concerning ordinary physical objects, they always seem to be reversible. Why is that? The apparent answer is that the assessment of such probabilities is based ultimately upon statistical induction. In evaluating prob(Axy/Rxy) by statistical induction, we observe a finite sample of pairs $\langle x,y \rangle$ satisfying R and see what proportion of them also satisfy A. We then estimate that prob(Axy/Rxy) is approximately equal to that proportion. Proportions in finite sets are just relative frequencies, and hence are always reversible. Thus the observed relative frequency of pairs satisfying R that also satisfy A is automatically the same as the observed relative frequency of pairs satisfying the converse of R that also satisfy the converse of A. In other words, insofar as induction gives us a reason for estimating that prob(Axy/Rxy) = r, it automatically also gives us a reason for estimating that prob(Ayx/Ryx) = r. Thus our grounds for evaluating contingent nomic probabilities automatically give us a reason for thinking that each such probability is reversible.

Of course, not all contingent nomic probabilities are evaluated by a direct application of statistical induction. Some are calculated on the basis of others. But all of the computational principles uncovered so far have the characteristic that when they are applied to reversible probabilities they yield reversible probabilities. Thus such computations cannot be a source of nonreversibility. It appears then that the only way we could ever be led to conclude that a contingent nomic probability is nonreversible is when some nonreversible *a priori* probability is used in its computation. That would be at least unusual.

My conclusion is that there should be no logical guarantee of reversibility for nomic probabilities. However, the epistemological characteristics of these probabilities make it natural to expect any concrete example of a contingent nomic probability to be reversible. That is the source of our ordinary expectation of reversibility. Such reversibility should not be forthcoming from our computational principles by themselves.

3. The Derivation of *(PFREQ)*

We can use the principle of translation invariance to relate probabilities and frequencies. Suppose the property G is such that, necessarily, freq$[F/G] = r$. prob(F/G) is supposed to be the proportion of physically possible G's that would be F's. As the proportion of actual G's that are F is the same in every possible world, that should also be the value of prob(F/G). Precisely:

(3.1)   If r is a nomically rigid designator of a a real number and
         $\Diamond_p \exists G$ and $\Box[\exists G \supset$ freq$[F/G] = r]$ then prob$(F/G) = r$.

Proof: Suppose $\langle w,x \rangle \in G$. Then w is a physically possible world and x exemplifies G at w. Hence at w it is true that freq$[F/G] = r$. By the frequency principle (5.2) of Chapter 2,

$$\wp\Big(\{x|\ \langle w,x \rangle \in G\ \&\ x \text{ is } F \text{ at } w\}\ /\ \{x|\ \langle w,x \rangle \in G\} \Big) = r.$$

Recalling that G_w is the w-projection of the binary relation G, this can be expressed equivalently as:

$$\wp((F \cap G)_w\ /\ G_w) = r.$$

Therefore, by the constancy principle (2.5), $\wp(F \cap G/G) = r$. But $\wp(F \cap G/G) = \wp(F/G) = $ prob(F/G).[3]

(PFREQ) is an immediate consequence of (6.20) of Chapter 2[4] and (3.1):

---

[3] It is worth noting that this proof requires only a special case of translation invariance. It assumes translation invariance for properties, i.e., binary relations whose domains are sets of possible worlds. It would be possible to adopt this special case of translation invariance as an axiom without assuming the general principle.

[4] (6.20) requires that r be a rigid designator rather than a nomically rigid designator, but inspection of the proof of (6.20) reveals that if $\Diamond_p \exists G$ then the theorem holds if r is a nomically rigid designator.

(*PFREQ*) If r is a nomically rigid designator of a real number
and $\Diamond_p\big[\exists G \ \& \ \text{freq}[F/G] = r\big]$ then $\text{prob}\big(F \ / \ G \ \&$
$\text{freq}[F/G] = r\big) = r.$

4. The Principle of Agreement

The Principle of Agreement has played a pivotal role in the
theory of nomic probability. I will now show that it can be
derived from the exotic theory of proportions. We first prove a
rather standard law of large number whose proof does not require
the use of exotic principles. Given any set B, let B^n be the set
of all ordered n-tuples of members of B, that is, $\{\langle x_1,\ldots,x_n\rangle \,|\,$
$x_1,\ldots,x_n{\in}B\}$. We have defined frequency in unordered sets, but
it is also convenient to define frequency in ordered sets. If σ is
an ordered n-tuple, let $\text{freq}[A/\sigma]$ be the frequency of places in the
n-tuple occupied by members of A. Let us say that $\langle x_1,\ldots,x_n\rangle$
is *nonrepeating* iff $x_i \neq x_j$ whenever $i \neq j$. Recall that $\ulcorner x \approx_\delta y\urcorner$
means $\ulcorner x$ is approximately equal to y, the difference being at most
$\delta\urcorner$, i.e., $\ulcorner x-y \leq \delta$ and $y-x \leq \delta\urcorner$. With this notation, the *Basic
Law of Large Numbers* can be formulated as follows:

(4.1)   If B is infinite then $\wp(A/B) = p$ iff for every $\delta > 0$,

$\lim_{n\to\infty}\wp\big(\text{freq}[A/\sigma] \approx_\delta p \ / \ \sigma{\in}B^n \ \& \ \sigma \text{ nonrepeating}\big) = 1.$

This principle is very similar to the standard Weak Law of Large
Numbers in mathematical probability theory, and its proof is also
similar to standard proofs of the Weak Law of Large Numbers.
There are a few minor differences, however, so I include the
proof here for the sake of completeness. The theorem is a
biconditional, and it is convenient to prove the two directions
separately:

(4.2)   If B is infinite and $\wp(A/B) = p$ then for every $\delta > 0$,

$\lim_{n\to\infty}\wp\big(\text{freq}[A/\sigma] \approx_\delta p \ / \ \sigma{\in}B^n \ \& \ \sigma \text{ nonrepeating}\big) = 1.$

Proof: Suppose B is infinite, $\wp(A/B) = p$, and > 0. By the concatenation principle (7.9) of Chapter 2:

$$\wp\left(x_1,\ldots,x_r{\in}A \ \& \ x_{r+1},\ldots,x_n{\notin}A \ / \ x_1,\ldots,x_n{\in}B\right) = p^r(1-p)^{n-r}.$$

Suppose $b{\in}B$. By the extended frequency principle, $\wp(b \neq y \ / \ y{\in}B) = 1$, so by the constancy principle $\wp(x \neq y \ / \ x,y{\in}B) = 1$. Analogously, for each n,

$$\wp(x_1,\ldots,x_n \text{ distinct } / \ x_1,\ldots,x_n{\in}B) = 1.$$

Therefore, by the probability calculus:

$$\wp\left(x_1,\ldots,x_r{\in}A \ \& \ x_{r+1},\ldots,x_n{\notin}A \ / \ x_1,\ldots,x_n{\in}B \ \& \ x_1,\ldots,x_n \text{ distinct}\right)$$
$$= p^r(1-p)^{n-r}.$$

By (7.11) of Chapter 2:

$$\wp\left(\text{freq}[A/\sigma] = \tfrac{r}{n} \ / \ \sigma{\in}B^n \ \& \ \sigma \text{ nonrepeating}\right)$$

$$= \wp\left(\text{freq}[A/\langle x_1,\ldots,x_n\rangle] = \tfrac{r}{n} \ / \ \langle x_1,\ldots,x_n\rangle{\in}B^n \ \& \ x_1,\ldots,x_n \text{ distinct}\right)$$

$$= \wp\left(\text{freq}[A/\langle x_1,\ldots,x_n\rangle] = \tfrac{r}{n} \ / \ x_1,\ldots,x_n{\in}B \ \& \ x_1,\ldots,x_n \text{ distinct}\right).$$

$\ulcorner\text{freq}[A/\langle x_1,\ldots,x_n\rangle] = \tfrac{r}{n}\urcorner$ is equivalent to a disjunction of $n!/r!(n-r)!$ mutually incompatible conjunctions of the form $\ulcorner x_1,\ldots,x_r{\in}A \ \& \ x_{r+1},\ldots,x_n{\notin}A\urcorner$. Consequently,

$$\wp\left(\text{freq}[A/\sigma] = \tfrac{r}{n} \ / \ \sigma{\in}B^n \ \& \ \sigma \text{ nonrepeating}\right)$$
$$= \frac{n!p^r(1-p)^{n-r}}{r!(n-r)!}.$$

This, of course, is the formula for the binomial distribution. $\ulcorner\text{freq}[A/\sigma] \approx_\delta p\urcorner$ is equivalent to $\ulcorner p-\delta \leq \text{freq}[A/\sigma] \leq p+\delta\urcorner$. Let $k(n)$ be the first integer greater than or equal to $n\cdot(p-\delta)$ and let $m(n)$ be the first integer less than or equal to $n\cdot(p+\delta)$. It then follows that

$$\wp\Big(\text{freq}[A/\sigma] \approx_\delta p \ / \ \sigma \in B^n \ \& \ \sigma \ \text{nonrepeating} \Big)$$

$$= \sum_{r=k(n)}^{m(n)} \frac{n! p^r (1-p)^{n-r}}{r!(n-r)!} \ .$$

But

$$\lim_{n \to \infty} \sum_{r=k(n)}^{m(n)} \frac{n! p^r (1-p)^{n-r}}{r!(n-r)!} = 1.$$

Therefore,

$$\lim_{n \to \infty} \wp\Big(\text{freq}[A/\sigma] \approx_\delta p \ / \ \sigma \in B^n \ \& \ \sigma \ \text{nonrepeating} \Big) = 1.$$

Conversely:

(4.3)   If B is infinite and for every $\delta > 0$,
$\lim_{n \to \infty} \wp\Big(\text{freq}[A/\sigma] \approx_\delta p \ / \ \sigma \in B^n \ \& \ \sigma \ \text{nonrepeating} \Big) = 1$,
then $\wp(A/B) = p$.

Proof: Suppose $\wp(A/B) = r \neq p$. Choose δ such that $0 < \delta < \frac{1}{2}|p-r|$.[5] By
(4.2), $\lim_{n \to \infty} \wp\Big(\text{freq}[A/\sigma] \approx_\delta r \ / \ \sigma \in B^n \Big) = 1$. But $\ulcorner\text{freq}[A/\sigma] \approx_\delta r \ \& \ \delta < \frac{1}{2}|p-r|\urcorner$
entails $\ulcorner\text{freq}[A/\sigma] \not\approx_\delta p\urcorner$, so $\lim_{n \to \infty} \wp\Big(\text{freq}[A/\sigma] \approx_\delta p \ / \ \sigma \in B^n \ \& \ \sigma$
nonrepeating$\Big) = 0 \neq 1$.

(4.2) and (4.3) jointly entail the Basic Law of Large Numbers
(4.1).

(4.1) is a close analogue of the standard Weak Law of Large
Numbers, but within the exotic theory of proportions it can be

---

[5] Recall that $|a|$ is the *absolute value* of a; that is, if $a \geq 0$ then $|a| = a$, and if $a < 0$ then $|a| = -a$.

used to prove a more powerful law of large numbers that has no analogue in standard probability theory. This is the Principle of Agreement for Proportions:

If B is infinite and $\wp(A/B) = p$ then for every $\delta > 0$,
$$\wp\Big(\wp(A/X) \approx_\delta p \,/\, X \subseteq B\Big) = 1.$$

Let us begin by proving the following lemma:

(4.4)   If B is infinite and $\wp(A/B) = p$ then for every $\delta > 0$,
$\lim_{n\to\infty}\wp\Big(\text{freq}[A/\sigma] \approx_\delta p \,/\, \sigma\in X^n$ & σ nonrepeating &
$X \subseteq B$ & X infinite$\Big) = 1$.

Proof: Suppose B is infinite and $\wp(A/B) = p$. Consider the proportion:

$$\wp\Big(x_1,\dots,x_r\in A \text{ \& } x_{r+1},\dots,x_n\notin A \,/\, x_1,\dots,x_n\in X \text{ \& } X \subseteq B \text{ \& }$$
$$X \text{ infinite \& } x_1,\dots,x_n \text{ distinct}\Big).$$

where we take the variables to be ordered alphabetically as follows: $x_1,\dots,x_n,X$. This proportion can be computed using the principle (2.8) of existential generalization. Consider first:

$$R = \{\langle\langle x_1,\dots,x_n\rangle,X\rangle \,|\, x_1,\dots,x_n\in X \text{ \& } X \subseteq B \text{ \& } X \text{ infinite}$$
$$\text{\& } x_1,\dots,x_n \text{ distinct}\}.$$

Observe that

$$(\exists y_1),\dots,(\exists y_n)[y_1,\dots,y_n\in X \text{ \& } X \subseteq B \text{ \& } X \text{ infinite \& } y_1,\dots,y_n$$
$$\text{distinct}]$$

is equivalent to $\ulcorner X \subseteq B$ & X infinite$\urcorner$. Therefore, by (7.14) of Chapter 2, if $a_1,\dots,a_n\in D(R)$:

$$\wp(R\langle x_1,\dots,x_n\rangle \,/\, R(R))$$

$$= \wp\Big(\{X|\ x_1,\dots,x_n\in X\} \,/\, \{X|\ (\exists y_1),\dots,(\exists y_n)[y_1,\dots,y_n\in X$$
$$\text{\& } X \subseteq B \text{ \& } X \text{ infinite \& } y_1,\dots,y_n \text{ distinct}]\}\Big)$$

$$= \wp(\{X|\ x_1,\ldots,x_n{\in}X\}\ /\ \{X|\ X \subseteq B\ \&\ X\ \text{infinite}\})$$

$$= 1/2^n.$$

By (2.8) it follows that:

$$\wp\Big(x_1,\ldots,x_r{\in}A\ \&\ x_{r+1},\ldots,x_n{\notin}A\ /\ x_1,\ldots,x_n{\in}X\ \&\ X \subseteq B\ \&$$
$$X\ \text{infinite}\ \&\ x_1,\ldots,x_n\ \text{distinct}\Big)$$

$$= \wp\Big(x_1,\ldots,x_r{\in}A\ \&\ x_{r+1},\ldots,x_n{\notin}A\ /\ (\exists Y)[x_1,\ldots,x_n{\in}Y\ \&$$
$$Y \subseteq B\ \&\ Y\ \text{infinite}\ \&\ x_1,\ldots,x_n\ \text{distinct}]\Big)$$

$$= \wp\Big(x_1,\ldots,x_r{\in}A\ \&\ x_{r+1},\ldots,x_n{\notin}A\ /\ x_1,\ldots,x_n{\in}B\ \&\ x_1,\ldots,x_n$$
$$\text{distinct}\Big)$$

$$= p^r(1-p)^{n-r}.$$

Note that it follows from (2.10) that the variables can be ordered either as $\langle\langle x_1,\ldots,x_n\rangle,X\rangle$ or $\langle X,\langle x_1,\ldots,x_n\rangle\rangle$. From this point, the proof is identical to the proof of (4.2).

The Principle of Agreement for Proportions is now proven as follows:

(4.5)   If B is infinite and $\wp(A/B) = p$ then for every $\delta > 0$:
$$\wp\Big(\wp(A/X) \approx_\delta p\ /\ X \subseteq B\Big) = 1.$$

Proof: Suppose B is infinite and $\wp(A/B) = p$. Consider any infinite subset C of B. By (4.1), for each $\delta,\epsilon > 0$ there is an N such that if $n > N$ and k is the first integer greater than or equal to $n\cdot(p-\delta)$ then:

$$\wp\Big(\text{freq}[A/\sigma] \approx_\delta \wp(A/C)\ /\ \sigma{\in}C^n\ \&\ \sigma\ \text{nonrepeating}\Big)$$

$$= \sum_{r=k(n)}^{m(n)} \frac{n!p^r(1-p)^{n-r}}{r!(n-r)!} \geq 1-\epsilon.$$

Taking the variables in the order $\langle X,\sigma \rangle$, it follows from the constancy principle (2.5) that:

$$\wp\Big(\text{freq}[A/\sigma] \approx_\delta \wp(A/X) \ / \ \sigma{\in}X^n \ \& \ \sigma \text{ nonrepeating } \&$$
$$X \subseteq B \ \& \ X \text{ infinite}\Big)$$

$$= \sum_{r=k(n)}^{m(n)} \frac{n!p^r(1-p)^{n-r}}{r!(n-r)!} \geq 1-\epsilon.$$

By theorem (4.4), for each $\delta,\epsilon > 0$ there is an M such that if $n > M$ then

$$\wp\Big(\text{freq}[A/\sigma] \approx_\delta p \ / \ \sigma{\in}X^n \ \& \ \sigma \text{ nonrepeating } \& \ X \subseteq B \ \&$$
$$X \text{ infinite}\Big) \geq 1-\epsilon.$$

It is a theorem of the classical probability calculus that if $\ulcorner\varphi \ \& \ \psi\urcorner$ entails χ and $\wp(\varphi/\theta) \geq 1-\epsilon$ and $\wp(\psi/\theta) \geq 1-\epsilon$ then $\wp(\chi/\theta) \geq 1-2\epsilon$. Therefore, if $n > N$ and $n > M$ then

$$\wp\Big(\wp(A/X) \approx_{2\delta} p \ / \ \sigma{\in}X^n \ \& \ \sigma \text{ nonrepeating } \& \ X \subseteq B \ \&$$
$$X \text{ infinite}\Big) \geq 1-2\epsilon.$$

Suppose $\eta{\in}B^n$ and η is nonrepeating. Then by (7.14) of Chapter 2

$$\wp\Big(\eta{\in}X^n \ / \ (\exists\sigma)[\sigma{\in}X^n \ \& \ \sigma \text{ nonrepeating } \& \ X \subseteq B \ \& \ X \text{ infinite}]\Big)$$

$$= \wp(\eta{\in}X^n \ / \ X \subseteq B \ \& \ X \text{ infinite})$$

$$= 1/2^n.$$

Therefore by existential generalization:

$$\wp\Big(\wp(A/X) \approx_{2\delta} p \ / \ \sigma{\in}X^n \ \& \ \sigma \text{ nonrepeating } \& \ B \ \&$$
$$X \text{ infinite}\Big)$$

$$= \wp\Big(\wp(A/X) \approx_{2\delta} p \ / \ (\exists \sigma)[\sigma \in X^n \ \& \ \sigma \text{ nonrepeating } \&$$
$$X \subseteq B \ \& \ X \text{ infinite}]\Big)$$

$$= \wp\Big(\wp(A/X) \approx_{2\delta} p \ / \ X \subseteq B \ \& \ X \text{ infinite}\Big).$$

Consequently, the latter is $\geq 1-2\epsilon$. This is true for every $\epsilon > 0$, so it follows that for every $\delta > 0$,

$$\wp\Big(\wp(A/X) \approx_{\delta} p \ / \ X \subseteq B \ \& \ X \text{ infinite}\Big) = 1.$$

As B is infinite, $\#\{X| \ X \subseteq B \ \& \ X \text{ finite}\} < \#\{X| \ X \subseteq B\}$. Thus by the extended frequency principle,

$$\wp(X \text{ infinite } / \ X \subseteq \mathrm{B}) = 1$$

and hence by (6.18) of Chapter 2,

$$\wp\Big(\wp(A/X) \approx_{\delta} p \ / \ X \subseteq B\Big)$$

$$= \wp\Big(\wp(A/X) \approx_{\delta} p \ / \ X \subseteq B \ \& \ X \text{ infinite}\Big) = 1.$$

The derivation of the principle of agreement makes very heavy use of the exotic axioms. Translation invariance and existential generalization play a prominent role. It is worth noting that these are relational principles, but the principle of agreement itself involves no relations—it concerns two one-place properties. This suggests that there may be another way of proving the principle of agreement that does not require these powerful relational principles. However, I have been unsuccessful in finding such a proof.

5. Summary

1. Exotic principles of proportions relate the measures of two-dimensional sets to the measures of their one-dimensional

projections.
2. Although these principles are highly intuitive, intuition leads to principles that are jointly inconsistent, so we can adopt only a subset of the intuitive axioms.
3. (*PFREQ*) and (*AGREE*) are theorems of the exotic theory of nomic probability that results from the exotic theory of proportions.

CHAPTER 8
ADVANCED TOPICS:
DEFEASIBLE REASONING

1. Introduction

The principal novelty this book brings to probability theory is a sophisticated epistemology accommodating defeasible reasoning. It is this that makes the theory of nomic probability possible. Earlier theories lacked the conceptual framework of prima facie reasons and defeaters, and hence were unable to adequately formulate principles of probabilistic reasoning. Thus far, the book has relied upon a loosely formulated account of the structure of defeasible reasoning, but that must be tightened up before the theory can be implemented. This chapter gives a more rigorous account of defeasible reasoning and compares the present theory with some related work in AI.[1]

2. Reasons

Reasoning begins from various kinds of inputs, which for convenience I will suppose to be encoded in beliefs.[2] Crudely put, reasoning proceeds in terms of reasons. Reasons are strung

---

[1] For a completely rigorous treatment, see Pollock [1988].

[2] This is simplistic. Perceptual states are nondoxastic states. For instance, something can look red to you without your having any belief to that effect. Furthermore, if you look around, you will form myriad beliefs about your surroundings but few if any beliefs about how things look to you. You do not usually attend to the way things look to you. (For an extended discussion of this point see Pollock [1986], 61ff.) Instead, you reason directly from the way things look to conclusions about their objective properties without forming intermediate beliefs about your own perceptual states. This means that if we are to describe this process in terms of reasoning, then we must acknowledge that reasons need not be beliefs. At the very least, perceptual states can be reasons. However, this sophistication makes no difference to the structure of probabilistic reasoning, so I propose to ignore it for now.

together into arguments and in this way the conclusions of the arguments become justified. The general notion of a reason can be defined as follows:

(2.1)   A set of propositions $\{P_1,\ldots,P_n\}$ is a *reason* for S to believe Q if and only if it is logically possible for S to be justified in believing Q on the basis of believing $P_1,\ldots,P_n$.

There are two kinds of reasons—*defeasible* and *nondefeasible*. Nondefeasible reasons are those reasons that logically entail their conclusions. For instance, $(P\&Q)$ is a nondefeasible reason for P. Such reasons are *conclusive reasons*. P is a defeasible reason for Q just in case P is a reason for Q, but it is possible to add additional information that undermines the justificatory connection. Such reasons are called 'prima facie reasons'. This notion can be defined more precisely as follows:

(2.2) P is a *prima facie reason* for S to believe Q if and only if P is a reason for S to believe Q and there is an R such that R is logically consistent with P but $(P\&R)$ is not a reason for S to believe Q.

(To keep this and subsequent definitions simple, they are just formulated for the case of a reason that is a single proposition rather than a set of propositions, but the general case is analogous.) The R's that defeat prima facie reasons are called 'defeaters':

(2.3) R is a *defeater* for P as a prima facie reason for Q if and only if P is a reason for S to believe Q and R is logically consistent with P but $(P\&R)$ is not a reason for S to believe Q.

There are two kinds of defeaters for prima facie reasons. Rebutting defeaters are reasons for denying the conclusion:

(2.4) R is a *rebutting defeater* for P as a prima facie reason for Q if and only if R is a defeater and R is a reason for believing $\sim Q$.

Undercutting defeaters attack the connection between the reason
and the conclusion rather than attacking the conclusion itself:

(2.5) R is an *undercutting defeater* for P as a prima facie reason
          for S to believe Q if and only if R is a defeater and R is
          a reason for denying that P wouldn't be true unless Q
          were true.

In (2.5), $\ulcorner P$ wouldn't be true unless Q were true$\urcorner$ is some kind of
conditional. It is symbolized here as $\ulcorner(P \gg Q)\urcorner$. This is clearly
not a material conditional, but beyond that it is unclear how it is
to be analyzed.

Defeaters are defeaters by virtue of being reasons for either
$\ulcorner{\sim}Q\urcorner$ or $\ulcorner{\sim}(P \gg Q)\urcorner$. They may be only defeasible reasons for
these conclusions, in which case their defeaters are "defeater
defeaters". There may similarly be defeater defeater defeaters,
and so on.

3. Warrant

In constructing a theory of reasoning, it is useful to begin by
considering the fiction of an ideal reasoner, or if you like, an ideal
intelligent machine with no limits on memory or computational
capacity. How should such a reasoner employ reasons and
defeaters in deciding what to believe? Let us say that a proposition is *warranted* in an epistemic situation iff an ideal reasoner
starting from that situation would be justified in believing the
proposition. This section is devoted to giving a more precise
account of the set of warranted propositions.

3.1 An Analysis of Warrant

Reasoning proceeds in terms of arguments. What arguments
can be constructed depends upon what input beliefs one has.
These constitute the *epistemic basis*. In the simplest case, an
argument is a finite sequence of propositions each of which is
either (1) contained in the epistemic basis or (2) inferred from a
set of earlier members of the sequence that constitutes a reason

for it. Such arguments are *linear arguments*. As we have seen, not all arguments are linear arguments. There are various kinds of "indirect" arguments that involve suppositional reasoning. Indirect arguments make suppositions that have not been established, using those as premises to obtain a conclusion, and then "discharge" the premises by using some rule like conditionalization or *reductio ad absurdum*.

We can take a line of an argument to be an ordered quadruple $\langle \Gamma, P, R, \{m, n, \ldots\} \rangle$ where Γ is a finite set of propositions (the *premise set of the line*, i.e., the set of suppositions from which the line has been inferred), P is a proposition (the proposition *supported* by the line), R is a rule of inference, and $\{m, n, \ldots\}$ is the set of line numbers of the previous lines to which the rule R appeals in justifying the present line. Arguments are finite sequences of such lines. If σ is an argument, σ_i is the proposition supported by its ith line, and $\wp(\sigma, i)$ is the premise set of the ith line. An argument *supports* a conclusion P iff P is supported by some line of the argument having an empty premise set.

Arguments are constructed in accordance with "rules of inference". These are just rules for argument formation. They are not necessarily rules of deductive inference, but they will usually be analogous to rules of deductive inference. I assume that the rules for linear argument formation include the following:

Rule F: *Foundations*
    If P is contained in the epistemic basis, and Γ is any finite set of propositions, $\langle \Gamma, P, F, \varnothing \rangle$ can be entered as any line of the argument.

Rule R: *Closure under reasons*
    If $\{P_1, \ldots, P_n\}$ is a reason for Q and $\langle \Gamma, P_1, \ldots \rangle, \ldots,$ $\langle \Gamma, P_n, \ldots \rangle$ occur as lines $i_1, \ldots, i_n$ of an argument, $\langle \Gamma, Q, R, \{i_1, \ldots, i_n\} \rangle$ can be entered on any later line.

In addition we will have at least the following two rules governing indirect arguments:

Rule **P**: *Premise introduction*
  For any finite set Γ and any P in Γ, $\langle\Gamma,P,\mathbf{P},\varnothing\rangle$ can be entered as any line of the argument.

Rule **C**: *Conditionalization*
  If $\langle\Gamma\cup\{P\},Q,\dots\rangle$ occurs as the ith line of an argument then $\langle\Gamma,(P\supset Q),\mathbf{C},\{i\}\rangle$ can be entered on any later line.

Conditionalization is a very pervasive form of inference. There are a number of different kinds of conditionals—material, indicative, subjunctive, and so on—and what is perhaps characteristic of conditionals is that they can all be inferred by some form of conditionalization. It can be shown that any conditional satisfying both rule **C** and *modus ponens* is equivalent to the material conditional,[3] but many kinds of conditionals satisfy weaker forms of conditionalization. In particular, I will assume the following weak form of conditionalization, related to the conditionals involved in undercutting defeaters:

Rule **WC**: *Weak conditionalization*
  If $\langle\{P\},Q,\dots\rangle$ occurs as the ith line of an argument then $\langle\varnothing,(P\gg Q),\mathbf{SC},\{i\}\rangle$ can be entered on any later line.

The difference between conditionalization and weak conditionalization is that the latter requires you to discharge all your assumptions at once. Given *modus ponens* and the principle of exportation:

$$[(P\&Q)\gg R]\supset[P\gg(Q\gg R)]$$

---

[3] The proof is simple. Given such a conditional '$\rightarrow$', suppose $\{P,(P\supset Q)\}$. By *modus ponens* for '$\supset$', we get Q, and then by strong conditionalization for '$\rightarrow$', $\ulcorner(P\rightarrow Q)\urcorner$ follows from the supposition $\{(P\supset Q)\}$. A second conditionalization (this time with respect to '$\supset$') gives us $\ulcorner[(P\supset Q)\supset(P\rightarrow Q)]\urcorner$. Conversely, using *modus ponens* for '$\rightarrow$' and strong conditionalization for '$\supset$', we get $\ulcorner[(P\rightarrow Q)\supset(P\supset Q)]\urcorner$. So $\ulcorner[(P\supset Q)\equiv(P\rightarrow Q)]\urcorner$ becomes a theorem.

conditionalization and weak conditionalization would be equivalent, but no conditional other than the material conditional seems to satisfy exportation.

3.2 Ultimately Undefeated Arguments

Warrant is always relative to an epistemic basis that provides the starting points for arguments. In the following, by 'argument' I always refer to arguments relative to some fixed epistemic basis. Merely having an argument for a proposition does not guarantee that the proposition is warranted, because one might also have arguments for defeaters for some of the steps in the first argument. Repeating the process, one argument might be defeated by a second, but then the second argument could be defeated by a third, thus reinstating the first, and so on. A proposition is warranted only if it ultimately emerges from this process undefeated.

The defeat of one argument always results from another argument supporting a defeater for some use of rule **R**. Suppose the jth line of an argument η is obtained in this way using some prima facie reason, and the ith line of an argument σ contains a defeater for this prima facie reason. A necessary condition for this to defeat η is that the ith line of σ and the jth line of η have the same premise set. What conclusions can be drawn from different suppositions is not, in general, relevant. For instance, if one of the suppositions of the ith line of σ is that η_j is false, then obtaining η_j in η is simply irrelevant to σ. Consequently, we must define:

(3.1) $\langle \eta, j \rangle$ *defeats* $\langle \sigma, i \rangle$ iff (1) $\langle \sigma, i \rangle$ is obtained by rule **R** using $\{P_1, \ldots, P_n\}$ as a prima facie reason for Q, (2) η_j is either $\sim Q$ or $\sim[(P_1 \& \ldots \& P_n) \gg Q]$, and (3) $\wp(\eta, j) = \wp(\sigma, i)$.

We can then capture interactions between argument as follows. Let all arguments be *level 0 arguments*. Then define:

(3.2) σ is a *level $n+1$ argument* iff σ is a level 0 argument and there is no level n argument η such that for some i and j, $\langle \eta, j \rangle$ defeats $\langle \sigma, i \rangle$.

A given level 0 argument may be defeated and reinstated many times by this alternating process. Only if we eventually reach a point where it stays undefeated can we say that it warrants its conclusion. An argument is *ultimately undefeated* iff there is some m such that the argument is a level n argument for every $n > m$. Epistemological warrant can be characterized in terms of arguments that are ultimately undefeated:

(3.3) P is warranted relative to an epistemic basis iff P is supported by some ultimately undefeated argument proceeding from that epistemic basis.

3.3 Collective Defeat

The analysis of warrant will have to be modified in one respect, but in order to appreciate the difficulty we must first note that the analysis entails the following form of *the principle of collective defeat*:

(3.4)   If R is warranted and there is a set Γ of propositions such that:
1. there are equally good defeasible arguments for believing each member of Γ;
2. for each P in Γ there is a finite subset Γ_P of Γ such that the conjunction of R with the members of Γ_P provides a deductive argument for $\sim P$ that is as strong as our initial argument is for P; and
3. none of these arguments is defeated except possibly by their interactions with one another;
then none of the propositions in Γ is warranted on the basis of these defeasible arguments.

Proof: Suppose we have such a set Γ and proposition R. For each P in Γ, combining the argument supporting R with the arguments supporting the members of Γ_P gives us an argument supporting $\sim P$. We have equally strong support for both P and $\sim P$, and hence we could not reasonably believe either on this basis, i.e., neither is warranted. This holds for each P in Γ, so none of them should be warranted. They collectively defeat one another. This is forthcoming from the analysis of warrant. We have level 0 arguments supporting each P. But these can be combined to generate level 0 arguments that also support rebutting defeaters for the argument for each P. Thus none of these are level 1 arguments. But this means that none of the defeating arguments are level 1 arguments either. Thus all of the arguments are level 2 arguments. But then they fail to be level 3 arguments. And so on. For each even number n, each P is supported by a level n argument, but that argument is not a level $n+1$ argument. Thus the P's are not supported by ultimately undefeated arguments, and hence are not warranted.

The most common instances of (3.4) occur when Γ is a minimal finite set of propositions deductively inconsistent with R. In that case, for each P in Γ, $\{R\}\cup(\Gamma-\{P\})$ gives us a deductive reason for $\sim P$. This is what occurs in the lottery paradox.

3.4 Two Paradoxes of Defeasible Reasoning

The simple account of warrant given in subsection 3.2 has some unacceptable consequences that will force its modification. This is illustrated by two apparent paradoxes of reasoning. First, consider the lottery paradox again. The lottery paradox is generated by supposing that we are warranted in believing a proposition R describing the lottery (it is a fair lottery, has 1 million tickets, and so on). Given that R is warranted, we get collective defeat for the proposition that any given ticket will not be drawn. But the present account makes it problematic how R can ever be warranted. Normally, we will believe R on the basis of being told that it is true. In such a case, our evidence for R is statistical, proceeding in accordance with the statistical syllogism (3.10). That is, we know inductively that most things we are told that fall within a certain broad range are true, and that gives us a prima facie reason for believing R. So we have only a defeasible reason for believing R. Let σ be the argument supporting R. Let T_i be the proposition that ticket i will be drawn. In accordance with the standard reasoning involved in the lottery paradox, we can extend σ to generate a longer argument η supporting $\sim R$. This is diagramed in Figure 1. The final step of the argument is justified by the observation that if none of the tickets is drawn then the lottery is not fair.

The difficulty is now that η defeats σ by (3.1). Thus σ and η defeat one another, with the result that neither is ultimately

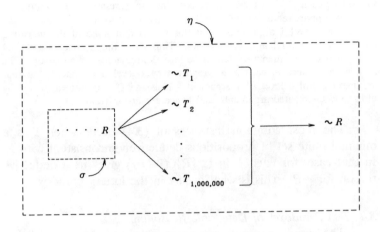

Figure 1. The lottery paradox.

undefeated. In other words, R and $\sim R$ are subject to collective defeat. This result is intuitively wrong. It should be possible for us to become warranted in believing R on the basis described.

Consider a second instance of paradoxical reasoning. Suppose we observe n A's r of which are B's, and then by statistical induction (3.13) we infer that $\text{prob}(Bx/Ax) \approx r/n$. Suppose that r/n is high. Then if we observe a further set of k A's without knowing whether they are B's, we can infer by statistical syllogism (3.10) that each one is a B. This gives us $n+k$ A's of which $r+k$ are B's. By (3.13), this in turn provides a reason for thinking that $\text{prob}(Bx/Ax) \approx (r+k)/(n+k)$. If k is large enough, $(r+k)/(n+k) \neq r/n$, and so we can infer that $\text{prob}(Bx/Ax) \neq r/n$, which contradicts our original conclusion and undermines all of the reasoning. Making this more precise, we have the two nested arguments diagrammed in Figure 2. Arguments σ and η defeat each other by (3.1), so we have a case of collective defeat. But this is intuitively wrong. All we actually have in this case is a reason for believing that $\text{prob}(Bx/Ax) \approx r/n$,

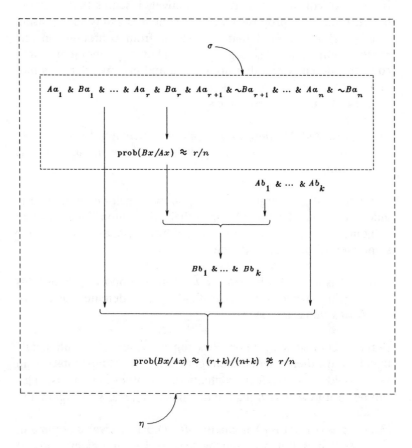

Figure 2. The paradox of statistical induction.

and a bunch of *A*'s regarding which we do not know whether they are *B*'s. The latter should have no effect on our warrant for the former. But by (3.1), it does. I will call this *the paradox of statistical induction*.

I suggest that these two paradoxes illustrate a single inadequacy in the preceding analysis of warrant. In each case we begin with an argument σ supporting a conclusion P, and then we extend σ to obtain an argument η supporting $\sim P$. By (3.1), this is a case of collective defeat, but intuitively it seems that P should be warranted. I propose that argument η is faulty all by itself. It is *self-defeating*, and that removes it from contention in any contest with conflicting arguments. Thus it cannot enter into collective defeat with other arguments, and in particular it cannot enter into collective defeat with σ.

Let us define more precisely:

(3.5) σ is *self-defeating* iff σ supports a defeater for one of its own defeasible steps, i.e., for some i and j, $\langle \sigma, i \rangle$ defeats $\langle \sigma, j \rangle$.

More generally, an argument σ can be made self-defeating by the independent justification of some other proposition P which, when conjoined with propositions supported by σ, implies a defeater for some step of σ. Let us define:

(3.6) σ is *self-defeating relative to* P iff σ supports propositions which when conjoined with P imply a defeater for one of σ's defeasible steps.

If the argument supporting P is unproblematic (i.e., ultimately undefeated), then σ should be regarded as undermining itself, and that should remove it from contention with other arguments. This can be captured by revising the definition of 'level $k+1$ argument':

(3.7) σ is a level $k+1$ argument iff (1) σ is a level 0 argument, (2) σ is not defeated by any level k argument, and (3) there is no proposition P supported by a level k argument such that σ is self-defeating relative to P.

Arguments that are self-defeating simpliciter fail to be level $k+1$ arguments for any k (this is because they are self-defeating relative to a tautology). Furthermore, if an argument μ supporting P is ultimately undefeated and σ is self-defeating relative to P, then σ will not be a level k argument for any k beyond the level at which μ becomes perpetually undefeated. This has the effect of removing self-defeating arguments from contention, and it resolves the two paradoxes of defeasible reasoning described above. This modification to the analysis of warrant will also provide one of the ingredients in the resolution of the paradox of the preface.

3.5 Logical Properties of Warrant

The analysis of warrant entails that the set of warranted propositions has a number of important properties. Let us symbolize $\ulcorner P$ is warranted relative to the epistemic basis $E\urcorner$ as $\ulcorner \models_E P\urcorner$. Let us also define *warranted consequence*:

(3.8) $\Gamma \models_E P$ iff there is an ultimately undefeated argument relative to E that contains a line of the form $\langle \Gamma, P, \ldots \rangle$.

A proposition P is a *deductive consequence* of a set Γ of propositions (symbolized $\ulcorner \Gamma \vdash P\urcorner$) iff there exists a deductive argument leading from members of Γ to the conclusion P. Deductive arguments are arguments containing no defeasible steps. Trivially:

(3.9)   If $\Gamma \vdash P$ then $\Gamma \models_E P$.

A set of propositions is *deductively consistent* iff it does not have an explicit contradiction as a deductive consequence. The set of warranted propositions must be deductively consistent. (It is assumed here and throughout that an epistemic basis must be consistent.) If a contradiction could be derived from it, then reasoning from some warranted propositions would lead to the denial (and hence defeat) of other warranted propositions, in which case they would not be warranted. More generally:

(3.10)   If Γ is deductively consistent so is $\{P \mid \Gamma \models_E P\}$.

The set of warranted propositions must also be closed under deductive consequence:

(3.11)   If for every P in Γ, $\models_E P$, and $\Gamma \vdash Q$, then $\models_E Q$.

Proof: Suppose $P_1, \ldots, P_n$ are warranted and Q is a deductive consequence of them. Then an argument supporting Q can be constructed by combining arguments for $P_1, \ldots, P_n$ and adding onto the end an argument deducing Q from $P_1, \ldots, P_n$. The last part of the argument consists only of deductive nondefeasible steps of reasoning. If Q is not warranted, there must be an argument defeating the argument supporting Q. There can be no defeaters for the final steps, which are nondefeasible, so such a defeater would have to be a defeater for an earlier step. But the earlier steps all occur in the arguments supporting $P_1, \ldots, P_n$, so one of those arguments would have to be defeated, which contradicts the assumption that $P_1, \ldots, P_n$ are warranted. Thus there can be no such defeater, and hence Q is warranted.

More generally:

(3.12) If for every P in Γ, $\Lambda \models_E P$, and $\Gamma \models_E Q$, then $\Lambda \models_E Q$.

We also have the following analogue of the standard deduction theorem in classical logic:

(3.13) If $\Gamma \cup \{P\} \models_E Q$ then $\Gamma \models_E (P \supset Q)$.

This follows immediately from rule **C**, and contrasts with standard nonmonotonic logics.

As in (3.6) of Chapter 3 we have a principle of constructive dilemma:

(3.14) If $\Gamma \cup \{P\} \models_E R$ and $\Gamma \cup \{Q\} \models_E R$ then $\Gamma \cup \{P \vee Q\} \models_E R$.

3.6 The Principle of Collective Defeat

The principle (3.4) of collective defeat remains true in our general theory of warrant, and its proof remains essentially unchanged. It also has an interesting analogue. Principle (3.4) is a principle of *collective rebutting defeat*. It only pertains to cases in which there are arguments supporting both P and $\sim P$. A *principle of collective undercutting defeat* can be obtained in precisely the same way:

(3.14) If R is warranted and there is a set Γ of propositions such that:
    1. there are equally good defeasible arguments for

believing each member of Γ;

2. for each P in Γ, the supporting argument involves a defeasible step proceeding from some premises $S_1, \ldots, S_n$ to a conclusion T, and there is a finite subset Γ_P of Γ such that the conjunction of R with the members of Γ_P provides a deductive argument for $\sim[(S_1, \ldots, S_n) \gg T]$ that is as strong as the initial argument is for P; and

3. none of these arguments is defeated except possibly by their interactions with one another;

then none of the propositions in Γ is warranted on the basis of these defeasible arguments.

A simple illustration of this principle will involve a pair of arguments having the structure diagramed in Figure 3. For instance, R might be 'People generally tell the truth'. Suppose P is 'Jones says Smith is unreliable' and Q is 'Smith says Jones is unreliable'. By statistical syllogism (3.10), $(P\&R)$ is a prima facie reason for believing S: 'Smith is unreliable', and $(Q\&R)$ is a prima facie reason for believing T: 'Jones is unreliable'. But S is an undercutting defeater for the reasoning from $(Q\&R)$ to T, and T is an undercutting defeater for the reasoning from $(P\&R)$ to S. Presented with this situation, what should we believe about

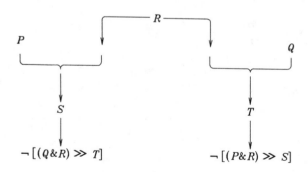

Figure 3. Collective undercutting defeat.

Smith and Jones? The intuitive answer is "Nothing". We have no basis for deciding that one rather than the other is unreliable. Under the circumstances, we should withhold belief regarding their reliability. And that is just what principle (3.14) tells us.

We have separate principles of collective defeat for rebutting defeaters and undercutting defeaters. Can there also be mixed cases of collective defeat involving both rebutting and undercutting defeaters? That does not seem to be possible. Such cases would have the form diagramed in Figure 4, or a generalization of that form, where all steps are defeasible. σ provides rebutting defeat for η, and η provides undercutting defeat for σ. But the relation 'provides rebutting defeat for' is symmetrical, so η also provides rebutting defeat for σ. Thus this is a simple case of collective rebutting defeat, and the last step of η is irrelevant.

If we have equally good reasons for P and Q, but these hypotheses conflict, then our reasoning is subject to collective defeat and we cannot conclude that either is true. But we are not left without a conclusion to draw. For instance, consider a doctor who has equally good evidence that his patient has disease$_1$ and that he has disease$_2$. He cannot conclude that the patient has disease$_1$ and he cannot conclude that the patient has

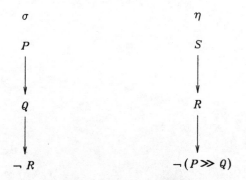

Figure 4. Mixing collective rebutting defeat and collective undercutting defeat.

disease$_2$, but he can reasonably conclude that his patient has one or the other. This seems intuitively right, and it is forthcoming from our theory of reasoning. The doctor has two arguments at his disposal—σ is an argument for the conclusion that the patient has disease$_1$, and η is an argument for the conclusion that the patient has disease$_2$. These arguments collectively defeat one another, but a third argument can be constructed that is not subject to collective defeat. This is the following conditional argument η^*:

Suppose the patient does not have disease$_1$.

η $\Big\{$ $\begin{array}{l} \vdots \\ \text{The patient has disease}_2. \end{array}$

If the patient does not have disease$_1$ then he has disease$_2$. Therefore, either the patient has disease$_1$ or he has disease$_2$.

This argument repeats argument η subject to the supposition that the patient does not have disease$_1$. Recall that an argument making a supposition can only be defeated by another argument if it makes the same supposition. Accordingly, η^* is not subject to collective defeat because we cannot similarly repeat η within that supposition. This is because given the supposition that the patient does not have disease$_1$, argument σ is automatically defeated by a rebutting defeater.

3.7 The Principle of Joint Defeat

It is natural to suppose that, except in cases of collective defeat, the only way a prima facie reason can be defeated is by our being warranted in believing a defeater for it. But that is simplistic in two different ways. First, to defeat a prima facie reason, it is not necessary that we justifiably believe a particular defeater. It is sufficient that we be warranted in believing that *there is* is a true defeater. Formally, this is justified by the following two theorems:

(3.15) If P is a prima facie reason for Q, and $R_1,\ldots,R_n$ are
       undercutting defeaters for this inference, none of the R_i
       is defeated, and the disjunction $(R_1 \vee \ldots \vee R_n)$ is war-
       ranted, then the inference from P to Q is defeated.

(3.16) If P is a prima facie reason for Q, and $R_1,\ldots,R_n$ are
       rebutting defeaters for this inference, none of the R_i is
       defeated, and the disjunction $(R_1 \vee \ldots \vee R_n)$ is warranted,
       then the inference from P to Q is defeated.

Proof: To establish (3.15), recall that what makes R_i an undercutting defeater
is that it is a reason for $\sim(P \succ Q)$. By (3.6) of Chapter 3, it follows that the
disjunction $(R_1 \vee \ldots \vee R_n)$ is also a reason for $\sim(P \succ Q)$. Thus if the latter
disjunction is warranted, and none of the defeaters is defeated, it follows that
$\sim(P \succ Q)$ is warranted, and that defeats the inference from P to Q. Principle
(3.16) can be established analogously.

This can be generalized. Suppose we have prima facie
reasons for each of a set of conclusions $Q_1,\ldots,Q_n$. Suppose
further that we are warranted in believing that there is a true
defeater for at least one of these prima facie reasons. That
makes the conjunction $(Q_1 \& \ldots \& Q_n)$ unwarranted. If each of
$Q_1,\ldots,Q_n$ were warranted, then the conjunction $(Q_1 \& \ldots \& Q_n)$
would be warranted by deductive reasoning, but the latter is not
warranted, so it follows that not all of $Q_1,\ldots,Q_n$ are warranted.
If there is no proper subset of the Q_i's such that we are warranted in believing that the true defeater is a defeater for a
member of that set, then there is nothing to favor one of the Q_i's
over any of the others, so *none* of the Q_i's is warranted. This is
summed up in *The Principle of Joint Defeat*:

(3.17) If Γ is a finite set of triples $\langle P_i, Q_i, \Lambda_i \rangle$ such that for each
       i, P_i is a prima facie reason for Q_i and Λ_i is a finite set
       of defeaters for P_i (either all undercutting defeaters or all
       rebutting defeaters), and we are warranted in believing
       that some member of some Λ_i is true, but there is no
       proper subset Γ_0 of Γ such that we are warranted in
       believing that some member of some Λ_i for $\langle P_i, Q_i, \Lambda_i \rangle$ in
       Γ_0 is true, then (prima facie) each of the P_i's is defeated.

To keep the argument simple, I will prove the following special case of (3.17):

(3.18) Suppose P is a prima facie reason for Q and R is a prima facie reason for S, and $T = \ulcorner\sim(P \succ Q)\urcorner$ and $U = \ulcorner\sim(R \succ S)\urcorner$. If $\ulcorner T \vee U\urcorner$ is warranted but neither T nor U is warranted, then no argument employing either P as a reason for Q or R as a reason for S (in accordance with rule **R**) is ultimately undefeated.

Proof: Suppose σ is an argument employing one of these reasons. For specificity, suppose σ uses R as a reason for S. Suppose μ is an ultimately undefeated argument for $\ulcorner T \vee U\urcorner$. Construct the argument η by adding the following lines to the end of μ:

(i)	$\langle\emptyset, \sim(P \succ Q) \vee \sim(R \succ S), \dots\rangle$	(last line of μ)
(i+1)	$\langle\{P\}, P, \mathbf{P}, \{i+1\}\rangle$	
(i+2)	$\langle\{P\}, Q, \mathbf{R}, \{i+1\}\rangle$	
(i+3)	$\langle\emptyset, P \succ Q, \mathbf{WC}, \{i+2\}\rangle$	
(i+4)	$\langle\emptyset, \sim(R \succ S), \text{deductive inference}, \{i, i+3\}\rangle$	

Similarly, we can construct an argument γ for $\langle\emptyset, \sim(P \succ Q), \dots\rangle$. η and γ defeat one another, so each fails to be a level n argument for odd n, but each is a level n argument for all even n. η defeats σ, so σ also fails to be a level n argument for any odd n. Thus σ is not ultimately undefeated.

The principle of joint defeat should not be confused with the principle of collective defeat. The two principles are similar in important respects, but there are also important differences. The principle of joint defeat has little role to play in most of epistemology, because we are rarely in a position of knowing that there is a true defeater for some member of a set of prima facie reasons without knowing that there is a true defeater for a specific member of that set. But we will see in the next chapter that we are frequently in that position in direct inference. This is because the probability calculus establishes logical ties between different direct inferences.

4. Nonmonotonic Logic

Defeasible reasoning has been a topic of major concern in AI. A number of theories have been proposed that aim at characterizing the logical structure of defeasible reasoning.[4] These are known as theories of *nonmonotonic logic*. They differ from the present theory of defeasible reasoning in a number of important respects, and I will briefly survey some of those differences here.

The species of nonmonotonic logic that looks most like the present theory of defeasible reasoning is Reiter's default logic [1980]. In default logic, inferences are mediated by *default rules*, which are written in the form: $\ulcorner P : Q / R \urcorner$. The intended meaning of such a rule is that if P is believed and $\sim Q$ is not, then R is to be inferred. A set of formulas E is *inferentially closed* under the default rule $\ulcorner P : Q / R \urcorner$ iff (1) E is closed under logical consequence and (2) if $P \in E$ and $\ulcorner \sim Q \urcorner \notin E$ then $R \in E$. A *minimal extension* of A (relative to some set of default rules) is any set E containing A that is inferentially closed under the default rules and has no inferentially closed proper subset containing A. On the orthodox construal of default logic, a set of beliefs is reasonable relative to a set of inputs and a set of default rules iff it is a minimal extension of the set of inputs.

An alternative, due to McDermott and Doyle [1980] and Moore [1983], is the modal approach, which formulates the same default rule as $\ulcorner (P \, \& \, \Diamond Q) \supset R \urcorner$. Konolige [1987] shows that Moore's autoepistemic logic is equivalent to Reiter's default logic. Another alternative is *circumscription*, due to McCarthy [1980].[5] The exact relationship between circumscription and other forms of nonmonotonic logic is unclear.[6]

Theories of nonmonotonic logic can be regarded as theories

---

[4] See Doyle [1979], Etherington [1983], Hanks and McDermott [1985], [1987], Konolige [1987], Lifschitz [1984], Loui [1987], McCarthy [1980], [1984], McDermott [1982], McDermott and Doyle [1980], Moore [1983], Nute [1986], [1987], [1988], Poole [1985], Reiter [1980].

[5] See also McCarthy [1984], Lifschitz [1985], [1987], and Etherington, Mercer, and Reiter [1985].

[6] But see Shoham [1987].

of warrant. Most such theories are semantical, proceeding in terms of models. My account of warrant is syntactical, proceeding instead in terms of arguments. Ginsberg [1987] has criticized truth maintenance systems for not having semantical foundations, and the same objection might be leveled at my theory of warrant. But I suggest that that is to put the cart before the horse. Warrant is defined as an idealization of justified belief, and the latter must ultimately be characterized procedurally, in terms of reasoning. The theory of reasoning must take priority. If a semantical characterization of warrant is incompatible with principles of reasoning, then the characterization of warrant is wrong. Furthermore, I will argue that all existing nonmonotonic logics fail in precisely this way. The structure of defeasible reasoning is subtle, and existing nonmonotonic logics do not do it justice.

To illustrate this point in connection with Reiter's default logic, note first that the notion of a minimal extension is defined semantically. This is because it is defined in terms of logical consequence, which is defined in terms of models. The first thing to observe is that this can only work for *complete* logical theories—those for which logical consequence coincides with deducibility. This is because Reiter's theory implies that any reasonable set of beliefs is closed under logical consequence. However, if P is a logical consequence of an ideal agent's beliefs but is not deducible from them, there is no reason that even an ideal agent should believe P. This cannot happen in either the propositional or predicate calculi, but it does happen in second-order logic, ω-order logic, set theory, and even in reasoning about the natural numbers (if one takes the relevant set of models to be the natural models). Thus, in general, minimal extensions should be defined in terms of deducibility rather than semantically. Such a change will make default logic more like my theory of defeasible warrant.

The next problem to be raised is the multiple extension problem.[7] There will frequently be more than one minimal

---

[7] See Hanks and McDermott [1987], McCarthy [1984], Reiter [1978].

extension of a given set of inputs, and according to Reiter's theory and other familiar nonmonotonic logics, each of those minimal extensions is a reasonable set of beliefs. But if we consider concrete examples, this is clearly wrong. Suppose Jones tells me that *P* is true and Smith tells me that *P* is false. I regard Jones and Smith as equally reliable, and I have no other evidence regarding *P*. What should I believe? According to Reiter's theory, I can take my choice and believe either *P* or ~*P*. But surely, in the face of such conflicting evidence, I should believe neither. I should *withhold belief* rather than randomly choosing what to believe.

Responding to a related criticism by Touretzky [1984], Etherington [1987] writes:

> These criticisms suggest that a (common) misapprehension about default logic has occurred. ... In fact, while extensions are all acceptable, the logic says nothing about preference of one to another. It has always been assumed that an agent would "choose" one extension or another, but nothing has been said about how such choices should be made.

But if rationality can dictate preference of one extension over its competitors, then what notion of "acceptable" can default logic possibly be analyzing? Certainly not epistemic warrant, which is my target.

Multiple extensions arise from conflicting default rules. The analogous problem arises for my theory of defeasible reasoning when we have equally good reasons for conflicting conclusions. In the theory of defeasible reasoning, that gives rise to collective defeat, with the result that none of the conflicting conclusions is warranted. As we have seen, this is an extremely important phenomenon in defeasible reasoning. We can alter default logic so that it gives the right answer in cases of collective defeat.[8] Instead of saying that any minimal extension is a reasonable set of beliefs, we should insist that a belief is reasonable for an ideal agent only if it is in every minimal extension. More precisely, we

---

[8] This is to endorse a "skeptical" theory of nonmonotonic reasoning, in the sense of Horty, Thomason, and Touretzky [1987].

can interpret default logic as proposing that a proposition is
warranted relative to a set of inputs and a set of default rules iff
the proposition is in the intersection of all minimal extensions of
that set of inputs.[9] Although the connection between default logic
and circumscription is not entirely clear,[10] circumscription theories
also encounter the multiple extension problem and fail to satisfy
the principle of collective defeat.   To that extent they are
inadequate analyses of warrant.

The analysis of warrant in terms of the intersection of all
default minimal extensions constitutes an elegant analysis if it
correct. To what extent can it capture the logical structure of
warrant and defeasible reasoning? "Normal" default rules are
those of the form $\ulcorner P : Q \: / \: Q \urcorner$. Reiter tried initially to handle all
defeasible reasoning using only normal default rules, but subse-
quently acknowledged the need for more complex default rules.[11]
Normal default rules correspond to prima facie reasons for which
the only defeaters are rebutting defeaters. There probably are no
such prima facie reasons. However, undercutting defeaters can be
accommodated by more complicated default rules. The second
element of a default rule is a conjunction of negations of defeaters, and we can handle prima facie reasons generally by taking
default rules to have the form $\ulcorner P : Q \: \& \: (P \geqslant Q) \: / \: Q \urcorner$. Treating
prima facie reasons in this way and taking the default conse-
quences of a set of premises to be the propositions in the
intersection of all minimal extensions of that set, default logic can
handle much defeasible reasoning.

Some difficulties remain, however. These stem from the
desire to define the default consequence relation semantically
rather than in terms of arguments. I have argued that we must
define extensions in terms of deducibility rather than logical
consequence, but even if we do that the resulting default logic

---

[9] It is noteworthy that this is the original proposal of McDermott and
Doyle [1980]. In this respect, their modal logic is not equivalent to Moore's
autoepistemic logic, a point that seems to have gotten lost in the literature.

[10] See Etherington, Mercer, and Reiter [1985] and McCarthy [1980] and
[1984].

[11] See Etherington and Reiter [1983].

does not delve into the actual structure of arguments. That turns out to be a mistake. The account is satisfactory only for linear reasoning. To illustrate, suppose we hear a small animal moving around in the next room, and we are trying to figure out what it is. Knowing that most birds can fly, we might reasonably assert, "If it is a bird, then it can fly". This cannot be accommodated by default logic. Our reasoning involves the use of prima facie reasons within subsidiary arguments and subsequent conditionalization. That is, we *suppose* that it is a bird, defeasibly infer that it can fly, and then discharge the supposition through conditionalization to infer that *if* it is a bird then it can fly. There is no way to do this in default logic. Circumscription seems equally unable to handle this sort of reasoning. The *only* way to handle this is to take argument structure seriously and investigate how defeasible reasoning can occur in both the main argument and in embedded subsidiary arguments. The theory of defeasible reasoning presented in this book accommodates this.

There is a related difficulty. If we have equally good prima facie reasons for two conflicting hypotheses, collective defeat makes it unreasonable to believe either, but it is still reasonable to believe their disjunction. For instance, if a doctor has equally good evidence that a patient has disease$_1$ and that he has disease$_2$, it is not reasonable to pick one of the diseases at random and conclude that the patient has that disease, but it is reasonable to conclude that the patient has *either* disease$_1$ or disease$_2$. As we saw in section 4, this is forthcoming from the theory of defeasible reasoning, but the way it is obtained is by using conditional reasoning, and hence this conclusion cannot be obtained in nonmonotonic logics that do not accommodate nonlinear reasoning.

As we have seen, there are other complexities in the theory of defeasible reasoning. For instance, there must be a stricture against self-defeating arguments. But nothing like this occurs in nonmonotonic logic, and it is unclear how such a stricture could be incorporated into existing nonmonotonic logics.

In conclusion, the standard theories of nonmonotonic logic are incapable of dealing with many of the more sophisticated aspects of defeasible reasoning. In order to construct a theory of such reasoning, we must examine the reasoning itself and attempt to model that directly. In this connection, AI cannot afford to ignore the work in epistemology. It might ultimately prove possible to give some abstract nonsyntactical account of a defeasible consequence relation, but the structure of defeasible reasoning is sufficiently complex that we are not apt to get it right without first looking closely at the reasoning itself. Only after we have a good characterization of such reasoning is there much point in trying to construct an abstract nonsyntactical characterization.

5. Summary

1. Paradoxical instances of defeasible reasoning force us to acknowledge that self-defeating arguments do not support their conclusions and cannot enter into collective defeat with other arguments.
2. A defeasible inference can be defeated by a disjunction of defeaters even when we do not know which disjunct is true.
3. By the principle of joint defeat, if we know that there is a true defeater for some member of a finite set of defeasible inferences but cannot narrow the set of possibly defeated inferences further, then all the inferences in the set are defeated.
4. Existing nonmonotonic logics in AI give inadequate characterizations of warrant because they fail to handle collective defeat correctly, and they can accommodate only linear reasoning.

CHAPTER 9
ADVANCED TOPICS:
ACCEPTANCE RULES

1. Introduction

There once was a man who wrote a book. He was very careful in his reasoning, and was confident of each claim that he made. With some display of pride, he showed the book to a friend (who happened to be a probability theorist). He was dismayed when the friend observed that any book that long and that interesting was almost certain to contain at least one falsehood. Thus it was not reasonable to believe that all of the claims made in the book were true. If it were reasonable to believe each claim then it would be reasonable to believe that the book contained no falsehoods, so it could not be reasonable to believe each claim. Furthermore, because there was no way to pick out some of the claims as being more problematic than others, there could be no reasonable way of withholding assent to some but not others. "Therefore," concluded his friend, "you are not justified in believing anything you asserted in the book."

This is the paradox of the preface (so named because in the original version the author confesses in the preface that his book probably contains a falsehood).[1] The paradox of the preface is more than a curiosity. It has been used by some philosophers to argue that the set of one's warranted beliefs need not be deductively consistent, and by others to argue that you should not befriend probability theorists. If $(A1)$ is to be a correct acceptance rule it must be capable of explaining what is involved in the paradox of the preface.

The lottery paradox and the paradox of the preface seem superficially similar, so it might be supposed that a resolution of one will automatically generate a resolution of the other in some trivial manner. But in fact, the opposite is true. It is the prin-

---

[1] The paradox of the preface originated with D. C. Makinson [1965].

ciple of collective defeat that makes possible the resolution of
the lottery paradox, but it is the principle of collective defeat that
is responsible for the creation of the paradox of the preface. The
reasoning involved in the latter turns upon the observation that
we have as good a reason for believing the book to contain a
falsehood as we do for believing the individual claims in the book,
and argues that because we have no way to choose between these
various conclusions, they are subject to collective defeat.

2. The Gambler's Fallacy

It will be argued here that the paradoxical reasoning involved in
the paradox of the preface is similar in important respects to the
gambler's fallacy, so let us consider the latter. Suppose a fair coin
is tossed six times and the first five tosses are heads.[2] We
observe that it is highly improbable for six consecutive tosses to
be heads, and this tempts us to conclude that the last toss will not
be a head. Our intuitive reasoning proceeds by noting that

$$(2.1) \quad \text{prob}\Big(Hx_1 \& \ldots \& Hx_6 \: / \: Tx_1 \& \ldots \& Tx_6 \& x_1,\ldots,x_6$$

$$\text{distinct}\Big)$$

$$= 1/64.$$

and using $(A1)$ and the fact that $t_1,\ldots,t_6$ are distinct tosses to
conclude $\ulcorner \sim(Ht_1 \& \ldots \& Ht_6)\urcorner$. Then because we know $\ulcorner Ht_1 \&$
$\ldots \& Ht_5\urcorner$ we infer $\ulcorner \sim Ht_6\urcorner$.

There is actually a very simple objection to this reasoning.
It is not licensed by $(A1)$ because of the projectibility constraint.
Although $\ulcorner Hx_1 \& \ldots \& Hx_6\urcorner$ is projectible (because projectibility
is closed under conjunction), its negation is not. Disjunctions of
projectible properties are not generally projectible, and $\ulcorner \sim(Hx_1$

---

[2] I will take a coin to be fair iff the probability of heads is ½ and getting
heads on one toss is statistically independent of the outcomes of any other
finite set of tosses.

& ... & Hx_6)$^\rceil$ is equivalent to the unprojectible property $\ulcorner{\sim}Hx_1$ $\vee$ ... $\vee$ $\sim Hx_6\urcorner$. But this is not the objection to the gambler's fallacy that I want to press. There is a more standard objection that is customarily levied at it. Because the coin is fair, we know that the tosses are independent, and hence:

(2.2)   prob$\Big(Hx_6 \,/\, Hx_1 \,\&\, \ldots \,\&\, Hx_5 \,\&\, Tx_1 \,\&\, \ldots \,\&\, Tx_6 \,\&\, x_1,\ldots,x_6$
          distinct$\Big) = \frac{1}{2}$.

By the probability calculus:

(2.3)   prob$\Big(Hx_1 \,\&\, \ldots \,\&\, Hx_6 \,/\, Hx_1 \,\&\, \ldots \,\&\, Hx_5 \,\&\, Tx_1 \,\&\, \ldots$
          $\&\, Tx_6 \,\&\, x_1,\ldots,x_6$ distinct$\Big)$
        $= $ prob$\Big(Hx_6 \,/\, Hx_1 \,\&\, \ldots \,\&\, Hx_5 \,\&\, Tx_1 \,\&\, \ldots \,\&\, Tx_6 \,\&\, x_1,\ldots,x_6$
          distinct$\Big)$

Therefore,

(2.4)   prob$\Big(Hx_1 \,\&\, \ldots \,\&\, Hx_6 \,/\, Hx_1 \,\&\, \ldots \,\&\, Hx_5 \,\&\, Tx_1 \,\&\, \ldots$
          $\&\, Tx_6 \,\&\, x_1,\ldots,x_6$ distinct$\Big)$
        $= \frac{1}{2}$.

Because we are warranted in believing $\ulcorner Ht_1 \,\&\, \ldots \,\&\, Ht_5\urcorner$ and $\frac{1}{2}$ $<$ $^{63}/_{64}$, (D1) provides a subset defeater for any inference from (2.1) in accordance with (A1). This is a reconstruction of the standard objection to the gambler's fallacy, and it is illuminating to see how it proceeds by appealing to (D1).

Next notice that we can reformulate the gambler's fallacy in a way that makes its resolution more difficult. Suppose that $t_1,\ldots,t_6$ are tosses of a particular fair coin c that it tossed just these six times and then melted down. Taking $\ulcorner Txy\urcorner$ to be $\ulcorner y$ is a toss of coin $x\urcorner$ and $\ulcorner Fx\urcorner$ to be $\ulcorner x$ is a fair coin$\urcorner$, we have:

(2.5) $\text{prob}\big((\exists y)(Txy \ \& \ \sim Hy) \ / \ Fx \ \& \ (\exists y_1)\ldots(\exists y_6)[y_1,\ldots,y_6 \text{ are}$

$\qquad$ distinct $\& \ (\forall z)(Tzx \equiv [z = y_1 \vee \ldots \vee z = y_6])])\big)$

$\qquad = \ ^{63}/_{64}.$

Because we also know

(2.6) $Fc \ \& \ t_1,\ldots,t_6 \text{ are distinct } \&$
$\qquad (\forall z)(Tzc \equiv [z = t_1 \vee \ldots \vee z = t_6])$

we are warranted in believing

(2.7) $Fc \ \& \ (\exists y_1)\ldots(\exists y_6)[y_1,\ldots,y_6 \text{ are distinct } \&$
$\qquad (\forall z)(Tzc \equiv [z = y_1 \vee \ldots \vee z = y_6])].$

Therefore, by (*A1*), we have a prima facie reason for believing

(2.8) $(\exists y)(Tcy \ \& \ \sim Hy).$

Having observed that $t_1,\ldots,t_5$ were heads, we can infer from (2.6) and (2.8) that t_6 is not a head. Thus we have the gambler's fallacy again in a more sophisticated guise.

It was remarked earlier that an inference from (2.1) could be blocked by the projectibility constraint in (*A1*). That is at least not clearly true for an inference from (2.5). It is unclear what effect existential quantifiers have on projectibility, so it is best not to try to resolve the problem in that way. Instead it seems intuitively that the same considerations that blocked the first version of the gambler's fallacy should also block this version. Different tosses of coin c are still independent of one another, so just as we had (2.2) before, now we have:

(2.9) $\text{prob}\big(Hy_6 \ / \ Fx \ \& \ Hy_1 \ \& \ \ldots \ \& \ Hy_5 \ \& \ y_1,\ldots,y_6 \text{ are}$

$\qquad$ distinct $\& \ (\forall z)(Tzx \equiv [z = y_1 \vee \ldots \vee z = y_6])\big)$

$\qquad = \ \frac{1}{2}.$

Intuitively, this ought to provide us with a defeater for the inference from (2.5). There is, however, a problem in explaining how this can be the case. Let

$$(2.10) \quad s = \text{prob}\big((\exists y)(Txy \ \& \sim Hy) \ / \ Fx \ \& \ (\exists y_1)\ldots(\exists y_6)[y_1,\ldots,y_6$$
$$\text{are distinct} \ \& \ Hy_1 \ \& \ \ldots \ \& \ Hy_5 \ \&$$
$$(\forall z)(Tzx \equiv [z = y_1 \vee \ldots \vee z = y_6])]\big).$$

This probability takes account of more information than does (2.5), so if it can be established that $s < {}^{63}/64$, this provides us with a defeater of the form $(D1)$ for an inference from (2.5). By the classical probability calculus and (2.9),

$$(2.11) \quad \text{prob}\big((\exists y)(Txy \ \& \sim Hy) \ / \ Fx \ \& \ y_1,\ldots,y_6 \text{ are distinct}$$
$$\& \ Hy_1 \ \& \ \ldots \ \& \ Hy_5 \ \&$$
$$(\forall z)(Tzx \equiv [z = y_1 \vee \ldots \vee z = y_6])\big)$$

$$= \text{prob}\big(\sim Hy_1 \vee \ldots \vee \sim Hy_6 \ / \ Fx \ \& \ y_1,\ldots,y_6 \text{ are distinct}$$
$$\& \ Hy_1 \ \& \ \ldots \ \& \ Hy_5 \ \& \ (\forall z)(Tzx \equiv [z = y_1 \vee \ldots$$
$$\vee z = y_6])\big)$$

$$= \text{prob}\big(\sim Hy_6 \ / \ Fx \ \& \ y_1,\ldots,y_6 \text{ are distinct} \ \& \ Hy_1 \ \& \ \ldots$$
$$\& \ Hy_5 \ \& \ (\forall z)(Tzx \equiv [z = y_1 \vee \ldots \vee z = y_6])\big)$$

$$= \tfrac{1}{2}.$$

Note, however, that the first formula in (2.11) is not the same as (2.10). The difference lies in the quantifiers. (2.10) is of the general form $\ulcorner \text{prob}(Ax \ / \ (\exists y)Rxy)\urcorner$, whereas the first formula in (2.11) is of the form $\ulcorner \text{prob}(Ax/Rxy)\urcorner$. What is the connection between these probabilities?

The classical probability calculus provides us with no connection between these two probabilities, but a connection is forthcoming from the exotic theory of nomic probabilities developed in Chapter 7. As I remarked there, when I first began thinking about this it seemed to me that these probabilities should always be equal, but there are simple counterexamples to that. The reason these probabilities are different is that there may be more y's related to some x's than to others and this gives some x's more weight than others in computing prob(Ax/Rxy). If it were a necessary truth that each x had the same number of y's related to it then each x would have the same weight and then these probabilities would be the same. Thus taking $\#X$ to be the number of members in a finite set X, the following principle should be true:

(2.12) If for some natural number n, $\Box(\forall x)[(\exists y)Rxy \supset \#\{y \mid Rxy\} = n]$, then prob($Ax/Rxy$) = prob($Ax / (\exists y)Rxy$).

This is a kind of probabilistic principle of existential instantiation, and it is a theorem of the calculus of nomic probability defended in Chapter 7.

(2.12) implies a somewhat more general principle, and it is this more general principle that is required for an adequate treatment of the gambler's fallacy. Let us say that an n-tuple $\langle y_1, \ldots, y_n \rangle$ is *nonrepeating* if no object occurs in it twice. Let $\ulcorner Sx\langle y_1, \ldots, y_n \rangle \urcorner$ be $\ulcorner y_1, \ldots, y_n$ distinct & Rxy_1 & ... & $Rxy_n \urcorner$. If $\#X = n$ then there are $n!$ nonrepeating n-tuples of members of X. Thus if the hypothesis of (2.12) holds then $\Box(\forall x)[(\exists \sigma)Sx\sigma \supset \#\{\sigma \mid Sx\sigma\} = n!]$, and hence by (2.12):

$$\text{prob}(Ax / y_1, \ldots, y_n \text{ distinct } \& \ Rxy_1 \ \& \ \ldots \ \& \ Rxy_n)$$

$$= \text{prob}(Ax / Sx\sigma)$$

$$= \text{prob}(Ax / (\exists \sigma)Sx\sigma)$$

$$= \text{prob}\Big(Ax / (\exists y_1) \ldots (\exists y_n)[y_1, \ldots, y_n \text{ distinct } \& \ Rxy_1 \ \& \ \ldots \ \& \ Rxy_n]\Big).$$

Consequently, (2.12) implies the somewhat more general theorem:

(2.13) If for some natural number n, $\Box(\forall x)[(\exists y)Rxy \supset \#\{y \mid Rxy\}$
$= n]$ then
prob($Ax / y_1,\ldots,y_n$ distinct & Rxy_1 & $\ldots$ & Rxy_n) =
prob$\Big(Ax / (\exists y_1)\ldots(\exists y_n)[y_1,\ldots,y_n$ distinct & Rxy_1 &
$\ldots$ & $Rxy_n]\Big)$.

It is (2.13) that enables us to dispose of the gambler's fallacy. Letting $\ulcorner Rxy \urcorner$ be $\ulcorner Hy$ & Tyx & $\#\{z \mid Tzx\} = 6 \urcorner$, it follows immediately from (2.13) that the probability in (2.10) is the same as that in (2.11), and hence $s = \frac{1}{2} < {}^{63}\!/_{64}$. Consequently, our intuitions are vindicated, and (2.11) provides a subset defeater for the reasoning involved in the gambler's fallacy. It is surprising just how much powerful machinery is required to thus dispose of the gambler's fallacy.

3. The Paradox of the Preface

Now let us return to the paradox of the preface. Consider a precise formulation of that paradox. We have a book b of the general description B, and we know that it is highly probable that a book of that general description makes at least one false claim. Letting T be the property of being true and Czy be the relation $\ulcorner z$ is a claim made in $y \urcorner$, we can express this probability as:

(3.1) prob$\Big((\exists z)(Czy$ & $\sim Tz) / By\Big) = r$.

Suppose N claims are made in the book—call them $p_1,\ldots,p_N$. Because r is high, $(A1)$ gives us a prima facie reason for believing $\ulcorner(\exists z)(Czb$ & $\sim Tz)\urcorner$. The paradox arises when we observe that the following is warranted:

(3.2) $(\forall z)[Czb \equiv (z = p_1 \vee \ldots \vee z = p_N)]$.

The set of propositions consisting of $\ulcorner(\exists z)(Czb \ \& \ \sim Tz)\urcorner$, and $\ulcorner Tp_1\urcorner,\ldots,\ulcorner Tp_N\urcorner$, is a minimal set deductively inconsistent with (3.2), so if r is large enough it seems to follow that the members of this set are subject to collective defeat, and hence that none of the p_i's is warranted. But this conclusion is obviously wrong. What will now be shown is that this paradox can be resolved in a manner formally similar to the above resolution of the gambler's fallacy.

Let

(3.3) $s = \mathrm{prob}\big((\exists z)(Czy \ \& \ \sim Tz) \ / \ By \ \& \ (\exists x_1)\ldots(\exists x_N)[x_1,\ldots,x_N$
           are distinct $\& \ (\forall z)(Czy \equiv (z = x_1 \vee \ldots \vee$
           $z = x_N)) \ \& \ Tx_1 \ \& \ \ldots \ \& \ Tx_{i\text{-}1} \ \& \ Tx_{i+1} \ \& \ \ldots$
           $\& \ Tx_N]\big)$.

This probability takes account of more information than does (3.1), so if it can be established that $s < r$, this provides part of what is needed for a subset defeater for the reasoning involved in the paradox. Let us turn then to the evaluation of s. By (2.13):

(3.4) $\mathrm{prob}\big((\exists z)(Czy \ \& \ \sim Tz) \ / \ By \ \& \ (\exists x_1)\ldots(\exists x_N)[x_1,\ldots,x_N$ are
           distinct $\& \ (\forall z)(Czy \equiv (z = x_1 \vee \ldots \vee z = x_N)) \ \&$
           $Tx_1 \ \& \ \ldots \ \& \ Tx_{i\text{-}1} \ \& \ Tx_{i+1} \ \& \ \ldots \ \& \ Tx_N\big)$

           $= \mathrm{prob}\big((\exists z)(Czy \ \& \ \sim Tz) \ / \ By \ \& \ x_1,\ldots,x_N$ are distinct $\&$
           $(\forall z)(Czy \equiv (z = x_1 \vee \ldots \vee z = x_N)) \ \& \ Tx_1 \ \& \ \ldots$
           $\& \ Tx_{i\text{-}1} \ \& \ Tx_{i+1} \ \& \ \ldots \ \& \ Tx_N\big)$.

By the classical probability calculus:

(3.5) $\text{prob}\big((\exists z)(Czy \ \& \ {\sim}Tz) \ / \ By \ \& \ x_1,\dots,x_N \text{ are distinct } \&$
$\qquad\qquad (\forall z)(Czy \equiv (z = x_1 \vee \dots \vee z = x_N)) \ \& \ Tx_1 \ \& \ \dots \ \&$
$\qquad\qquad Tx_{i\text{-}1} \ \& \ Tx_{i+1} \ \& \ \dots \ \& \ Tx_N\big)$

$\qquad = \text{prob}\big({\sim}Tx_1 \vee \dots \vee {\sim}Tx_N \ / \ By \ \& \ x_1,\dots,x_N \text{ are distinct}$
$\qquad\qquad \& \ (\forall z)(Czy \equiv (z = x_1 \vee \dots \vee z = x_N)) \ \&$
$\qquad\qquad Tx_1 \ \& \ \dots \ \& \ Tx_{i\text{-}1} \ \& \ Tx_{i+1} \ \& \ \dots \ \& \ Tx_N\big)$

$\qquad = \text{prob}\big({\sim}Tx_i \ / \ By \ \& \ x_1,\dots,x_N \text{ are distinct } \& \ (\forall z)(Czy \equiv$
$\qquad\qquad (z = x_1 \vee \dots \vee z = x_N)) \ \& \ Tx_1 \ \& \ \dots \ \& \ Tx_{i\text{-}1}$
$\qquad\qquad \& \ Tx_{i+1} \ \& \ \dots \ \& \ Tx_N\big)$

The gambler's fallacy is avoided by noting that we take the different tosses of a fair coin to be statistically independent of one another. In contrast, we are not apt to regard the disparate claims in a single book as being statistically independent of one another, but the important thing to realize is that insofar as they are statistically relevant to one another, they *support* one another. In other words, they are not negatively relevant to one another:

(3.6) $\text{prob}\big(Tx_i \ / \ By \ \& \ x_1,\dots,x_N \text{ are distinct } \& \ (\forall z)(Czy \equiv$
$\qquad\qquad (z = x_1 \vee \dots \vee z = x_N)) \ \& \ Tx_1 \ \& \ \dots \ \& \ Tx_{i\text{-}1}$
$\qquad\qquad \& \ Tx_{i+1} \ \& \ \dots \ \& \ Tx_N\big)$

$\qquad \geq \text{prob}\big(Tx_i \ / \ By \ \& \ x_1,\dots,x_N \text{ are distinct } \& \ (\forall z)(Czy \equiv$
$\qquad\qquad (z = x_1 \vee \dots \vee z = x_N))\big).$

We may not know the value of the latter probability, but we can be confident that it is not low. In particular, it is much larger than the very small probability $1\text{-}r$. By the probability calculus it then follows that:

(3.7) $\text{prob}\big(\sim Tx_i \,/\, By \,\&\, x_1,\ldots,x_N \text{ are distinct } \&\, (\forall z)(Czy \equiv$

$\qquad (z = x_1 \lor \ldots \lor z = x_N)) \,\&\, Tx_1 \,\&\, \ldots \,\&\, Tx_{i\text{-}1}$

$\qquad \&\, Tx_{i+1} \,\&\, \ldots \,\&\, Tx_N\big)$

$\quad < r.$

Therefore, by (3.4) and (3.5) it follows that $s < r$.

The reasoning in the paradox of the preface proceeds by using $(A1)$ in connection with (3.1). It follows that

(3.8) $s < r \,\&\, Tp_1 \,\&\, \ldots \,\&\, Tp_{i\text{-}1} \,\&\, Tp_{i+1} \,\&\, \ldots \,\&\, Tp_N$

is a defeater for that reasoning. But this is not yet to say that the reasoning is defeated. In order for this to defeat the reasoning in a straightforward way, not only $\ulcorner s < r \urcorner$ but also $\ulcorner Tp_1 \,\&\, \ldots \,\&\, Tp_{i\text{-}1} \,\&\, Tp_{i+1} \,\&\, \ldots \,\&\, Tp_N \urcorner$ would have to be warranted. It is precisely the warrant of the $\ulcorner Tp_j \urcorner$'s that is called into question by the paradox of the preface. However, it will now be argued that the reasoning is defeated in a more complicated way.

The putative collective defeat for each Tp_i proceeds by constructing an argument σ_i for $\sim Tp_i$. σ_i is diagramed in Figure 1. However, $\ulcorner s < r \urcorner$ is warranted, and when it is conjoined with $Tp_1 \,\&\, \ldots \,\&\, Tp_{i\text{-}1} \,\&\, Tp_{i+1} \,\&\, \ldots \,\&\, Tp_n$ this produces a defeater for the very first diagramed inference in σ_i. This means that σ_i is self-defeating relative to $\ulcorner s < r \urcorner$. Thus the arguments that appear to enter into collective defeat in the paradox of the preface are all self-defeating. There is no collective defeat and all of the propositions Tp_i remain warranted. Hence $(A1)$ leads to a satisfactory resolution of the paradox of the preface.

4. Defeasible Defeaters

Thus far the only epistemological principles required in the theory of nomic probability have been the acceptance rules $(A1)$-$(A3)$ and the associated subset defeaters $(D1)$-$(D3)$. But now it must be acknowledged that some details have been ignored that require

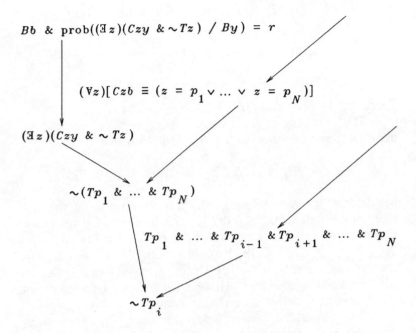

Figure 1. The counterargument for $\sim Tp_i$.

the account to be made more complicated. These are discussed in this section and the next.

First, consider a difficulty pertaining to (*D2*) directly, and (*D1*) and (*D3*) indirectly. Recall that (*A2*) and (*D2*) are formulated as follows:

(*A2*)   If F is projectible with respect to G and $r > .5$, then $\lceil \text{prob}(F/G) \geq r \ \& \ \sim Fc \rceil$ is a prima facie reason for $\lceil \sim Gc \rceil$, the strength of the reason depending upon the value of r.

(*D2*)   If F is projectible with respect to H then $\lceil Hc \ \& \ \text{prob}(F/G\&H) < \text{prob}(F/G) \rceil$ is an undercutting defeater for $\lceil \sim Fc \ \& \ \text{prob}(F/G) \geq r \rceil$ as a prima facie reason for $\lceil \sim Gc \rceil$.

Consider an instance of (*A2*). Suppose we know that *c* is an electrical circuit connecting a light with a power source via a switch, and we know it to be highly probable that in such a circuit the light will be on if the switch is closed. Knowing that the light is off, it is reasonable to infer, in accordance with (*A2*), that the switch is open. (*D2*) works correctly in most cases. For instance, if we discover that the wire from the power source is broken, this defeats the inference. This defeat can be accommodated by (*D2*), because it is improbable that in such a circuit the light will be on if the switch is closed but the wire broken. But there are also apparent counterexamples to (*D2*). For example, let *H* be $\sim F$. $\text{prob}(F/G \& \sim F) = 0$, so by (*D2*), $\ulcorner \sim Fc \urcorner$ should be a defeater for an application of (*A2*). But this is obviously wrong—we have $\ulcorner \sim Fc \urcorner$ in all cases of (*A2*). Similarly, let $\ulcorner Hc \urcorner$ be our reason for believing $\ulcorner \sim Fc \urcorner$; for instance, 'it looks like the light is off'. Again, $\text{prob}(F/G \& H)$ will normally be low, but this should not be a defeater for the use of (*A2*). Apparently, some choices of H that satisfy (*D2*) should not yield a defeater for (*A2*). These *H*'s are characterized by the fact that we believe $\ulcorner \sim Fc \urcorner$ on the basis of some argument that supports $\ulcorner Hc \urcorner$ in the process of getting to $\ulcorner \sim Fc \urcorner$. More generally, $\ulcorner Hc \urcorner$ might be a deductive consequence of propositions supported by the argument. Let us say that such an $\ulcorner Hc \urcorner$ is *presupposed by* $\ulcorner \sim Fc \urcorner$. Notice that $\ulcorner \sim Fc \urcorner$ itself is automatically presupposed by $\ulcorner \sim Fc \urcorner$.

The principles of defeat for (*A2*) must preclude appeal to $\ulcorner Hc \urcorner$'s that are presupposed by $\ulcorner \sim Fc \urcorner$. There are two ways of doing this. We might build a qualification into (*D2*) requiring that $\ulcorner Hc \urcorner$ not to be presupposed by $\ulcorner \sim Fc \urcorner$. Alternatively, we might leave (*D2*) as it is, but introduce a defeater defeater whose function is to rule out an instance of (*D2*) when the $\ulcorner Hc \urcorner$ involved is found to be presupposed by $\ulcorner \sim Fc \urcorner$. The difference between these two ways of handling defeat is that the second proposal results in (*A2*) being defeated by such an $\ulcorner Hc \urcorner$ *until* it is discovered that it is presupposed by $\ulcorner \sim Fc \urcorner$. In some cases it may be obvious that $\ulcorner Hc \urcorner$ is presupposed, but in other cases it may require a fair amount of reflection on one's reasoning to discover that $\ulcorner Hc \urcorner$ is presupposed. For this reason, qualifying (*D2*) rather than making it defeasible puts too heavy a burden on

the epistemological agent. It is unreasonable to require the agent
to continually monitor his past reasoning. As in the case of all
defeasible reasoning, the reasoning is presumed correct until some
difficulty is noted. So my proposal is that (D2) is correct as it
stands, but it is a defeasible defeater, the defeater being '⌜Hc⌝ is
presupposed by ⌜~Fc⌝'.

This must also be a defeater for (D3) and (D1). This is
because all applications of (A2) can be regarded as derivative
from (A3), or derivative from (D1) via conditional reasoning. So
my final proposal is:

(DD)   '⌜~Fc⌝ is warranted, and ⌜Hc⌝ is a deductive consequence
       of propositions supported by the argument on the basis of
       which ⌜~Fc⌝ is warranted' is an undercutting defeater for
       each of (D1), (D2), and (D3).

In normal uses of (A1) and (D1) this defeater cannot arise, but
in complex cases of conditional reasoning it can arise.

5. Domination Defeaters

Even with the addition of (DD), the preceding account of
acceptance rules is incomplete. Specifically, inferences proceeding
in accordance with (A1)–(A3) can be defeated in more ways than
can be accommodated by (D1)–(D3). The need for additional
defeaters is illustrated by the following example. Suppose that in
the course of investigating a certain disease, Roderick's syndrome,
it is discovered that 98% of the people having enzyme E in their
blood have the disease. This becomes a powerful tool for
diagnosis. It is used in accordance with (A1). In statistical
investigations of diseases, it is typically found that some factors
are statistically irrelevant. For instance, it may be discovered that
the color of one's hair is statistically irrelevant to the reliability of
this diagnostic technique. Thus, for example, it is also true that
98% of all redheads having enzyme E in their blood have the
disease. We may also discover, however, that there are specifi-

able circumstances in which the diagnostic technique is unreliable. For instance, it may be found that of patients undergoing radiation therapy, only 32% of those with enzyme E in their blood have Roderick's syndrome. Note that as we have found hair color to be irrelevant to the reliability of the diagnostic technique, we would not ordinarily go on to collect data about the effect of radiation therapy specifically on redheads. Now consider Jerome, who is redheaded, undergoing radiation therapy, and is found to have enzyme E in his blood. Should we conclude that he has Roderick's syndrome? Intuitively, we should not, but this cannot be accommodated by $(A1)$ and $(D1)$. We have statistical knowledge about the reference properties E = *person with enzyme E in his blood*, R = *redheaded person with enzyme E in his blood*, and T = *person with enzyme E in his blood who is undergoing radiation therapy*. The entailment relations between these properties can be diagramed as in Figure 2. Letting S be the property of having Roderick's syndrome, we know that:

(5.1) Ej & $\text{prob}(S/E) = .98$.

(5.2) Rj & $\text{prob}(S/R) = .98$.

(5.3) Tj & $\text{prob}(S/T) = .32$.

By $(A1)$, both (5.1) and (5.2) constitute prima facie reasons for concluding that Jerome has Roderick's syndrome. By $(D1)$, (5.3) defeats the inference from (5.1), but it does not defeat the inference from (5.2). Thus it should be reasonable to infer that because Jerome is a redhead and most redheads with enzyme E

Figure 2. Entailment relations.

in their blood have Roderick's syndrome, Jerome has Roderick's syndrome. The fact that Jerome is undergoing radiation therapy should not be relevant, because that is not more specific information than the fact that Jerome has red hair. But, obviously, this is wrong. We regard Jerome's having red hair as irrelevant. The important inference is from the fact that most people with enzyme E in their blood have Roderick's syndrome to the conclusion that Jerome has Roderick's syndrome, and *that* inference is defeated.

We should not be allowed to infer $\ulcorner Sj \urcorner$ from (5.2) unless we can also infer it from (5.1), because (5.2) results from simply conjoining statistically irrelevant considerations to (5.1). We regard $\ulcorner \text{prob}(S/R) = .98 \urcorner$ to be true only *because* $\ulcorner \text{prob}(S/E) = .98 \urcorner$ is true, and hence any application of ($A1$) should appeal to the latter rather than the former. Accordingly, in a situation of the sort diagramed in Figure 3, the inference to $\ulcorner Ac \urcorner$ should be defeated. This can be captured as follows:

(5.4)   If A is projectible with respect to both B and D then $\ulcorner \text{prob}(A/B) = \text{prob}(A/C)$ & $C \preccurlyeq B \urcorner$ is an undercutting defeater for the inference from $\ulcorner Cc$ & $\text{prob}(A/C) \geq r \urcorner$ to $\ulcorner Ac \urcorner$.

Defeaters of the form of (5.4) will be called *domination defeaters*. B "dominates" C, with the result that we are unable to make inferences from $\text{prob}(A/C)$ in accordance with ($A1$). Instead, if we know $\ulcorner Cc \urcorner$, then we must infer $\ulcorner Bc \urcorner$ and make our inference from $\text{prob}(A/B)$. If the latter inference is defeated, we cannot fall back on an inference from $\text{prob}(A/C)$. In light of the intercon-

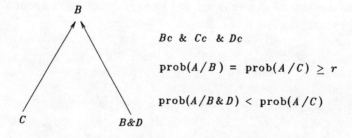

Figure 3. The structure of domination defeaters.

nections between $(A1)-(A3)$, this must be extended to yield defeaters for each:

(DOM)   If A is projectible with respect to both B and D then $\ulcorner \text{prob}(A/B) = \text{prob}(A/C) \ \& \ C \preccurlyeq B \urcorner$ is an undercutting defeater for each of the following inferences:
(a) from $\ulcorner Cc \ \& \ \text{prob}(A/C) \geq r \urcorner$ to $\ulcorner Ac \urcorner$ by ($A1$);
(b) from $\ulcorner {\sim}Ac \ \& \ \text{prob}(A/C) \geq r \urcorner$ to $\ulcorner {\sim}Cc \urcorner$ by ($A2$);
(c) from $\ulcorner \text{prob}(A/C) \geq r \urcorner$ to $\ulcorner Cc \supset Ac \urcorner$ by ($A3$).

It is worth emphasizing how important domination defeaters are for applications of the statistical syllogism. The preceding sort of situation is typical of rich scientific contexts in which we have a lot of statistical information regarding not only relevant factors but also irrelevant ones. Without domination defeaters, we would be precluded from ever using the statistical syllogism in such cases.

It is also worth noting that domination defeaters would be largely useless if the statistical syllogism had to employ frequencies rather than subjunctive probabilities. This is because in cases like the above, we evaluate the probabilities by statistical induction. We observe the relative frequency of Roderick's syndrome in rather small samples and use that to estimate the values of the probabilities. If the observed frequency of Roderick's syndrome among redheads with enzyme E in their blood is approximately the same as the observed frequency among all people with enzyme E in their blood, then we may judge that the probability of Roderick's syndrome is equal in the two cases. But we do not judge that the relative frequency of Roderick's syndrome in the whole set of people with enzyme E in their blood is precisely equal to the relative frequency of Roderick's syndrome in the whole set of redheads with enzyme E in their blood — only that they are approximately equal. Equal probabilities only give us a reason for thinking that frequencies are approximately equal. Accordingly, if statistical syllogism had to proceed from frequencies, then in cases like the above we would not have reason to believe the requisite frequencies to be equal and so could not employ domination defeaters and could not defeat the inference to the conclusion that Jerome has Roderick's syndrome.

Defeaters can themselves be defeasible. We have seen that $(D1)-(D3)$ are defeasible, and it turns out that domination defeaters are as well. The purpose of domination defeaters is to make irrelevant strengthenings of reference properties irrelevant to the use of $(A1)$, but such irrelevant strengthenings can also work in the opposite direction to generate spurious domination defeaters. For example, suppose we know that the probability of a person dying within the next hour (D) given that he has taken a large dose of arsenic (R) is .98. However, the probability that he will die in the next hour given that he has also taken an antidote (A) is only .001. On the other hand, if he takes a large dose of arsenic, the antidote, and also strychnine (S) the probability is .98 again. However, the probability of a person dying within the next hour given that he takes a large dose of arsenic but has also been eating small doses of arsenic regularly to build up a tolerance (T) is only .5. Now consider Frank, who takes arsenic as a party trick. He has been eating small doses for weeks to build up a tolerance, but after he takes the large dose he panics and drinks the antidote. Unknown to Frank, his wife has laced the arsenic antidote with strychnine. What should we conclude about poor Frank? Obviously, that he is not long for this world. But now consider the various reference properties and the corresponding probabilities of dying within the next hour. They are related as in Figure 4. The inference from R is blocked by our knowing that Frank has taken the antidote, but we should be able to infer that he will die because he has also taken strychnine. However, because $\mathrm{prob}(D/R) = \mathrm{prob}(D/R\&A\&S)$, an inference from the latter is blocked by a domination defeater deriving from the fact that $\mathrm{prob}(D/R\&A\&S) \neq \mathrm{prob}(D/R)$. This results in our getting the wrong answer.

What is happening here is that the domination defeater works on the presumption that if $\mathrm{prob}(D/R) = \mathrm{prob}(D/R\&A\&S)$ then $(A\&S)$ is an irrelevant strengthening of R and the high value of $\mathrm{prob}(D/R\&A\&S)$ is due simply to the fact that the value of $\mathrm{prob}(D/R)$ is high. But this sense of irrelevance is stronger than just statistical irrelevance. It is required that if $(R\&A\&S) \leqslant C \leqslant$

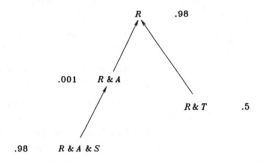

Figure 4. Defeated domination defeater.

R then we also have prob(D/C) = prob(D/R). This fails in the present example if we let $C = A$. Discovering this shows that A and S are not irrelevant strengthenings of R. They are relevant strengthenings that just happen to cancel out. Accordingly, there is no reason why a defeater for the inference from the high value of prob(D/R) should also defeat an inference from the high value of prob($D/R\&A\&S$).

In general, in a situation having the form of Figure 5, the inference to $\ulcorner Ac\urcorner$ should not be defeated. This can be captured by adopting the following defeater defeater:

(*DOM-D*) If A is projectible with respect to E then $\ulcorner C \preceq E \preceq B$ & prob(A/E) $\neq$ prob(A/B)$\urcorner$ is a defeater for (*DOM*).

This is getting complicated, but there remain still more wheels within wheels. As things stand, we could almost always circumvent domination defeaters by purely logical manipulations. Consider the first example above, in which (5.2) is defeated by a domination defeater. Because it is necessarily true that prob($R\&S/R$) = prob(S/R), we can generate another prima facie reason that is not defeated:

(5.5) Rj & prob($R\&S/R$) = .98.

We do not have a domination defeater for (5.5) because we have

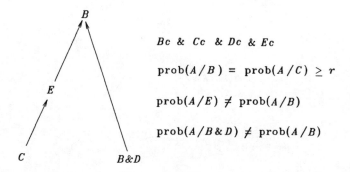

Figure 5. The structure of (*DOM-D*).

no reason to expect that $\text{prob}(R\&S/R) = \text{prob}(R\&S/E)$. But this is just a logical trick. The statistical information embodied in (5.5) is equivalent to that embodied in (5.2). In effect, logical manipulations enable us to construct an end run around the domination defeater, as diagramed in Figure 6.

The same sort of logical trick can be performed using physical necessities rather than logical necessities. For example, if we know that $(R \Rightarrow U)$, it follows that $\text{prob}(U\&S/R) = \text{prob}(S/R)$. Then an inference from

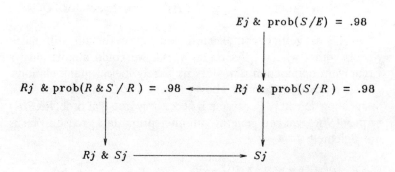

Figure 6. End run around a domination defeater.

(5.6) Rj & prob($U\&S/R$) = .98.

gives us a reason for believing ⌜Uj & Sj⌝ and hence ⌜Sj⌝. This inference must be defeated.

If we turn to multiplace properties (relations), similar tricks can be performed in another way. Suppose we have:

(5.7) Bbc & prob(Axy/Bxy) ≥ r.

and

(5.8) Cbc & prob(Axy/Bxy & Cxy) < prob(Axy/Bxy).

(5.8) should defeat an inference to ⌜Abc⌝ from (5.7). But suppose we also know that b is a "typical" object, in the sense that:

(5.9)   prob(Axy/Bxy & $x = b$) = prob(Axy/Bxy).

This should not change things. The inference to ⌜Abc⌝ should still be defeated. The difficulty is that (5.9) introduces a new reason of the form of ($A1$) for ⌜Abc⌝, namely:

(5.10) Bbc & $b = b$ & prob(Axy/Bxy & $x = b$) ≥ r.

Although (5.8) does not provide a defeater of the form ($D1$) for the inference from (5.10), we do have a domination defeater. However, the domination defeater can also be circumvented, this time by using the principle (IND) of Chapter 2. That principle tells us that prob(Axy/Bxy & $x = b$) = prob(Aby/Bby). Thus we can infer from (5.7) and (5.9):

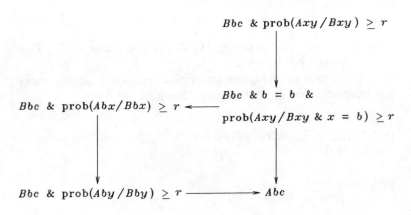

Figure 7. End run around a domination defeater.

(5.11) Bbc & prob(Aby/Bby) $\geq r$.

(5.11) provides a new reason for inferring $\ulcorner Abc \urcorner$, and we have no domination defeater for it. The resulting end run can be diagramed as in Figure 7.

Basically, what we have here are three different logical tricks for constructing end runs to circumvent domination defeaters. There must be some general kind of defeater that blocks these logical tricks. I suggest that these tricks can be characterized by the fact that they use logical or nomic facts to transform one probability prob($Ax_1 \ldots x_n/Bx_1 \ldots x_n$) into another probability prob($A^*x_1 \ldots x_k/B^*x_1 \ldots x_k$) and appeal to (possibly different) tuples $b_1, \ldots, b_n$ and $c_1, \ldots, c_k$ where those same logical or nomic facts guarantee that $\ulcorner Bb_1 \ldots b_n \urcorner$ is nomically equivalent to $\ulcorner B^*c_1 \ldots c_k \urcorner$ (i.e., these express equivalent information about the world) and either of the latter nomically implies $\ulcorner Ab_1 \ldots b_n \equiv A^*c_1 \ldots c_k \urcorner$ (i.e., the conclusions drawn from either by (AI) express equivalent information about the world). We can think of these transformations as functions transforming triples $\langle X, Y, \sigma \rangle$ into $\langle X^*, Y^*, \sigma^* \rangle$, where X, X^*, Y, and Y^* are properties and σ and σ^* are their lists of arguments. For instance, in the first example, f is chosen so that $Y^* = Y$, $X^* = X\&Y$, and $\sigma^* = \sigma$. In the second example, we can let f be a function such that $Y^* = Y$, $X^* = U\&X$ if $X \Rightarrow U$ and $X^* = X$ otherwise, and $\sigma^* = \sigma$. In the third example, f is such that if Y has the form $(Zxy \ \& \ x = b)$ then $Y^* = Yby$, $X^*xy = Xby$, and if $\sigma = \langle x,y \rangle$ then $\sigma^* = \langle y \rangle$. In each case f has the property that $\square_p(\text{prob}(X\sigma/Y\sigma) = \text{prob}(X^*\sigma^*/Y^*\sigma^*))$. Let us call such f's *nomic transforms*. Then the requisite defeaters can be captured by the following principle:

(*E-DOM*) Consider the following:
    (a) $Bb_1, \ldots, b_n$ & $\text{prob}(A/B) \geq r$;
    (b) $B^*c_1 \ldots c_k$ & $\text{prob}(A^*/B^*) \geq r$;
    If P is a domination defeater for (a) as a prima facie reason for $\ulcorner Ab_1 \ldots b_n \urcorner$, then
    (c) f is a nomic transform such that $f(\langle A,B,\langle b_1, \ldots, b_n \rangle \rangle)$
        $= \langle A^*,B^*,\langle c_1, \ldots, c_k \rangle \rangle$, and P & $(Bb_1 \ldots b_n \Leftrightarrow B^*c_1 \ldots c_k)$
        & $[Bb_1 \ldots b_n \Rightarrow (Ab_1 \ldots b_n \equiv A^*c_1 \ldots c_k)]$
    is a defeater for (b) as a prima facie reason for $\ulcorner A^*c_1 \ldots c_k \urcorner$. Defeaters of the form (*DOM-D*) for P are also defeaters for (c).

I will call such defeaters *extended domination defeaters*. We must extend the domination defeaters for (*A2*) and (*A3*) in precisely the same way. With the addition of (*E-DOM*), I believe that our account of acceptance rules is now complete.

6. Summary

1. It is more difficult to show that the gambler's fallacy is fallacious than is generally appreciated, but it can be done using a principle from the exotic theory of nomic probabilities.
2. The paradox of the preface is resolved by showing that

reasoning of the same form as that involved in the gambler's fallacy shows that crucial arguments underlying the paradox of the preface are self-defeating.

3. Defeaters of the form $(D1)-(D3)$ are defeasible, the defeater defeater being described by (DD).

4. A correct account of acceptance rules must also add domination defeaters and extended domination defeaters.

CHAPTER 10
ADVANCED TOPICS:
DIRECT INFERENCE

1. Propensities

It was argued in Chapter 1 that various kinds of propensities make intuitive sense, but existing propensity theories do little to clarify their nature. We are now in a position to give precise definitions for several different kinds of propensities in terms of nomic probabilities. The characteristics of these propensities can then be deduced from the theory of nomic probability. The technique for defining propensities in terms of nomic probabilities is modeled on the definitions of objective and physical/epistemic definite probabilities proposed in Chapter 4.

1.1 Counterfactual Probability

We often want to know how probable it is that P *would be* true if Q *were* true. This is a kind of *counterfactual probability*, and I will symbolize it as $\lceil prob(P/Q) \rceil$.[1] Counterfactual conditionals constitute the limiting case of counterfactual probability, in the sense that if $(Q > P)$ obtains then $prob(P/Q) = 1$. $prob(P/Q)$ is to be distinguished from $prob(P/Q)$ and $\mathsf{PROB}(P/Q)$, either of which can be regarded as an "indicative" probability that P *is* true if Q *is* true. Recall that $(Q > P)$ obtains iff P obtains at every nearest world at which Q obtains. In other words, where $\mathbf{M}(Q)$ is the set of nearest Q-worlds, $(Q > P)$ obtains iff $\mathbf{M}(Q) \subseteq |P|$. Analogously, $prob(P/Q)$ can be regarded as a measure of the proportion of nearest Q-worlds that are also P-worlds. This has the immediate result that if $(Q > P)$ obtains then $prob(P/Q) =$

---

[1] I introduced counterfactual probability in Pollock [1976], where I called it 'simple subjunctive probability'. Prior to that time it was almost entirely overlooked. Even those philosophers who tried to weld an intimate connection between counterfactuals and probability did so by considering only indicative probabilities, and usually only subjective probabilities.

1 (but not conversely). This sort of heuristic description of counterfactual probability enables us to investigate its formal properties, but if the notion is to be of any real use we must do more than wave our hands and talk about an unspecified measure on $\mathbf{M}(Q)$. We are now in a position to accurately define counterfactual probability. Let $\mathbf{C}_Q$ be the conjunction of all of the counterfactual consequences of Q, that is, the conjunction of all states of affairs R such that $(Q > R)$ obtains. We can then define:

(1.1) $prob(P/Q) = \mathrm{prob}(P/\mathbf{C}_Q)$.

This is completely analogous to the definition of physical/epistemic probability. The following analogues of (4.5)–(4.9) hold for counterfactual probability:

(1.2) $0 \leq prob(P/Q) \leq 1$.

(1.3)   If $(Q \Rightarrow P)$ then $prob(P/Q) = 1$.

(1.4)   If $(Q \Leftrightarrow R)$ then $prob(P/Q) = prob(P/R)$.

(1.5)   If $\Diamond R$ and $[R \Rightarrow \sim(P \& Q)]$ then $prob(P \lor Q/R) = prob(P/R) + prob(Q/R)$.

(1.6)   If $[(P \& R) \Rightarrow Q]$ then $prob(P/R) \leq prob(Q/R)$.

As was pointed out in Pollock [1976], the multiplicative axiom fails for counterfactual probability, even in the following weak form:

(1.7)   If $\Diamond_p R$ then $prob(P \& Q/R) = prob(P/R) \cdot prob(Q/P \& R)$.

The reason this fails is that $\mathbf{M}(R)$ and $\mathbf{M}(P \& R)$ will normally be different sets of possible worlds, so these probabilities are measuring proportions in different sets, and there is no way to combine the proportions in any meaningful way.

The logic *SS* of subjunctive conditionals was discussed in Chapter 2. It is being assumed that the following are necessarily true:

(1.8)   If $(P > Q)$ and $(P > R)$ then $[P > (Q\&R)]$.

(1.9)   If $(P > R)$ and $(Q > R)$ then $[(P \lor Q) > R]$.

(1.10) If $(P > Q)$ and $(P > R)$ then $[(P\&Q) > R]$.

(1.11) If $(P\&Q)$ then $(P > Q)$.

(1.12) If $(P > Q)$ then $(P \supset Q)$.

(1.13) If $\Box(Q \supset R)$ and $(P > Q)$ then $(P > R)$.

(1.14) If $\Box(P \equiv Q)$ then $[(P > R) \equiv (Q > R)]$.

(1.15) If $(P \Rightarrow Q)$ then $(P > Q)$.

The Generalized Consequence Principle is also being assumed:

(1.16) If Γ is a set of counterfactual consequences of P, and Γ entails Q, then $(P > Q)$ obtains.

Let us define $\ulcorner P \between Q \urcorner$ to mean $\ulcorner(P > Q) \& (Q > P)\urcorner$. The following is a theorem of SS:

(1.17) If $(P \between Q)$ then $[(P > R) \equiv (Q > R)]$.

Given these principles it follows easily that the axioms for counterfactual probability can be strengthened as follows:

(1.18) If $(Q > P)$ then $prob(P/Q) = 1$.

(1.19) If $(R \between S)$ then $prob(P/R) = prob(R/S)$.

(1.20) If $(R > \sim(P\&Q))$ and $\Diamond_p R$ then $prob(P \lor Q/R) = prob(P/R) + prob(Q/R)$.

(1.21) If $(R > (P \supset Q))$ then $prob(P/R) \leq prob(Q/R)$.

Using the model of the reduction of classical direct inference to nonclassical direct inference, a theory of counterfactual direct inference for *prob* can be derived from the theory of nonclassical direct inference. The basis of the theory is a *Rule of Counterfactual Direct Inference* that is formally parallel to our rule of conditional classical direct inference:

(1.22) If F is projectible with respect to $(G\&H)$ then $\ulcorner(Ga > Ha)$ & $(Q > (P \equiv Fa))$ & $(Q \ s\ Ga)\urcorner$ is a prima facie reason for $\ulcorner prob(P/Q) = \text{prob}(F/G\&H)\urcorner$.

Proof: Suppose we are justified in believing

$$(Ga > Ha) \ \& \ (Q > (P \equiv Fa)) \ \& \ (Q \ s \ Ga).$$

Nonclassical direct inference provides a prima facie reason for believing that

$$\text{prob}(Fx \ / \ Gx \ \& \ Hx \ \& \ \mathbf{C}_{Ga} \ \& \ x = a) = \text{prob}(F/G\&H)$$

which by definition is equivalent to:

$$\text{prob}(Fa \ / \ Ga \ \& \ Ha \ \& \ \mathbf{C}_{Ga}) = \text{prob}(F/G\&H),$$

and hence also to:

$$prob(Fa/Ga\&Ha) = \text{prob}(F/G\&H).$$

By (1.19) and (1.21), $prob(P/Q) = prob(Fa/Ga)$. As $(Ga > Ha)$, it follows from SS that $(Ga \ s \ (Ga\&Ha))$, so by (1.19), $prob(P/Q) = prob(Fa/Ga\&Ha)$. Hence we have a prima facie reason for believing that $prob(P/Q) = \text{prob}(F/G\&H)$.

In a similar fashion, we can justify the following subset defeaters for counterfactual direct inference:

(1.23) If F is projectible with respect to $(G\&H\&J)$ then $\ulcorner\text{prob}(F/G\&H\&J) \neq \text{prob}(F/G\&H)$ & $(Q > Ja)\urcorner$ is an undercutting defeater for (1.22).

1.2 Hard Propensities

The propensities discussed by Ronald Giere and James Fetzer are supposed to be purely physical propensities, untinged with any epistemic element.[2] Furthermore, they are supposed to be nonconditional probabilities. Quantum mechanics is often taken to provide the paradigms of such propensities. For example, we can talk about the propensity of a particular radium atom to decay within a certain amount of time. These propensities are supposed to represent purely physical facts about the objects involved and reflect genuine nomic uncertainty in the world.[3] I will call them *hard propensities*.

It is frequently asserted that hard propensities are the probabilities that the quantum physicist deals with. I cannot undertake an extended discussion of the philosophy of quantum mechanics here, but I seriously doubt that this claim is true. Most of the discussion of probability in quantum mechanics completely overlooks the distinction between definite and indefinite probabilities. Quantum mechanical laws are, *a fortiori*, laws. As such, the probabilistic conclusions derived from them are statistical laws, and hence nomic probabilities. When we use quantum mechanics to deduce the probability distribution expected in a scattering experiment, we are deducing indefinite probabilities pertaining to a *kind* of physical situation. To apply these to a concrete situation we must employ some kind of direct inference. Hard propensities must be the result of the direct inference, not the original calculation from laws. I am not sure that the resulting hard propensities play any significant role in actual scientific practice, but be that as it may, it is still of interest to try to make sense of them.

In trying to analyze hard propensities, the first thought that may occur to one is that they are derived from counterfactual probabilities. Hard propensities are nonconditional probabilities, whereas counterfactual probabilities are conditional probabilities, but it is always possible to define a nonconditional probability in terms of a conditional probability, and it might be supposed that that is what hard propensities are. The standard way of getting

---

[2] See Giere [1973] and [1976]; and Fetzer [1971], [1977], and [1981].

[3] Giere ([1973], p. 475) explicitly acknowledges that such propensities exist only (or better, only have values different from 0 and 1) in nondeterministic universes.

nonconditional probabilities from conditional probabilities is by identifying them with probabilities conditional on tautologies:

(1.24) $prob(P) = prob(P/Q \lor \sim Q)$.

However, a simple theorem establishes that these nonconditional probabilities cannot be identified with hard propensities:

(1.25)   If P obtains then $prob(P) = 1$, and if P does not obtain then $prob(P) = 0$.

Proof: If P obtains then $((Q \lor \sim Q)\&P)$ obtains.   Then by the centering principle (1.11), $((Q \lor \sim Q) > P)$ obtains, and hence by (1.18), $prob(P) = 1$. By the same reasoning, if P does not obtain then $prob(\sim P) = 1$ and hence $prob(P) = 0$.

Thus hard propensities cannot be defined in terms of counterfactual probabilities.   A related observation is that (conditional) counterfactual probabilities can take values intermediate between 1 and 0 even in a deterministic world, but that is not supposed to be true for hard propensities.

There is another way of making sense of hard propensities. Consider the radium atom again.   The probability (= hard propensity) of its undergoing radioactive decay during a specified interval of time should depend upon whether that interval is in the past or the future.   If the interval is past then it is already determined whether the atom decayed during that interval.   Thus propensities should be relativized to times.   Because of relativistic problems concerning simultaneity, if they are to be relativized to times then they must also be relativized to places.   Thus I suggest that we explicitly relativize them to space time points $\langle x,y,z,t \rangle$, writing $PROB_{\langle x,y,z,t \rangle}(P)$ for the hard propensity of P at $\langle x,y,z,t \rangle$. My suggestion is then that this is the objective probability of P's being true given the entire state of the universe earlier than or simultaneous with $\langle x,y,z,t \rangle$.   To make this precise we must have absolute temporal relations.   Just how they should be defined depends upon what physical theories are true, but within current physics they might be defined in terms of the past light cone:

$\langle x_0,y_0,z_0,t_0 \rangle$ is earlier than or simultaneous with $\langle x,y,z,t \rangle$ iff $\langle x_0,y_0,z_0,t_0 \rangle$ occurs in the past light cone of $\langle x,y,z,t \rangle$.[4] The state of the universe earlier than or simultaneous with $\langle x,y,z,t \rangle$ consists of the values of "physical quantities" at all space-time points earlier than or simultaneous with $\langle x,y,z,t \rangle$. Again, what we take to be the relevant physical quantities will depend upon our physical theory, so this is not so much a definition of hard propensities as a definition schema to be completed in the light of physical theory. But however it is to be filled out, let $S_{\langle x,y,z,t \rangle}$ be the state of affairs describing, in this sense, the entire state of the universe earlier than or simultaneous with $\langle x,y,z,t \rangle$. My proposal is then;

(1.26) $PROB_{\langle x,y,z,t \rangle}(P) = \text{prob}(P/S_{\langle x,y,z,t \rangle})$.

If the universe is deterministic then $S_{\langle x,y,z,t \rangle}$ will either nomically imply P or nomically imply $\sim P$, in which case $PROB_{\langle x,y,z,t \rangle}(P)$ will be either 1 or 0, but if the world is genuinely indeterministic then $PROB_{\langle x,y,z,t \rangle}(P)$ can have values intermediate between 1 and 0.

The evaluation of hard propensities will proceed by direct inference from nomic probabilities. The following principles of direct inference can be justified in the same way as the corresponding principles of direct inference for counterfactual probabilities:

(1.27) If F is projectible with respect to G then $\ulcorner \Box(S_{\langle x,y,z,t \rangle} \supset Ga) \urcorner$ is a prima facie reason for $\ulcorner PROB_{\langle x,y,z,t \rangle}(Fa) = \text{prob}(F/G) \urcorner$.

(1.28) If F is projectible with respect to $(G\&H)$ then $\ulcorner \text{prob}(F/G\&H) \neq \text{prob}(F/G) \ \& \ \Box(S_{\langle x,y,z,t \rangle} \supset Ha) \urcorner$ is an undercutting defeater for (1.27).

---

[4] Of course, we cannot expect this relation to be connected, that is, it is not automatically true that given any two space-time points, one is earlier than or simultaneous with the other.

For example, in calculating the probability of a certain result in a scattering experiment in particle physics we might observe that there is a proton with such-and-such an energy and momentum in a potential field of a certain shape, and then using quantum mechanics we can compute the ψ-function and arrive at a probability. But this is a nomic probability—not a hard propensity. By (1.27) we can make a direct inference to the value of the hard propensity, but if it is pointed out that there are other factors we overlooked in calculating the ψ-function, that constitutes a subset defeater for the direct inference.

The upshot of this is that by beginning with nomic probabilities we can make perfectly good sense of hard propensities. The difficulties encountered in earlier theories of hard propensities had their origin in the fact that those theories attempted to make hard propensities the most basic kind of probability rather than defining them in terms of more fundamental kinds of probability as I have done here. The present alternative avoids all the difficulties discussed in Chapter 1.

1.3 Weak Indefinite Propensities

In Chapter 1, the distinction between definite and indefinite propensities was formulated, and it was noted that propensity theorists tend to confuse that distinction. Counterfactual probabilities and hard propensities are definite propensities, but there are also indefinite propensities in which we are interested. For example, suppose I have a coin I believe to be a fair coin. This means that I regard the indefinite probability of a flip of this particular coin landing either heads or tails to be .5. This is not nomic probability. Nomic probabilities only take account of physical laws, but we would expect this probability to have a different value if the coin were bent, and bending a coin does not alter physical laws. In other words, the probability of a flip of this coin landing heads is sensitive to nomically contingent facts about the coin. This makes it a propensity rather than a nomic probability.

The simplest kind of indefinite propensity is what I will call a *weak indefinite propensity*. Suppose the coin in the preceding example has never been flipped. The weak propensity of a flip of this coin landing heads is the nomic probability of a flip landing heads given all the things that would obtain if there were a flip of this coin. Recalling that $C_{(\exists x)Gx}$ is the conjunction of all states of affairs that would obtain if there were a G, we can define:

(1.29) $prob_w(Fx/Gx) = prob(Fx \ / \ Gx \ \& \ C_{(\exists x)Gx})$.

Assuming it is true that the coin would remain unbent and in an ordinary setting if it were flipped, the weak propensity of a flip landing heads is .5.

Weak propensities satisfy the following unremarkable axioms:

(1.30) If $(G \Rightarrow F)$ then $prob_w(F/G) = 1$.

(1.31) If $(G \Leftrightarrow H)$ then $prob_w(F/G) = prob_w(F/H)$.

(1.32) If $\Diamond H$ and $[H \Rightarrow \sim(F \& G)]$ then $prob_w(F \lor G/H) = prob_w(F/H) + prob_w(G/H)$.

(1.33) If $[(F \& H) \Rightarrow G]$ then $prob_w(F/H) \leq prob_w(G/H)$.

The multiplicative axiom fails for familiar reasons.

Weak propensity is a well-behaved kind of probability, but it has a surprising feature:

(1.34) If there are finitely many G's then $prob_w(F/G)$
        $= freq[F/G]$.

Proof: Suppose there are finitely many G's. Then $(\exists x)Gx$ obtains, so by (1.11) and (1.12), for any state of affairs Q, $((\exists x)Gx > Q)$ obtains iff Q obtains.[5] Consequently, $C_{(\exists x)Gx}$ is the conjunction of all states of affairs that actually obtain. As there are finitely many G's, $freq[F/G]$ exists and has some value r. It follows that $C_{(\exists x)Gx}$ entails $\ulcorner freq[F/G] = r \urcorner$. The principle (PFREQ) of Chapter 2 tells us for any nomically rigid designator r:

---

[5] In the logic of counterfactuals, this is known as the 'centering principle'. It amounts to the principle that if P obtains then the only nearest possible P-world is the actual world.

If $\lozenge_p\Big[(\exists x)(Gx \ \& \ C_{(\exists x)Gx}) \ \& \ freq[F/G\&C_{(\exists x)Gx}] = r\Big]$ then

$prob\Big(Fx \ / \ Gx \ \& \ C_{(\exists x)Gx} \ \& \ freq[F/G\&C_{(\exists x)Gx}] = r\Big) = r.$

If P is any state of affairs that obtains then $freq[F/G\&P] = freq[F/G]$. In other words, $\square[P \supset freq[F/G] = freq[F/G\&P]]$. As $C_{(\exists x)Gx}$ entails $\ulcorner freq[F/G] = r\urcorner$, it also entails $\ulcorner freq[F/G\&C_{(\exists x)Gx}] = r\urcorner$. Therefore, by the probability calculus:

$$prob_w(F/G) =$$

$$prob(F \ / \ G \ \& \ C_{(\exists x)Gx}) =$$

$$prob\Big(F \ / \ G \ \& \ C_{(\exists x)Gx} \ \& \ freq[F/G\&C_{(\exists x)Gx}] = r\Big)$$

$$= r.$$

Theorem (1.34) might be regarded as either a desirable or an untoward result. It is undeniable that for some purposes we want a stronger kind of propensity for which (1.34) fails. We will construct such strong propensities below. But I think that we also employ weak propensities in our reasoning and we make use of the fact that (1.34) holds. It was remarked in Chapter 1, where nomic probability was introduced for the first time, that we are sometimes more interested in actual relative frequencies than in nomic probabilities. This is particularly true in betting. If we know that the actual relative frequency is different from the nomic probability, we will calculate our odds on the basis of the relative frequency. That relative frequency is the same thing as the weak propensity. But weak propensities have an important property not shared by relative frequencies, and that is that they exist and are well defined even when the reference class is either empty or infinite.

1.4 Strong Indefinite Propensities

Although weak indefinite propensities are of some use, we are generally more interested in a stronger kind of propensity for which (1.34) fails. Consider the coin of the preceding example again. Suppose it has been flipped only once, landing heads on that single occasion. Despite the fact that the frequency and hence weak propensity of a toss of this coin landing heads is 1, we may still regard it as a fair coin. This is to say that the propensity, in some stronger sense, of a toss landing heads is .5. Such propensities are not directly about the actual tosses. Instead, they are about the probability of a toss landing heads even if the actual tosses were quite different than they are. I suggest that the strong indefinite propensity can be identified with the nomic probability given that *there is a different toss*, that is, a toss different from the tosses there actually are. Where w is the actual world, this counterfactual hypothesis is the hypothesis that there is a toss that is not a toss at w. I propose that we define:

(1.35) If w is the actual world then $prob(F/G) =$

$$prob\Big(Fx \ / \ Gx \ \& \ \mathbf{C}_{(\exists x)(Gx \ but \ x \ is \ not \ G \ at \ w)}\Big).$$

The nearest worlds at which $\ulcorner(\exists x)(Gx$ *but* x *is not* G *at* $w)\urcorner$ obtains are nearest worlds w^* for which there is an x such that x is G at w^* but x is not G at w. The state of affairs

$$\mathbf{C}_{(\exists x)(Gx \ but \ x \ is \ not \ G \ at \ w)}$$

is the conjunction of all states of affairs obtaining at all such worlds w^*.

I suspect that when we talk about indefinite propensities we ordinarily have strong indefinite propensities in mind, so I will henceforth drop the adjective 'strong'. Indefinite propensities satisfy the following analogues of $(1.30)-(1.33)$:

(1.36) If $(G \Rightarrow F)$ then $prob(F/G) = 1$.

(1.37) If $(G \Leftrightarrow H)$ then $prob(F/G) = prob(F/H)$.

(1.38) If $\Diamond H$ and $[H \Rightarrow {\sim}(F\&G)]$ then $prob(F{\vee}G/H) = prob(F/H) + prob(G/H)$.

(1.39) If $[(F\&H) \Rightarrow G]$ then $prob(F/H) \leq prob(G/H)$.

Again, the multiplicative axiom fails.

To recapitulate, it is possible to make precise sense of a number of different kinds of propensities by defining them in terms of nomic probability. Propensities need no longer be regarded as creatures of darkness. They are on precisely the same footing as the physical/epistemic definite probabilities evaluated in classical direct inference. It is the theory of nonclassical direct inference that makes it possible to analyze propensities in this way, because nonclassical direct inference enables us to understand how the resulting propensities can be evaluated. Without nonclassical direct inference we could still give the definitions, but the resulting probabilities would be as mysterious as ever because we would have no way to discover their values.

2. Joint Subset Defeaters

Direct inference is of considerable practical importance. It underlies any use of probability in decision theory, and perhaps in other arenas as well. The theory of direct inference presented in this book is based upon relatively simple principles, but its application is often difficult. The source of the difficulty is twofold. On the one hand, the rich mathematical structure of probabilities introduces complex interrelations that provide sometimes surprising sources of defeat for the prima facie reasons upon which direct inference is based. On the other hand, the epistemological framework governing the way in which prima facie reasons and defeaters interact is complicated in ways we have not yet explored, and that too introduces unexpected but very important sources of defeat. The present section is devoted to clarifying some of these complexities and putting the reader in a position to make more practical use of the theory of direct inference.

Thus far the theory of direct inference consists of two rules — a rule of direct inference and a rule of subset defeat. But it appears that a satisfactory theory of direct inference requires more defeaters than just subset defeaters.[6] Paradoxically, the more

---

[6] I attribute this discovery to Kyburg [1974]. Kyburg does not agree with

defeaters we have, the more direct inferences we license. This is because we rarely lack reasons for direct inference. The problem is almost always that we have too many reasons. They support conflicting conclusions and hence enter into collective defeat with one another. By adding defeaters, we defeat some of the conflicting reasons, allowing the others to avoid collective defeat and hence support warranted direct inferences. In this section and the next I describe three additional kinds of defeaters. In each case, I defend the description of the defeater by appealing to intuitive examples of direct inference. Then I go on to show that the defeaters can be derived from the rest of the theory.

Subset defeaters defeat a direct inference by showing that an appeal to more complete information leads to a conflicting direct inference. As such, subset defeaters constitute a partial articulation of the traditional "total evidence" requirement. But they can be generalized. There are other ways in which the appeal to more complete evidence can lead to conflicting direct inferences. Let c be a counter in a shipment of oddly shaped counters 50% of which are red. If we want to know the probability that c is red, and we do not know anything else about c, we will judge that $\text{PROB}(Rc) = .5$. But suppose we also know that c was in a particular crate in the shipment, and 40% of the counters in the crate were both red and square. We also know that c was subsequently used in a particular game of chance (where all the counters were from the shipment containing c), and that 30% of the counters used in that game were red but not square. Then we have prima facie reasons for concluding that $\text{PROB}(Rc\&Sc) = .4$ and $\text{PROB}(Rc\&{\sim}Sc) = .3$, and these jointly entail that $\text{PROB}(Rc) = .7$, which conflicts with the original direct inference. The latter two direct inferences were both based upon more specific information about c than was the first direct inference, so it seems intuitively reasonable that they should take precedence and we should conclude that $\text{PROB}(Rc) = .3$.

me regarding what additional defeaters are required, but he appears to have been the first person to observe that additional defeaters of some kind are required, and it was by reflecting upon his theory of direct inference that I was led to the theory of nomic probability.

Consider another example. Syndrome X is a genetic abnormality linked to heart disease. Suppose we know that the probability of a person having both syndrome X and heart disease is .005. But the probability of a person of Slavic descent having syndrome X is .1, and the probability of a male with syndrome X having heart disease is .3. Boris is a male of Slavic descent. We would judge that the probability of his having both syndrome X and heart disease is .03, not .005. This is because, although we have a reason for judging that PROB($Xb\&Hb$) = .005, we have reasons based upon more specific information for judging that PROB((Xb) = .1 and PROB(Hb/Xb) = .3, and these jointly entail that PROB($Xb\&Hb$) = .03.

These examples illustrate what I call *joint subset defeaters*. They are like subset defeaters in defeating a direct inference by appealing to more specific information, but they are more general than subset defeaters in that they proceed by combining several different direct inferences each based upon more specific information to defeat a single direct inference based on less specific information. Rather than describe these as a special kind of defeater, I will show that they arise from the application of the principle of joint defeat (principle (3.17) of Chapter 8) to ordinary subset defeaters.

In the first example we had prob(R/U) = .5, prob($R\&S/C$) = .4, prob($R\&{\sim}S/G$) = .3, $C \leqslant U$, $G \leqslant U$, and W($Cc\&Gc$). We want to conclude that PROB($Rc\&Sc$) = .4 and PROB($Rc\&{\sim}Sc$) = .3, and hence PROB(Rc) = .3. In order to draw this conclusion, the conflicting direct inference from $\ulcorner$prob(R/U) = .5$\urcorner$ must somehow be defeated. This comes about through joint defeat. As prob(R/U) $\neq$ prob($R\&S/C$)+prob($R\&{\sim}S/G$), one of the following is false:

(2.1)   prob($R\&S/C$) = prob($R\&S/U$);

(2.2)   prob($R\&{\sim}S/G$) = prob($R\&{\sim}S/U$).

Consider:

(2.3)   prob($R\&{\sim}S/U$) = prob($R\&{\sim}S/C$);

(2.4) $\text{prob}(R/U) = \text{prob}(R/C)$;

(2.5) $\text{prob}(R\&S/U) = \text{prob}(R\&S/G)$;

(2.6) $\text{prob}(R/U) = \text{prob}(R/G)$.

If (2.3)–(2.6) were all true then (2.1) and (2.2) would both be true, so it follows that at least one of (2.3)–(2.6) is false. But this means that there is a true subset defeater for the direct inference to at least one of the following:

(2.7) $\text{PROB}(Rc\&\sim Sc) = \text{prob}(R\&\sim S/U)$;

(2.8) $\text{PROB}(Rc) = \text{prob}(R/U) = .5$;

(2.9) $\text{PROB}(Rc\&Sc) = \text{prob}(R\&S/U)$.

The principle of joint defeat then implies that all three of these direct inferences are defeated. In particular, (2.8) is defeated. Thus there is nothing to conflict with the direct inferences to the conclusions that $\text{PROB}((Rc\&Sc) = .4$ and $\text{PROB}(Rc\&\sim Sc) = .3$, and so we can conclude, in accordance with our initial intuitions, that $\text{PROB}(Rc) = .3$.

The second example is handled similarly. We have $\text{prob}(X\&H/P) = .005$, $\text{prob}(X/P\&S) = .1$, $\text{prob}(H/X\&M\&P) = .3$, and $W(Pb\&Mb\&Sb)$. We want to conclude that $\text{PROB}(Xb) = .1$ and $\text{PROB}(Hb/Xb) = .3$, and hence $\text{PROB}(Xb\&Hb) = .03$. In order to draw this conclusion, the conflicting direct inference from $\ulcorner\text{prob}(X\&H/P) = .005\urcorner$ must be defeated. This results very simply from joint defeat. We know that one of the following must be false:

(2.10) $\text{prob}(H/X\&P\&S) = \text{prob}(H/X\&P)$;

(2.11) $\text{prob}(X\&H/P\&S) = \text{prob}(X\&H/P)$;

(2.12) $\text{prob}(H/X\&P\&M) = \text{prob}(H/X\&P)$;

(2.13) $\text{prob}(X\&H/P\&M) = \text{prob}(X\&H/P)$.

Thus we know that there is a true subset defeater for the direct inference to at least one of the following conclusions:

(2.14) PROB(Hb/Xb) = prob($H/X\&P$);

(2.15) PROB($Xb\&Hb$) = prob($X\&H/P$) = .005.

Thus by the principle of joint defeat they are both defeated. In particular, (2.15) is defeated, and that is all we need to allow us to conclude that PROB($Xb\&Hb$) = .03.

The upshot of all this is that there are such things as joint subset defeaters in direct inference, but they need not be posited as primitive defeaters. Instead, they follow from the principle of joint defeat.

3. Domination Defeaters

3.1 Description of Domination Defeaters

The defeasible inferences constituting direct inference result from applications of ($A1$). Subproperty defeaters for ($A1$) generate subset defeaters for direct inference. It turns out that domination defeaters for ($A1$) also generate analogous domination defeaters for direct inference. The need for domination defeaters for direct inference is illustrated by the following example. Typically, in a statistical investigation of the cause of some kind of event (for instance, a person's getting cancer), scientists discover many factors to be irrelevant. For example, the color of one's hair is irrelevant to the probability of getting lung cancer. More precisely, the incidence of lung cancer among residents of the United States is the same as the incidence of lung cancer among redheaded residents of the United States: prob(C/U) = prob(C/R) = .1. It has also been found that the incidence of lung cancer among residents of the United States who smoke is much greater: prob(C/S) = .3. If we know that Charles is a redheaded resident of the United States who smokes, we will estimate his chances of getting lung cancer to be .3 rather than .1. But this cannot be justified in terms of the rules of direct inference endorsed so far. We have two different prima facie reasons for concluding that PROB(Cc) = .1:

(3.1)   prob(C/R) = .1 & $\mathbf{W}(Rc)$

and

(3.2)   prob(C/U) = .1 & $\mathbf{W}(Uc)$;

and one prima facie reason for concluding that PROB(Cc) = .3:

(3.3)   prob(C/S) = .3 & $\mathbf{W}(Sc)$.

Because $S \preccurlyeq U$, (3.3) provides a subset defeater for (3.2), but we do not have a subset defeater for (3.1). Accordingly, (3.1) and (3.3) support conflicting direct inferences and hence rebut one another, leaving us with no undefeated direct inference.

We should be able to make a direct inference to the conclusion that PROB(Cc) = .3, ignoring (3.1). What justifies this intuitively is that we have found that being redheaded is irrelevant to the probability of getting lung cancer. Because it is irrelevant, we regard (3.1) as true only *because* (3.2) is true, and hence take any defeater for a direct inference from (3.2) to be a defeater for a direct inference from (3.1) as well. This can be captured by precluding a direct inference from (3.1) and requiring any direct inference to be made from (3.2):

(3.4)   If C is projectible with respect to U then $\ulcorner$prob(C/R) = prob(C/U) & $R \preccurlyeq U\urcorner$ is an undercutting defeater for the direct inference from $\ulcorner$prob(C/R) = r & $\mathbf{W}(Rc)\urcorner$ to $\ulcorner$PROB(Cc)$\urcorner$.

Defeaters of the form of (3.4) are analogous to domination defeaters for the statistical syllogism, so I will call them 'domination defeaters' as well.

Domination defeaters are as important for classical direct inference as they are for the statistical syllogism. It was remarked in Chapter 9 that the preceding sort of situation is typical of rich scientific contexts in which we have a lot of statistical information regarding not only relevant factors but also irrelevant ones. Without domination defeaters, we would be precluded from making classical direct inferences in such cases.

We can derive domination defeaters for nonclassical direct inference from domination defeaters for classical direct inference in the same way we derived subset defeaters for nonclassical direct inference from subset defeaters for classical direct inference. Suppose we have reference properties arranged as in Figure 1, where F is projectible with respect to G and $r^* \neq r$. By the principle of nonclassical direct inference:

(3.5) $\ulcorner H \preceq G$ & $\mathrm{prob}(F/G) = r \urcorner$ is a prima facie reason for $\ulcorner \mathrm{prob}(F/H) = r \urcorner$.

But this inference should be defeated. This is described by the following *domination defeater* for this nonclassical direct inference:

(3.6)   If F is projectible with respect to U then $\ulcorner G \preceq U$ & $\mathrm{prob}(F/G) = \mathrm{prob}(F/U) \urcorner$ is an undercutting defeater for (3.5).

Conversely, domination defeaters for classical direct inference can be derived from domination defeaters for nonclassical direct inference by way of our reduction of classical direct inference to nonclassical direct inference.

The next thing to be established is that principle (3.6) is a theorem following from the reduction of direct inference to (*A1*). The derivation given in Chapter 4 derived (3.5) from the following instance of (*A1*):

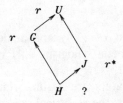

Figure 1. Domination defeater.

(3.7) $\ulcorner H^0 \ll G$ & $\text{prob}\big(\text{prob}(F/X) \approx_\delta r \,/\, X \ll G\big) = 1\urcorner$ is a
        prima facie reason for $\ulcorner \text{prob}(F/H^0) \approx_\delta r \urcorner$.

Given that $\text{prob}(F/G) = \text{prob}(F/U)$, it follows that

$$\text{prob}\big(\text{prob}(F/X) \approx_\delta r^* \,/\, X \ll G\big)$$
$$= \text{prob}\big(\text{prob}(F/X) \approx_\delta r^* \,/\, X \ll U\big).$$

As $G \lesssim U$, it follows that the property expressed by $\ulcorner X \ll G \urcorner$ is a subproperty of the property expressed by $\ulcorner X \ll U \urcorner$. Consequently, by (DOM) we have a domination defeater for (3.7). Thus domination defeaters for direct inference can be derived from domination defeaters for the statistical syllogism.

Domination defeaters are extremely important for nonclassical direct inference. Without domination defeaters, various kinds of logical difficulties would arise that would make nonclassical direct inference completely unworkable. For example, suppose we know that $\text{prob}(A/B\&C) = r$ and $\text{prob}(A/B) = r^*$, where $r \neq r^*$. If we want to know the value of $\text{prob}(A/B\&C\&D)$, and those are the only relevant probabilities whose values we know, then it seems clear that we should conclude, prima facie, that $\text{prob}(A/B\&C\&D) = r$. It might seem that this is guaranteed by the operation of subset defeaters. $\ulcorner \text{prob}(A/B\&C) = r \urcorner$ undercuts the direct inference from $\ulcorner \text{prob}(A/B) = r^* \urcorner$ to $\ulcorner \text{prob}(A/B\&C\&D) = r^* \urcorner$, but nothing undercuts the direct inference from $\ulcorner \text{prob}(A/B\&C) = r \urcorner$ to $\ulcorner \text{prob}(A/B\&C\&D) = r \urcorner$, so the latter conclusion is warranted. But there is a surprising twist that makes this reasoning incorrect. If we take the reference property $B\&D$ into account as well, we have four reference properties and they are related as in Figure 2. Thus we have the prima facie direct inferences diagramed in Figure 3. None of these inferences is undercut by a subset defeater. Hence (1) and (3) will rebut one another and both will undergo collective defeat. But this seems clearly wrong. Intuitively, (1) should be undefeated and (3) defeated. That is precisely what a domination defeater ensures.

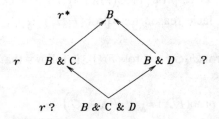

Figure 2. Domination defeater for nonclassical direct inference.

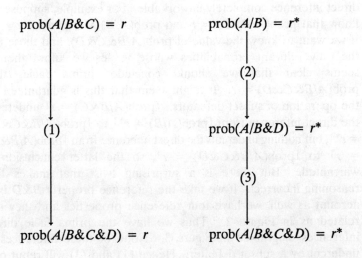

Figure 3. End run in nonclassical direct inference.

Given (2), we have $\lceil prob(A/B\&D) = prob(A/B)\rceil$, so we have a domination defeater for (3).

Domination defeaters provide the solution to another problem. Consider any intuitively legitimate nonclassical direct inference. For example, let B be the property of being a smoker, let C be the property of being a redheaded smoker, and let A be the property of contracting lung cancer. Then from

(3.8) $C \preceq B \ \& \ prob(A/B) = r$

we can infer

(3.9) $prob(A/C) = r$.

Now choose any property D such that

(3.10) $\diamond_p \exists (C\&D) \ \& \ prob(A/B) = r^*$

where $r^* \neq r$. There will always be such a D. In this example, we might let D be the property of being an asbestos worker. If all else failed we could let D be $\lceil E \ \& \ freq[A/E] = r^*\rceil$ for some property E. The entailment relations between the properties B, C, D, and $C\&D$ can be diagrammed as in Figure 4. By (DI), we have prima facie reasons for concluding that $prob(A/B) = prob(A/C)$, $prob(A/C) = prob(A/C\&D)$, and $prob(A/C\&D) = prob(A/D)$. But these conclusions are collectively inconsistent with the fact that $prob(A/B) \neq prob(A/D)$. Within the theory of nonclassical direct inference comprising (DI) and (SD), we have no reason to favor any of these nonclassical direct inferences over any of the others, so it follows that they are all collectively defeated. In particular, the inference from (3.8) to (3.9) is defeated. But by hypothesis, the inference from (3.8) to (3.9) is a legitimate nonclassical direct inference.

It seems intuitively clear that we would make the direct inference from (3.8) to (3.9), concluding that the probability of a redheaded smoker contracting lung cancer is the same as the probability of a smoker contracting lung cancer, but we would resist either of the other direct inferences, refraining from

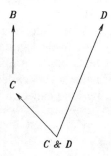

Figure 4. A problem for nonclassical direct inference.

concluding that the probability of lung cancer in a redheaded smoker who is also an asbestos worker is the same as either the probability in redheaded smokers in general or the probability in asbestos workers in general. This is explained by observing that in the following argument:

(1)   prob$(A/B) = r$
(2)   prob$(A/C) = r$
(3)   prob$(A/C\&D) = r$

lines (1) and (2) provide a domination defeater for the inference from (2) to (3). Accordingly, the argument is self-defeating. Thus we are precluded from inferring the conclusion ⌜prob$(A/C\&D) = r$⌝ from (2). We *can* infer this directly from (1), and that *does* enter into collective defeat with an inference from (3.10) to the conclusion that prob$(A/C\&D) = r^*$. Thus we are left with the undefeated conclusion that prob$(A/C) = r$, but we can conclude nothing about prob$(A/C\&D)$. This accords with our intuitions.

3.2 Defeaters for Domination Defeaters

Domination defeaters for $(A1)$ are defeasible, and that generates defeaters for domination defeaters in direct inference. For example, consider Albert, the smoking and jogging octogenarian. Suppose we know that although the probability of a smoker getting lung cancer is much higher than the probability of a person in general getting lung cancer, the probability of a smoking jogger getting lung cancer is the same as the probability of a person in general getting lung cancer: prob($C/S\&J$) = prob(C/P) = .1. But we also learn that the probability of an octogenarian getting lung cancer is twice as high as normal: prob(C/O) = .2. Under these circumstances, we would refrain from making any direct inference about the probability of Albert's getting lung cancer. This is not explicable in terms of the rules of direct inference adopted so far. We have four prima facie reasons for direct inferences:

(3.11) prob(C/P) = .1 & $\mathbf{W}(Pa)$;

(3.12) prob(C/S) = .3 & $\mathbf{W}(Sa)$;

(3.13) prob($C/S\&J$) = .1 & $\mathbf{W}(Sa\&Ja)$;

(3.14) prob(C/O) = .2 & $\mathbf{W}(Oa)$.

Noting the probabilities beside them, these reference properties are related as in Figure 5. What *should* happen is that we are unable to draw any conclusion because (3.13) and (3.14) collectively defeat one another. However, because prob($C/S\&J$) = prob(C/P), we have a domination defeater for (3.13), which

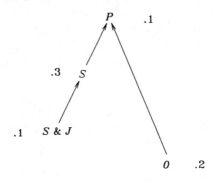

Figure 5. Defeated domination defeater.

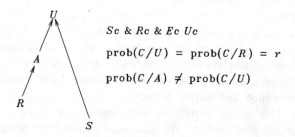

Figure 6. Defeater for domination defeaters.

removes it from competition with (3.14). Thus (3.14) is undefeated. But that is intuitively wrong.

What is happening here is precisely the same thing that led
to defeaters for domination defeaters for statistical syllogism. The
domination defeater works on the presumption that if $(S\&J) \leq P$
and $\text{prob}(C/S\&J) = \text{prob}(C/P)$ then $S\&J$ results from strengthening P in ways that are irrelevant to C. The sense of irrelevance
involved requires that if $(S\&J) \leq A \leq P$ then we also have $\text{prob}(C/A)$
$= \text{prob}(C/P)$. This fails in the present example if we let $A = S$.
Hence S and J are not irrelevant strengthenings of P. They are
relevant strengthenings that just happen to cancel out. Accordingly, we have no reason to think that they will be irrelevant or
that they will cancel out for octogenarians, and so (3.14) should
not undercut the direct inference from (3.13). In general, in a
situation having the form diagramed in Figure 6, the inference to
$\ulcorner \text{PROB}(Cc) = r \urcorner$ should not be defeated. This can be captured by
adopting the following defeater for domination defeaters:

(3.15) If C is projectible with respect to A then $\ulcorner R \leq A \leq U \&$
         $\text{prob}(C/A) \neq \text{prob}(C/U) \urcorner$ defeats the defeater in (3.4).

Domination defeaters for nonclassical direct inference must
be subject to analogous defeaters:

(3.16) If F is projectible with respect to L then $\ulcorner G \preccurlyeq L \preccurlyeq U$ & $\mathrm{prob}(F/L) \neq \mathrm{prob}(F/U)\urcorner$ is a defeater for the defeater in (3.6).

Principle (3.15) is derivable from (3.16) in the standard way. In turn, (3.16) can be derived from $(DOM\text{-}D)$ and the reduction of nonclassical direct inference to $(A1)$. The derivation is an immediate consequence of the fact that if $G \preccurlyeq L \preccurlyeq U$ then the property expressed by $\ulcorner X \ll G \urcorner$ is a subproperty of that expressed by $\ulcorner X \ll L \urcorner$, which is a subproperty of that expressed by $\ulcorner X \ll U \urcorner$, and the fact that if $\mathrm{prob}(F/L) = r^* \neq r$, then for every $\delta > 0$,

$$\mathrm{prob}\Big(\mathrm{prob}(F/X) \approx_\delta r \ / \ X \ll L\Big) = 0.$$

3.3 Extended Domination Defeaters

Just as in the case of domination defeaters for $(A1)$, the introduction of domination defeaters for direct inference requires the addition of extended domination defeaters. The same logical tricks that led to end runs around the domination defeaters for $(A1)$ produce end runs around domination defeaters for direct inference. For instance, consider the example above in which (3.1) is defeated by a domination defeater. We can generate another prima facie reason that is not defeated:

(3.17) $\mathrm{prob}(R\&C/R) = .1$ & $\mathbf{W}(Rc)$ & $\mathbf{W}(Cc \equiv (Rc\&Cc))$.

We do not have a domination defeater for (3.17) because we have no reason to expect that $\mathrm{prob}(R\&C/R) = \mathrm{prob}(R\&C/U)$. But this is just a logical trick. The statistical information embodied in (3.17) is equivalent to that embodied in (3.1) because it is necessarily true that $\mathrm{prob}(R\&C/R) = \mathrm{prob}(C/R)$ and that if $\mathbf{W}(Rc)$ then $\mathbf{W}(Cc \equiv (Rc\&Cc))$. As before, the same sort of logical trick can be performed using physical necessities rather than logical necessities.

The same problem arises for nonclassical direct inference.

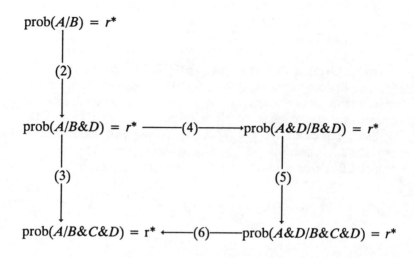

Figure 7. End run for nonclassical direct inference.

For example, in Figure 3, although (3) is defeated by a domination defeater, we can circumvent the domination defeater by reasoning as in Figure 7. Inferences (4) and (6) are entailments and hence nondefeasible, and although inference (5) is only a prima facie direct inference we have no defeater for it. Thus although (3) is defeated, it can be replaced by an end run for which we have no domination defeater. There must be some general kind of defeater that blocks (5). The nature of this defeater can be elicited by noting that in Figure 7, the end run for domination defeaters for (*DI*) corresponds to the end run for domination defeaters for (*A1*) diagramed in Figure 8.

Generalizing this, we obtain the following characterization of these defeaters, which follows from the characterization of extended domination defeaters for (*A1*):

(3.18) Consider the following:
(a) $H \preccurlyeq G$ & prob$(F/G) = r$;
(b) prob$(F/H) = r$;
and
(a*) $H^* \preccurlyeq G^*$ & prob$(F^*/G^*) = r$;
(b*) prob$(F^*/H^*) = r$;
If P is a domination defeater for (a*) as a prima facie reason for (b*), then $\ulcorner P$ & $(G \Leftrightarrow G^*)$ & $[G \Rightarrow (F \Leftrightarrow F^*)]$ & $(H \Leftrightarrow H^*)\urcorner$ is a defeater for (a) as a prima facie reason for (b).

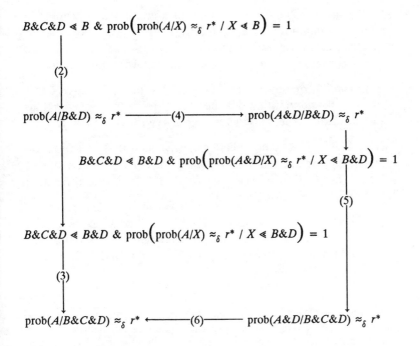

Figure 8. End run.

I will call such defeaters *extended domination defeaters for (DI)*.[7]
Given the reduction of classical direct inference to nonclassical
direct inference, (3.18) entails the existence of analogous extended
domination defeaters for classical direct inference:

(3.19) Consider the following:
    (a) $W(Gc)$ & $\mathrm{prob}(F/G) = r$;
    (b) $\mathsf{PROB}(Fc) = r$;
    and
    (a*) $W(G^*c)$ & $\mathrm{prob}(F^*/G^*) = r$;
    (b*) $\mathsf{PROB}(F^*c) = r$;
    If P is a domination defeater for (a*) as a prima facie
    reason for (b*), then $\ulcorner P$ & $(G \Leftrightarrow G^*)$ & $[G \Rightarrow (F \Leftrightarrow F^*)]\urcorner$ is a defeater for (a) as a prima facie reason for (b).

We saw in Chapter 9 that there is a another way generating
extended domination defeaters for (*A1*). This proceeds in terms
of (*IND*), and the same construction works for (*DI*), as in Figure
9. To accommodate such examples, the following must hold:

(3.20) $\ulcorner \mathrm{prob}(Aby/Bby) = \mathrm{prob}(Axy/Bxy) \neq \mathrm{prob}(Axy/Bxy\&Cxy)\urcorner$
    is an undercutting defeater for $\ulcorner \mathrm{prob}(Aby/Bby) = r^*\urcorner$ as
    a prima facie reason for $\ulcorner \mathrm{prob}(Aby/Bby\&Cby) = r^*\urcorner$.

Such defeaters might be called *projection defeaters*.[8] Projection
defeaters can be derived from the analogous extended domination
defeaters for (*A1*). In turn, they imply the following projection
defeaters for classical direct inference:

(3.21) $\ulcorner WCbc$ & $\mathrm{prob}(Aby/Bby) = \mathrm{prob}(Axy/Bxy) \neq \mathrm{prob}(Axy/Bxy\&Cxy)\urcorner$ is an undercutting defeater for
    $\ulcorner WBbc$ & $\mathrm{prob}(Aby/Bby) = r^*\urcorner$ as a prima facie reason
    for $\ulcorner \mathsf{PROB}(Abc) = r^*\urcorner$.

---

[7] In Pollock [1984], I gave a different justification for extended domination
defeaters, basing them on a general epistemological principle that I called 'the
no end runs principle'. I have since become convinced that that principle is
false, thanks in part to a counterexample due to Ian Pratt.

[8] Projection defeaters are a generalization of the *product defeaters* of
Kyburg [1974].

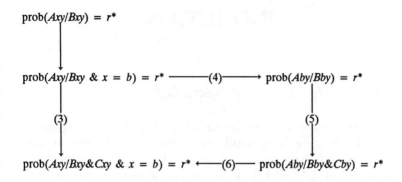

$\text{prob}(Axy/Bxy) = r^*$

$\text{prob}(Axy/Bxy \ \& \ x = b) = r^* \longrightarrow (4) \longrightarrow \text{prob}(Aby/Bby) = r^*$

(3)   (5)

$\text{prob}(Axy/Bxy\&Cxy \ \& \ x = b) = r^* \longleftarrow (6) \longrightarrow \text{prob}(Aby/Bby\&Cby) = r^*$

Figure 9. End run generated by (*IND*).

4. Summary

1. Several different kinds of propensities can be defined in terms of nomic probability on analogy to the way mixed physical/epistemic probabilities were defined, and in each case a theory of direct inference results that enables us to establish the numerical values of the propensities in actual cases.
2. The foundations for direct inference established in Chapter 4 lead us to the derivation of a number of additional defeaters for direct inference. These include joint subset defeater, domination defeaters, extended domination defeaters, and projection defeaters. These are important for the practical use of the theory of direct inference, because without them most inferences based upon the principles of direct inference would undergo collective defeat.

CHAPTER 11
ADVANCED TOPICS: INDUCTION

1. Projectibility

It is well known, since Goodman [1955], that principles of induction require a projectibility constraint. On the present account, such a constraint is inherited from the projectibility constraint on $(A1)-(A3)$. It remains to be shown, however, that this derived constraint is the intuitively correct constraint. Let us define:

(1.1)   A concept B (or the corresponding property) is *inductively projectible with respect to* a concept A (or the corresponding property) iff ⌜X is a set of A's, and all the members of X are also B's⌝ is a prima facie reason for ⌜$A \Rightarrow B$⌝, and this prima facie reason would not be defeated by learning that there are non-B's. A nomic generalization is projectible iff its consequent is inductively projectible with respect to its antecedent.[1]

What is needed is an argument to show that inductive projectibility is the same thing as projectibility. To make this plausible, I will argue that inductive projectibility has the same closure properties as those defended for projectibility in Chapter 3.

---

[1] The reason for the requirement that the reason would not be defeated by the observation of non-B's is this: If B is inductively projectible then observation of B's gives us a prima facie reason for believing that anything would be a B, and hence for any concept A we have a reason for thinking that any A would be a B. We do not want this to make B inductively projectible with respect to every concept. We avoid this untoward consequence by observing that this sort of reason for believing that any A would be a B would be defeated by the observation of non-B's (regardless of whether they are A's).

Goodman introduced inductive projectibility with examples of nonprojectible concepts like *grue* and *bleen*, and the impression has remained that only a few peculiar concepts fail to be inductively projectible. That, however, is a mistake. It is not difficult to show that most concepts fail to be inductively projectible, inductive projectibility being the exception rather than the rule. This results from the fact that, just like projectibility, the set of inductively projectible concepts is not closed under most logical operations. In particular, I will argue below that although inductive projectibility is closed under conjunction, it is not closed under either disjunction or negation. That is, negations or disjunctions of inductively projectible concepts are not automatically inductively projectible.

Just as for projectibility, we can argue fairly conclusively that inductive projectibility is closed under conjunction. More precisely, the following two principles hold:

(1.2)   If A and B are inductively projectible with respect to C, then $(A\&B)$ is inductively projectible with respect to C.

(1.3)   If A is inductively projectible with respect to both B and C, then A is inductively projectible with respect to $(B\&C)$.

The proof of (1.2) is the same as the proof of (2.3) in Chapter 3. Suppose A and B are inductively projectible with respect to C, and we have a sample of C's all of which are both A and B. This gives us prima facie reasons for both $\ulcorner C \Rightarrow A \urcorner$ and $\ulcorner C \Rightarrow B \urcorner$, and these entail $\ulcorner C \Rightarrow (A\&B) \urcorner$. Thus we have a prima facie reason for believing the latter, and hence $(A\&B)$ is inductively projectible with respect to C.

As in the case of (2.4) of Chapter 3, we cannot establish (1.3) quite so conclusively, but there is a fairly compelling reason for thinking that (1.3) is true. Suppose A is inductively projectible with respect to B. Then a scientist may set about trying to confirm $\ulcorner B \Rightarrow A \urcorner$. If his investigations turn up a counterexample to this generalization, he does not usually give up the investigation. Rather, he tries to confirm a less adventurous generalization. He restricts the antecedent and tries to confirm a generalization of the form $\ulcorner (B\&C) \Rightarrow A \urcorner$. If this is to be legitimate, A must be inductively projectible with respect to $(B\&C)$.

Like projectibility, inductive projectibility fails to be closed under disjunction. That is, neither of the following principles is true:

(1.4) If C is inductively projectible with respect to both A and B, then C is inductively projectible with respect to $(A \lor B)$.

(1.5) If A and B are both inductively projectible with respect to C, then $(A \lor B)$ is inductively projectible with respect to C.[2]

It is quite easy to see that (1.4) cannot be true in general. For example, *having whiskers* is inductively projectible with respect to both *cat* and *elephant*. That is, we could confirm that all cats have whiskers by observing whiskered cats, and we could confirm that all elephants have whiskers by observing whiskered elephants. But we cannot automatically confirm that anything that is either a cat or an elephant would have whiskers by observing a sample of things having the property of being either a cat or an elephant and seeing that all of its members are whiskered. A sample of things having that disjunctive property could consist entirely of cats, but observing a sample consisting of whiskered cats can give us no reason at all for thinking that any elephant would be whiskered, and the latter is entailed by the proposition that anything that is either a cat or an elephant would be whiskered. It is, of course, possible to confirm in a more complicated way that anything that is either a cat or an elephant would be whiskered, but this cannot be done just by observing a sample of cats-or-elephants and seeing that all of its members are whiskered. Instead, we must observe separate samples of cats and elephants, confirm the separate generalizations that any cat would be whiskered and that any elephant would be whiskered, and then

---

[2] I originally pointed out the failure of these principles in Pollock [1972] and [1974]. The failure of (1.4) has been observed by several authors, but as far as I know, the failure of (1.5) has been entirely overlooked.

infer deductively that anything that is either a cat or an elephant would be whiskered.

Turning to (1.5), it seems clear that *has a liver* and *has a kidney* are both inductively projectible with respect to *is an animal having B-endorphins in its blood*. Suppose now that we set about trying to confirm that any animal with B-endorphins in its blood has either a liver or a kidney. We observe a sample of animals with B-endorphins in their blood, and we note that many of them have livers, and many do not. We also observe that many have kidneys and many do not. But we never check whether any of those that fail to have livers have kidneys, or whether any of those that fail to have kidneys have livers. Obviously, these observations give us no reason at all to believe that every animal with B-endorphins in its blood has either a liver or a kidney. But notice that we have observed a sample of animals with B-endorphins in their blood that have either livers or kidneys (i.e., the sample consists of all of those that we ascertained either to have livers or to have kidneys), and we have not observed any that have neither. The latter is because when we found an animal in our sample to lack a kidney, we did not go on to check whether it lacked a liver, and vice versa. Thus if *has a liver or a kidney* were inductively projectible with respect to *is an animal with B-endorphins in its blood*, this sample would have to confirm that any animal with B-endorphins in its blood has either a liver or a kidney. As it does not, (1.5) is false. Note that from the fact that inductive projectibility is closed under conjunction but not under disjunction, it follows that inductive projectibility is not closed under negation.

This example provides an intuitive illustration of the failure of (1.5), but it does not explain the reason for this failure. The reason is actually a simple one. If we were allowed to make free use of disjunctions in induction then every sample would give us prima facie reasons for mutually inconsistent generalizations, with the result that we would not be justified in believing either of the generalizations on the basis of the sample. This would have the consequence that induction could never give us an undefeated reason for believing a generalization. This results as follows. Although inductive projectibility is not closed under negation, it does seem to be the case that the negations of logically simple inductively projectible concepts are inductively projectible (e.g., both *red* and *non-red* are inductively projectible). Suppose both B and $\sim B$ are inductively projectible with respect to A,[3] and we are attempting to confirm $\ulcorner A \Rightarrow B \urcorner$ by appealing to a sample X of A's all of which are B's. Given any finite sample X there will be properties C possessed by all the members of X and such that (1) C and $\sim C$ are both inductively projectible with respect to A and (2) we have reason to believe that there are A's that are not C's. An example of such a C might be *resident of the twentieth century* or *object not witnessed by Bertrand Russell*. As $(\exists x)(Ax \ \& \sim Cx)$, it follows that $\Diamond_p(\exists x)(Ax \ \& \ \sim Cx)$. Because inductive projectibility is closed under conjunction, both $(B\&C)$ and $(\sim B\&\sim C)$ are inductively projectible with respect to A. If inductive projectibility were also closed under disjunction it would follow that $[(B\&C) \vee (\sim B\&\sim C)]$ is also inductively projectible with respect to A. X is a sample of A's all of which have the latter disjunctive property, so observation of X would give us an equally good prima facie reason for believing

(1.6) $A \Rightarrow [(B\&C) \vee (\sim B\&\sim C)]$.

(1.6) is not counterlegal, so $\Box_p(\forall x)\{Ax \supset [(Bx \ \& \ Cx) \vee (\sim Bx \ \& \sim Cx)]\}$. Given that $\Diamond_p(\exists x)(Ax \ \& \sim Cx)$, it follows that $\sim\Box_p(\forall x)(Ax \supset Bx)$. We know that $\Diamond_p(\exists x)Ax$ (our sample consists of such objects), so it follows from (3.3) of Chapter 2 that $\sim(A \Rightarrow B)$. Consequently, our sample gives us equally good inductive reasons for affirming $\ulcorner A \Rightarrow B \urcorner$ and denying it, with the result that it does not justify the intuitively legitimate inductive inference. The only step of the argument that is doubtful is the assumption that $[(B\&C) \vee (\sim B\&\sim C)]$ is inductively projectible with respect to A, so it must be concluded that it is not. This argument is completely general, so what follows is that if inductive projectibility were

---

[3] The argument can also be made to work without the assumption that $\sim B$ is inductively projectible with respect to A if we assume instead that there is a concept D logically incompatible with B that is inductively projectible with respect to A, but I leave the details to the reader.

closed under disjunction then all inductive reasoning would be defeated.

Although generalizations involving disjunctions of inductively projectible concepts cannot usually be confirmed directly by enumerative induction, they can be confirmed indirectly by making use of entailments between nomic generalizations. For example, how, intuitively, would we confirm that any animal with B-endorphins in its blood has either a liver or a kidney? We might examine a number of such animals, look to see whether they have livers, and those that lack livers would be examined to see whether they have kidneys. If they do, that confirms the generalization. Similarly, we might examine a number of animals with B-endorphins in their blood to see whether they have kidneys, and then further examine those lacking kidneys to see whether they have livers. What we are doing here is obtaining direct inductive confirmation for the inductively projectible generalizations 'Any animal with B-endorphins in its blood but lacking a liver has a kidney' and 'Any animal with B-endorphins in its blood but lacking a kidney has a liver'. Because the non-counterlegal nomic generalizations $\ulcorner(B\&\sim K) \Rightarrow L\urcorner$ and $\ulcorner(B\&\sim L) \Rightarrow K\urcorner$ are both equivalent to $\ulcorner B \Rightarrow (K \lor L)\urcorner$, we are obtaining indirect confirmation for the latter.

It seems apparent that inductive projectibility behaves just like projectibility. Both are closed under conjunction, but fail to be closed under disjunction, negation, or other simple logical operations. In light of the derivation of principles of induction from the statistical syllogism, it is inescapable that projectibility and inductive projectibility are one and the same thing.

The objective logical nature of the preceding reasoning about projectibility deserves to be emphasized. Some philosophers feel that the concept of projectibility is irredeemably fuzzy, and that to build projectibility into an account of induction is to make the whole account suspect. It cannot be denied that to complete the account of induction we need a theory of projectibility, but the precise objective arguments we can give regarding projectibility show that there is nothing fuzzy or illegitimate about the concept. Its only failing is that it stands in need of philosophical analysis, but that is a failing it shares with almost all concepts.

Perhaps much of the resistance to projectibility is really resistance to Goodman's analysis in terms of entrenchment, so it should be emphasized that in talking about projectibility I do not mean to endorse that account. On the contrary, the possibility of the kind of objective arguments given above seems to rule out any analysis of projectibility in terms of something so ephemeral as entrenchment.

Before leaving the topic of projectibility, it is of interest to note that although the enumerative induction argument imposes the expected projectibility constraint on enumerative induction, the statistical induction argument imposes an additional projectibility constraint on statistical induction that, to the best of my knowledge, has been previously overlooked. Stage I of the statistical induction argument proceeds by applying nonclassical direct inference to the following probabilities:

(1.7) $\mathrm{prob}\Big(Ax_i \; / \; Ax_{i+1} \; \& \; \ldots \; \& \; Ax_r \; \& \; {\sim}Ax_{r+1} \; \& \; \ldots \; \& \; {\sim}Ax_n$

$\& \; x_1,\ldots,x_n \text{ are distinct} \; \& \; Bx_1 \; \& \; \ldots \; \& \; Bx_n \; \&$

$\mathrm{prob}(Ax/Bx) = p\Big)$

and

(1.8) $\mathrm{prob}\Big({\sim}Ax_i \; / \; {\sim}Ax_{r+1} \; \& \; \ldots \; \& \; {\sim}Ax_n \; \& \; x_1,\ldots,x_n \text{ are}$

$\text{distinct} \; \& \; Bx_1 \; \& \; \ldots \; \& \; Bx_n \; \& \; \mathrm{prob}(Ax/Bx) = p\Big).$

In order to apply direct inference to (1.8), it is not sufficient for A to be projectible with respect to B. ${\sim}A$ must also be projectible with respect to B. This requirement is easily illustrated by examples related to those demonstrating that projectibility is not closed under disjunction. Suppose A is the conjunction $C\&D$, where C and D are projectible with respect to B but $({\sim}C \vee {\sim}D)$ is not. Suppose we are testing samples of a substance for the properties C and D. We find some samples of the substance to have the property C and some to lack it, and we find some samples to have the property D and some to lack it. But when we find a sample to possess one of these properties, we never go on to test it for the other property. As a result, we know of many samples of the substance lacking the property A (i.e., all of those we know to lack one or the other of C and D), but we do not know of any possessing the property A. Thus taking X to consist of all samples of the substance regarding which we know whether they possess the property A, freq$[A/X] = 0$. But it would clearly be unreasonable for us to conclude that prob(A/B) is approximately 0.

It was observed in Chapter 4 that in enumerative induction, generalizations of the form $\ulcorner B \Rightarrow (C \lor D) \urcorner$ cannot be confirmed directly. Instead, we typically seek confirmation for $\ulcorner (B \& \sim C) \Rightarrow D \urcorner$ and $\ulcorner (B \& \sim D) \Rightarrow C \urcorner$, each of which is logically equivalent to $\ulcorner B \Rightarrow (C \lor D) \urcorner$. Similarly, prob$(C \& D/B)$ (or prob$(\sim C \lor \sim D/B)$) cannot be evaluated in the manner described in the preceding example. We must go on to examine samples having the property C and determine what proportion of them lack the property D; and we must examine samples having the property D and determine what proportion of them lack the property C. In this way we obtain evidence that allows us, by statistical induction, to evaluate prob(C/B), prob$(D/B \& C)$, prob(D/B), and prob$(C/B \& D)$. Then we can compute prob$(C \& D/B) = $ prob$(C/B) \cdot$ prob$(D/B \& C)$ $=$ prob$(D/B) \cdot$ prob$(C/B \& D)$.

The requirement that $\sim A$ be projectible with respect to B marks an important difference between statistical induction and enumerative induction. The reason it does not arise in enumerative induction is that in that case $r = n$ and hence (1.8) need not be evaluated by direct inference. As far as I know, the requirement of statistical induction that $\sim A$ be projectible with respect to B has previously gone unnoticed. Its prediction by the statistical induction argument constitutes further confirmation of the present account of statistical induction.

2. Subsamples

A satisfactory theory of statistical induction must be able to resolve the subsample problem. Suppose our sample is X, and upon observing that freq$[A/X] = \nu$, we wish to conclude that prob$(A/B) \approx \nu$ (where '$\approx$' means 'approximately equal'). By virtue of having observed the sample X, we have automatically observed other smaller samples that provide prima facie reasons for conflicting estimates of prob(A/B). For example, let X_0 be the set of all A's in X. Then freq$[A/X_0] = 1$, and this gives us a prima facie reason for concluding that prob$(A/B) \approx 1$. This and other inferences from subsamples of X must somehow be defeated. Intuitively, there should be a constraint on statistical induction requiring us to base our estimate of prob(A/B) on our total evidence, that is, on the largest sample to which we have access. But this constraint cannot be imposed by fiat, because the claim is that statistical induction can be derived from more general epistemic principles. This requires that those principles automatically generate some congenial way of dealing with the subsample problem. That has proven to be an apparently insuperable problem for some attempts to reconstruct statistical induction.[4]

The statistical induction argument generates a natural resolution of the subsample problem. Suppose the sample X consists of n B's r of which are A's, and consider a subsample Y of X consisting of k B's m of which are A's. Applying the statistical induction argument to the subsample, we first infer, as in (3.4) of Chapter 5:

$$(2.1) \quad \mathrm{prob}\Big(Ax_1 \ \& \dots \& \ Ax_m \ \& \ {\sim}Ax_{m+1} \ \& \dots \& \ {\sim}Ax_k \ / \ x_1, \dots, x_k$$

$$\text{are distinct} \ \& \ Bx_1 \ \& \dots \& \ Bx_k \ \& \ \mathrm{prob}(A/B) = p\Big)$$

$$= p^m(1\text{-}p)^{k\text{-}m}.$$

Then we compute:

---

[4] For example, in Kyburg's reconstruction of the Fiducial Argument (in Kyburg [1974]), this has come to be known as 'the Putnam subsample problem'. There is no apparent solution to the problem within Kyburg's [1974] theory, although many philosophers have worked hard at finding a solution. See Kyburg [1982]. Kyburg [1983] modifies his earlier theory in an attempt to handle this and other problems.

(2.2) $\text{prob}\Big(\text{freq}[A \ / \ \{x_1,\dots,x_n\}] = \frac{m}{k} \ / \ x_1,\dots,x_k$ are distinct &

   $Bx_1 \ \& \ \dots \ \& \ Bx_k \ \& \ \text{prob}(A/B) = p\Big)$

   $= k!p^m(1-p)^{k-m}m!(k-m)!$

and apply (*A5*). *However,* in the case of the subsample there are defeaters that block the application of (*A5*). The following frequency can be computed by familiar combinatorial mathematics:

(2.3) $\text{freq}\Big[Ax_1 \ \& \dots \& \ Ax_m \ \& \ {\sim}Ax_{m+1} \ \& \dots \& \ {\sim}Ax_k \ / \ x_1,\dots,x_k$
   are distinct & $(\forall x)(x{\in}X \supset Bx) \ \& \ x_1,\dots,x_k{\in}X \ \&$
   $\#(X) = n \ \& \ \text{freq}[A/X] = \frac{r}{n}\Big]$

   $= \dfrac{r!(n\text{-}r)!(n\text{-}k)!k!}{(r\text{-}m)!(n{+}m{-}r{-}k)!n!m!(k\text{-}m)!}.$

Let us abbreviate (2.3) as $\ulcorner\text{freq}[\psi/\xi] = \gamma\urcorner$. (2.3) is a necessary truth, so by the probability calculus,

   $\text{prob}\Big(\psi \ / \ \xi \ \& \ \text{prob}(A/B) = p\Big) =$

   $\text{prob}\Big(\psi \ / \ \xi \ \& \ \text{prob}(A/B) = p \ \& \ \text{freq}[\psi/\xi] = \gamma\Big).$

If *P* is any true proposition then $\text{freq}[\psi \ / \ \xi \ \& \ P] = \text{freq}[\psi/\xi]$. Thus for any proposition *P* we have the following logical equivalence:

(2.4) $\Box\Big[(P \ \& \ \text{freq}[\psi/\xi] = \gamma) \equiv (P \ \& \ \text{freq}[\psi \ / \ \xi \ \& \ P] = \gamma)\Big].$

Therefore, by the probability calculus and the principle (*PFREQ*) of Chapter 2:

$$(2.5) \quad \text{prob}\Big(\psi \,/\, \xi \,\&\, \text{prob}(A/B) = p\Big)$$
$$= \text{prob}\Big(\psi \,/\, \xi \,\&\, \text{prob}(A/B) = p \,\&\, \text{freq}[\psi/\xi] = \gamma\Big)$$
$$= \text{prob}\Big(\psi \,/\, \xi \,\&\, \text{prob}(A/B) = p \,\&$$
$$\text{freq}[\psi \,/\, \xi \,\&\, \text{prob}(A/B) = p] = \gamma\Big) = \gamma.$$

As in the statistical induction argument, we can then compute:

$$(2.6) \quad \text{prob}\Big(\text{freq}[A \,/\, \{x_1,\ldots,x_n\}] = m/k \,/\, x_1,\ldots,x_k \text{ are distinct}$$
$$\&\, Bx_1 \,\&\, \ldots \,\&\, Bx_k \,\&\, x_1,\ldots,x_k{\in}X \,\&\, \#(X) = n \,\&$$
$$\text{freq}[A/X] = \tfrac{r}{n} \,\&\, \text{prob}(A/B) = p\Big)$$

$$= \frac{r!(n{-}r)!(n{-}k)!k!}{(r{-}m)!(n{+}m{-}r{-}k)!n!m!(k{-}m)!}.$$

(*A5*) functions by applying (*A2*) to the different probabilities p thereby providing prima facie reasons, for each p, for believing that $\text{prob}(A/B) \neq p$. (*A5*) then weighs these prima facie reasons, and rejects the weakest of them. If any of these applications of (*A2*) are defeated, that defeats the use of (*A5*). The reference predicate in (2.2) is entailed by that in (2.6), so if

$$(2.7) \quad \frac{k!p^m(1{-}p)^{k-m}}{m!(k{-}m)!} \neq \frac{r!(n{-}r)!(n{-}k)!k!}{(r{-}m)!(n{+}m{-}r{-}k)!n!m!(k{-}m)!}$$

then we have a defeater of the form (*D2*) for the application of (*A2*) to the probability p. On the other hand, if (2.7) fails then the sample and subsample are in agreement and there is no need to defeat the inference from the subsample. The subsample problem is thereby resolved.

3. Fair Sample Defeaters for Statistical Induction

The inferences generated by the statistical induction argument are defeasible, and hence they are subject to both rebutting and undercutting defeat. In induction, undercutting defeaters can be called *fair sample defeaters*. These are reasons for suspecting that the inductive sample may not be a reliable guide to the value of the probability. For example, suppose we are attempting to ascertain the probability of a smoker contracting lung cancer. We examine a large sample of smokers and find that 50% of them have lung cancer. But if it is pointed out that the sample consists entirely of asbestos workers, that will defeat a statistical induction based upon the sample. It has proven very difficult to give a general characterization of fair sample defeaters. However, if the reconstruction of statistical induction given in this book is satisfactory, it should be possible to derive an account of fair sample defeaters from it.

There are several kinds of fair sample defeaters for statistical induction. The simplest is as follows. Given a sample X of B's, suppose we observe that

(3.1) $\text{freq}[A/X] = \nu$

and on this basis, for some particular $\alpha, \beta > 0$, we conclude by the statistical induction argument that

(3.2) $\text{prob}(A/B) \in [\nu\text{-}\alpha, \nu + \beta]$.

But suppose C is some property with respect to which A is projectible, and we observe that all the members of X are C's. As X is a sample of C's, and the statistical induction argument provides a prima facie reason for believing that $\text{prob}(A/B\&C) \in [\nu\text{-}\alpha, \nu + \beta]$. Any reason for believing that

$$|\text{prob}(A/B) - \text{prob}(A/B\&C)| > \alpha + \beta$$

constitutes a reason for believing that these conclusions cannot both be true, and so barring a reason for preferring one of the conclusions to the other, it defeats them both:

(3.3)   If A is projectible with respect to both B and C, $\ulcorner(\forall x)(x \in X \supset Cx)$ & $|\text{prob}(A/B\&C) - \text{prob}(A/B)| > \alpha+\beta\urcorner$ is a (defeasible) defeater for (3.1) as a reason for inferring (3.2).

Suppose A is also projectible with respect to some property D for which we know that $B \leqslant D$. Then nonclassical direct inference provides a prima facie reason for believing that $\text{prob}(C/A\&B) = \text{prob}(C/A\&D)$ and $\text{prob}(C/B) = \text{prob}(C/D)$. It is a theorem of the probability calculus that if these two equalities hold then

(3.4) $$\frac{\text{prob}(A/B\&C)}{\text{prob}(A/B)} = \frac{\text{prob}(A/D\&C)}{\text{prob}(A/D)}.$$

Applying this to the example of the smoking asbestos workers, let A be the property of getting lung cancer, B the property of smoking, C the property of being an asbestos worker, and D the property of being a person. (3.4) constitutes a reason for believing that the degree to which being an asbestos worker is positively relevant to contracting lung cancer is the same for smokers as for people in general. Given (3.4), it follows from the probability calculus that

(3.5) $\text{prob}(A/B\&C) - \text{prob}(A/B)$

$= \dfrac{\text{prob}(A/B)}{\text{prob}(A/D)} [\text{prob}(A/D\&C) - \text{prob}(A/D)].$

Given that $\text{prob}(A/B) > \text{prob}(A/D)$ (i.e., smoking is positively relevant to a person's contracting lung cancer),

(3.6) $\text{prob}(A/B\&C) - \text{prob}(A/B) > \text{prob}(A/D\&C) - \text{prob}(A/D).$

If we know that $\text{prob}(A/D\&C) - \text{prob}(A/B) \geq \alpha+\beta$ (i.e., being an asbestos worker makes it much more probable that one will get lung cancer), it follows that $\text{prob}(A/B\&C) - \text{prob}(A/B) \geq \alpha+\beta$. This is a defeater of the form of (3.3). Thus the example of the smoking asbestos workers illustrates the following fair sample defeater:

(3.7)   If A is projectible with respect to C and D then $\ulcorner(\forall x)(x \in X$ $\supset Cx)$ & $B \leq D$ & $|\text{prob}(A/D\&C) - \text{prob}(A/D)| > \alpha+\beta$ & $\text{prob}(A/B) > \text{prob}(A/D)\urcorner$ is a (defeasible) defeater for (3.1) as a reason for inferring (3.2).

Defeaters of the form of (3.7) employ "external" considerations to conclude that the sample is biased. A second kind of fair sample defeater employs fewer external considerations and more internal considerations. Here, instead of observing that $(\forall x)(x \in X$ $\supset Cx)$, we observe that the relative frequency of C's in X is not even approximately the same as $\text{prob}(C/B)$, i.e., $\text{freq}[C/X] \neq \text{prob}(C/B)$. If it is also true that $\text{freq}[A/X\&{\sim}C] \neq \text{freq}[A/X\&C]$ then we can conclude (defeasibly) that the sample is biased, and this defeats the induction. The reason for this is that by the probability calculus:

(3.8) $\text{prob}(A/B) = \text{prob}(A/B\&C) \cdot \text{prob}(C/B)$
$\qquad\qquad\qquad + \text{prob}(A/B\&{\sim}C) \cdot [1 - \text{prob}(C/B)]$

and

(3.9) $\text{freq}[A/X] = \text{freq}[A/X\&C] \cdot \text{freq}[C/X]$
$\qquad\qquad\qquad + \text{freq}[A/X\&{\sim}C] \cdot (1 - \text{freq}[C/X])$.

By the statistical induction argument we have reason to believe that $\text{prob}(A/B\&C) \approx \text{freq}[A/X\&C]$, $\text{prob}(A/B\&{\sim}C) \approx \text{freq}[A/X\&{\sim}C]$, and $\text{prob}(A/B) \approx \text{freq}[A/X]$. If those approximate equalities all hold and $\text{freq}[A/X\&C] \neq \text{freq}[A/X\&{\sim}C]$ then it follows, contrary to supposition, that $\text{prob}(C/B) \approx \text{freq}[C/X]$. As we know the latter to be false, we can conclude that one of the inductive inferences has a false conclusion, and not knowing which to disbelieve, they are collectively defeated. Thus we have the following fair sample defeater:

(3.10) If A is projectible with respect to both C and $\sim C$ then
⌜freq$[C/X] \neq$ prob(C/B) & freq$[A/X\&C] \neq$ freq$[A/X\&\sim C]$⌝
is a defeater for (3.1) as a reason for believing (3.2).

The defeater in (3.10) will itself be defeated by discovering either that prob$(A/B\&C) \neq$ freq$[A/X\&C]$ or that prob$(A/B\&\sim C) \neq$ freq$[A/X\&\sim C]$. To illustrate (3.10), suppose once more that we are attempting to ascertain the probability of a smoker contracting lung cancer. If we observe that our sample contains a disproportionate number of asbestos workers, and that the frequency of cancer among asbestos workers in the sample is markedly different from the frequency of cancer among non-asbestos workers in the sample, this gives us reason to distrust the sample and defeats the statistical induction based upon it.

To make (3.10) mathematically useful, we must take into account the accuracy of the approximations. Let $\nu = $ freq$[A/X]$, $\mu = $ freq$[A/X\&C]$, and $\eta = $ freq$[A/X\&\sim C]$. Then freq$[C/X] = \nu$-η/μ-η. By the probability calculus,

$$(3.11) \quad \text{prob}(C/B) = \frac{\text{prob}(A/B) - \text{prob}(A/B\&\sim C)}{\text{prob}(A/B\&C) - \text{prob}(A/B\&\sim C)}.$$

For some γ, δ, κ, and τ, statistical induction provides prima facie reasons for believing that prob$(A/B)\in[\nu$-$\alpha,\nu+\beta]$, prob$(A/B\&C)\in[\mu$-$\gamma,\mu+\delta]$, and prob$(A/B\&\sim C)\in[\eta$-$\kappa,\eta+\tau]$. From this and (3.11) we can conclude that

$$\frac{(\nu\text{-}\eta) + (\alpha+\tau)}{(\mu\text{-}\eta) + (\delta+\kappa)} \leq \text{prob}(C/B) \leq \frac{(\nu\text{-}\eta) + (\beta+\kappa)}{(\mu\text{-}\eta) - (\gamma\text{-}\tau)}.$$

Let

$$\alpha = \frac{(\nu\text{-}\eta) + (\beta+\kappa)}{(\mu\text{-}\eta) - (\gamma\text{-}\tau)} - \frac{\nu\text{-}\eta}{\mu\text{-}\eta}$$

and

$$\chi = \frac{\nu\text{-}\eta}{\mu\text{-}\eta} - \frac{(\nu\text{-}\eta) - (\alpha+\tau)}{(\mu\text{-}\eta) + (\delta+\kappa)}.$$

Then we can conclude that $\text{prob}(C/B) - \text{freq}[C/X] \le \alpha$ and $\text{freq}[C/X] - \text{prob}(C/B) \le \chi$. If either of these inequalities fails, that constitutes a defeater for the three defeasible steps of the argument. Thus (3.10) is made precise as follows:

(3.12) If A is projectible with respect to both C and $\sim C$ then ⌜Either $\text{prob}(C/B) - \text{freq}[C/X] > \alpha$ or $\text{freq}[C/X] - \text{prob}(C/B) > \chi$⌝ is a defeater for (3.1) as a reason for believing (3.2).

It is instructive to work out an example of (3.12). Suppose the size of the sample is such as to provide warrant at the .01 level for believing that $\text{prob}(A/B) \in [.78, .82]$, $\text{prob}(A/B\&C) \in [.84, .86]$, and $\text{prob}(A/B\&\sim C) \in [.74, .76]$, where $\text{freq}[C/X] = .5$. Then calculation reveals that $\alpha = 1/2$ and $\chi = 1/3$. Thus in order for (3.12) to yield a defeater for the inductive inference, we must know either that $\text{prob}(C/B) - \text{freq}[C/X] > 1/3$ or that $\text{freq}[C/X] - \text{prob}(C/B) > 1/2$. This illustrates that the divergence between $\text{freq}[C/X]$ and $\text{prob}(C/B)$ must be quite large before it biases the sample sufficiently to defeat the inductive inference, even in cases in which our inductive estimates of probabilities are quite precise. As we know that $\text{freq}[C/X] = 1/2$, the first condition is equivalent to requiring that $\text{prob}(C/B) > 5/6$, and the second is equivalent to requiring that $\text{prob}(C/B) < 0$ (which is impossible).

The upshot of this is that the proposed reconstruction of statistical induction accommodates fair sample defeaters quite easily and does so in a plausible way. No claim is made that the preceding constitutes a complete catalog of fair sample defeaters, but that is not a problem for the present account. If induction were treated as a primitive epistemic principle, then it would be necessary to give a complete catalog of the possible defeaters. But if, as here, it is only claimed that the rules of induction follow from certain other principles, then all that is required is that the defeaters for induction also follow from those principles. That is plausible even without our being in a position to give a complete list of those defeaters.

4. Fair Sample Defeaters for Enumerative Induction

It has proven just as difficult to describe fair sample defeaters for enumerative induction as for statistical induction. Most of the kinds of fair sample defeaters that apply to statistical induction are inapplicable when $r = n$ and hence inapplicable to enumerative induction, but the reconstruction of enumerative induction in terms of the enumerative induction argument make possible the description of another kind of fair sample defeater that is applicable to enumerative induction. Suppose we have a sample B_0 of B's all of which are A's. This provides a prima facie reason, in accordance with the enumerative induction argument, for believing that $B \Rightarrow A$. But now suppose we observe that B_0 has some property S guaranteeing that it could not but have consisted exclusively of A's. For example, we might have employed a sampling technique that proceeded by first looking for A's, then choosing B's from among those A's, and then checking to see whether all the B's so chosen are A's. Clearly, no inductive inference is justified in such a case. More precisely, $\ulcorner S(B_0)$ & $[S(Y) \Rightarrow (Y \subseteq A)]\urcorner$ should be a defeater provided $\ulcorner S(Y)\urcorner$ is projectible with respect to $\ulcorner \#Y = n$ & $Y \subseteq B\urcorner$.[5] This can be verified by noting that under these circumstances, $\ulcorner S(B_0)$ & $[S(Y) \Rightarrow (Y \subseteq A)]\urcorner$ defeats the enumerative induction argument by defeating Stage I of the embedded statistical induction argument. It follows from the earlier account of entailment relations between nomic generalizations that

(4.1) $\#B_0 = n$ & $B_0 \subseteq B$ & $\text{prob}(A/B) = p$ & $S(B_0)$ & $[S(Y) \Rightarrow (Y \subseteq A)]$

entails

---

[5] I am abbreviating $\ulcorner (\forall x)(x \in B_0 \supset Ax)\urcorner$ as $\ulcorner B_0 \subseteq A\urcorner$.

(4.2) $\Big(\#Z = n \ \& \ Z \subseteq B \ \& \ \text{prob}(A/B) = p \ \& \ S(Z) \ \&$

$[S(Y) \Rightarrow (Y \subseteq A)] \Big) \Rightarrow (Z \subseteq A)$

which in turn entails

(4.3) $\text{prob}\Big(Z \subseteq A \ / \ \#Z = n \ \& \ Z \subseteq B \ \& \ \text{prob}(A/B) = p \ \& \ S(Z)$

$\& \ [S(Y) \Rightarrow (Y \subseteq A)] \Big) = 1.$

By (*PPROB*) of Chapter 2:

(4.4) $\text{prob}\Big(X \subseteq A \ / \ \#X = n \ \& \ X \subseteq B \ \& \ \text{prob}(A/B) = p \ \& \ S(X)$

$\& \ [S(Y) \Rightarrow (Y \subseteq A)] \ \&$

$\text{prob}(Z \subseteq A \ / \ \#Z = n \ \& \ Z \subseteq B \ \& \ \text{prob}(A/B) = p$

$\& \ S(Z) \ \& \ [S(Y) \Rightarrow (Y \subseteq A)]) = 1 \Big) = 1.$

We then calculate, as in the statistical induction argument:

(4.5) $\text{prob}\Big(\text{freq}[A/X] \neq 1 \ / \ \#X = n \ \& \ X \subseteq B \ \&$

$\text{prob}(A/B) = p \ \& \ S(X) \ \& \ [S(Y) \Rightarrow (Y \subseteq A)] \ \&$

$\text{prob}(Z \subseteq A \ / \ \#Z = n \ \& \ Z \subseteq B \ \& \ \text{prob}(A/B) = p$

$\& \ S(Z) \ \& \ [S(Y) \Rightarrow (Y \subseteq A)]) = 1 \Big) = 0.$

The statistical induction argument proceeds by applying (*A2*) and (*A5*) to (3.6) of Chapter 5, and in this case (3.6) amounts to:

(4.6) $\text{prob}\Big(\text{freq}[A/X] \neq 1 \ / \ \#X = n \ \& \ X \subseteq B \ \&$

$\text{prob}(A/B) = p \Big) = 1 - p^n.$

The reference property of (4.5) entails that of (4.6), so (4.5) constitutes a subproperty defeater for the application of (*A2*), and this defeats the application of (*A5*) as well. Thus the statistical induction argument is blocked, and hence the enumerative induction is defeated. To summarize:

(4.7)   If $\ulcorner S(Y) \urcorner$ is projectible with respect to $\ulcorner \#Y = n \ \& \ Y \subseteq B \urcorner$ then $\ulcorner S(B_0) \ \& \ [S(Y) \Rightarrow (Y \subseteq A)] \urcorner$ is a defeater for the enumerative induction to the conclusion $\ulcorner B \Rightarrow A \urcorner$ based upon the sample B_0.

There are what initially appear to be counterexamples to (4.7). For example, what if we let $\ulcorner S(X) \urcorner$ be $\ulcorner X \subseteq A \urcorner$? By (4.7), that should give us a defeater, but obviously this should not be sufficient to defeat the enumerative induction. If it did, all cases of enumerative induction would be defeated. The reason this does not defeat the inductive inference is that $(D2)$ is a defeasible defeater. We saw in Chapter 8 that (DD) is a defeater for $(D2)$. In the present context, (DD) tells us that

(4.8) $\ulcorner$'$S(B_0) \ \& \ [S(Y) \Rightarrow (Y \subseteq A)]$' is a deductive consequence of the argument on the basis of which 'freq$[A/B_0] = 1$' is warranted$\urcorner$ is an undercutting defeater for (4.7).

Applying (4.8) to the case of $\ulcorner S(X) \urcorner = \ulcorner X \subseteq A \urcorner$ we have:

(4.9) $\ulcorner$'$B_0 \subseteq A \ \& \ [(Y \subseteq A) \Rightarrow (Y \subseteq A)]$' is a deductive consequence of the argument on the basis of which 'freq$[A/B_0] = 1$' is warranted$\urcorner$ is an undercutting defeater for (4.7).

But $\ulcorner B_0 \subseteq A \ \& \ [(Y \subseteq A) \Rightarrow (Y \subseteq A)] \urcorner$ is deductively equivalent to $\ulcorner$freq$[A/B_0] = 1 \urcorner$, so the problematic instance of (4.7) is defeated.

To take a more realistic example, suppose we are trying to confirm that no bluebirds are yellow. We do this by collecting a sample of bluebirds, noting of each that it is not yellow, and then concluding that no bluebirds are yellow. This induction should be legitimate. However, it is presumably the case that each member of the sample was blue, and that fact nomically implies that each member of the sample was nonyellow, so (4.7) provides a defeater for the inductive inference. But again, (4.8) provides a defeater defeater. Letting B be 'is a bluebird', A be 'is nonyellow', and

C be 'is blue', $\ulcorner S(X) \urcorner$ becomes $\ulcorner X \subseteq C \urcorner$. $\Box \forall (C \supset A)$, so $\ulcorner X \ll A \urcorner$ entails $\ulcorner X \ll C \urcorner$. $\ulcorner B_0 \subseteq C \ \& \ [(X \subseteq C) \Rightarrow (X \subseteq A)] \urcorner$ is a deductive consequence of $\ulcorner \text{freq}[A/B_0] = 1 \urcorner$. Thus we have a defeater defeater of the form of (4.8), and the inductive reasoning is reinstated.

On the other hand, consider a case in which the inductive reasoning ought to be defeated. Let $\ulcorner S(X) \urcorner$ describe some sampling procedure that is explicitly designed to select only A's. In that case, there is no reason to expect that $\ulcorner S(B_0) \urcorner$ will be a deductive consequence of the argument on the basis of which $\ulcorner \text{freq}[A/B_0] = 1 \urcorner$ is warranted.

Many defeaters for enumerative induction can be handled in terms of (4.7) and (4.8), but the most sophisticated defeaters for enumerative induction involve a different phenomenon, to which I turn next.

5. Diversity of the Sample

Suppose we are attempting to confirm a nomic generalization about all mammals. We collect a sample of mammals, verify that the generalization holds for the sample, and take that to confirm the generalization. But suppose the sample consists exclusively of puppy dogs and panda bears. Surely that should make us suspicious of extending the generalization to all mammals. Perhaps we are warranted in believing that the generalization holds for all puppy dogs and panda bears, but it seems that our warrant for believing that it holds for all mammals is much weaker. In general, in inductive confirmation we seek to make our sample as diverse as possible. The less homogeneous that sample, the greater we take the confirmation to be. In what does this desired diversity consist? There is a temptation to suppose it requires the different elements of the sample to have no properties in common, or as few common properties as possible. But it does not take much reflection to see that that cannot be right. In any finite set of objects, there will be infinitely many properties shared by all of them. We cannot assess the diversity of a sample just by counting the number of shared properties.

Most shared properties are irrelevant. For example, in a sample of mammals, all members of the sample will have been born on earth, and all will weigh less than 100 tons, but we would not regard these properties as relevant to the degree of confirmation.

By attending to the details of the statistical induction argument, we can make sense of this demand for diversity in our sample. In the instance of the statistical induction argument that is embedded in the enumerative induction argument, nonclassical direct inference is used to conclude that

$$(5.1) \quad \text{prob}\left(Ax_i \mid Ax_{i+1} \ \& \ \ldots \ \& \ Ax_n \ \& \ \theta_p\right)$$
$$= \text{prob}\left(Ax_i \mid Bx_i \ \& \ \text{prob}(A/B) = p\right).$$

This enables us to compute that

$$(5.2) \quad \text{prob}\left(\text{freq}[A/X] \neq 1 \mid \#X = n \ \& \ X \subseteq B \ \& \right.$$
$$\left. \text{prob}(A/B) = p\right) = 1 - p^n.$$

(5.1) amounts to the assumption that the different members of the sample are independent of one another with respect to whether they are A's, i.e., knowing that one is an A does not make it any more probable that another is an A. Nonclassical direct inference creates a presumption of independence. But this presumption of independence can be defeated. We might have an inductive reason for believing that if x and y are both C's and y is an A, that makes it more likely that x is an A. Suppose we know that for each $n > 1$,

$$(5.3) \quad \text{prob}\left(Ax_n \mid Ax_1 \ \& \ \ldots \ \& \ Ax_{n-1} \ \& \ Bx_1 \ \& \ \ldots \ \& \ Bx_n \ \& \right.$$
$$\left. Cx_1 \ \& \ \ldots \ \& \ Cx_n\right) = r.$$

By (6.21) of Chapter 2,

(5.4) $\text{prob}\big(Ax_n \: / \: Ax_1 \: \& \: ... \: \& \: Ax_{n-1} \: \& \: Bx_1 \: \& \: ... \: \& \: Bx_n \: \&$

$Cx_1 \: \& \: ... \: \& \: Cx_n\big) = r \: \& \: \text{prob}(A/B) = p$

entails

(5.5) $\text{prob}\big(Ax_n \: / \: Ax_1 \: \& \: ... \: \& \: Ax_{n-1} \: \& \: Bx_1 \: \& \: ... \: \& \: Bx_n \: \&$

$Cx_1 \: \& \: ... \: \& \: Cx_n \: \& \: \text{prob}(A/B) = p\big) = r.$

Then just as in Stage I of the statistical induction argument we acquire a prima facie reason for:

(5.6) $\text{prob}\big(\text{freq}[A/X] \neq 1 \: / \: \#X = n \: \& \: X \subseteq B \: \& \: X \subseteq C \: \&$

$\text{prob}(A/B) = p\big) = 1 - p \cdot r^{n-1}.$

If $r \neq p$, this gives us a defeater of the form (D2) for the use of (5.2) in Stage II of the statistical induction argument. Furthermore, this is a case of (D2) that will not ordinarily be defeated by (DD).

If $r < 1$, the preceding observations do not block the statistical induction argument altogether. The observations raise the probability of getting the data that we have, and hence lower the degree of confirmation (i.e., raise the acceptance level). This constitutes a "diminisher" rather than a defeater for the enumerative induction argument. I take it that this explains our intuitions regarding the desirability of diversity in our sample. The reason we are reluctant to draw conclusions about all mammals on the basis of a sample consisting entirely of puppy dogs and panda bears is that we do not regard the elements of the sample as independent. We think it more likely that different puppy dogs will be alike and different panda bears will be alike than that mammals of diverse species will be alike. Accordingly, our observations give us some confirmation for the generalization about all mammals, but less confirmation than we would expect from simply counting the number of confirming instances in the sample.

Now consider a special case in which a lack of diversity diminishes our degree of confirmation to the point of defeat. Suppose we note that our sample consists exclusively of C's, despite the fact that $\sim(B \Rightarrow C)$, and suppose we know that $C \Rightarrow A$. It seems intuitively that this should defeat the confirmation of $\ulcorner B \Rightarrow A \urcorner$. For example, suppose we are attempting to confirm that all mammals have kidneys, but we do this by examining only elephants, and we already know that all elephants have kidneys. Such a sample should not be taken to confirm the generalization that all mammals have kidneys because there was never any chance of its containing mammals without kidneys—we already knew that all elephants have kidneys and that is all the sample contains. We might try to explain this intuitive reasoning by appealing to the fair sample defeaters of section 4, noting that just as in (4.5), we have:

(5.7) $\text{prob}\Big(\text{freq}[A/X] \neq 1 \ / \ \#X = n \ \& \ X \subseteq B \ \&$

$$\text{prob}(A/B) = p \ \& \ X \subseteq C \ \& \ (C \Rightarrow A))] \Big) = 0.$$

By $(D2)$, this does provide us with a defeater. But just as in the example of the bluebirds in section 4, this defeater is defeated by (DD). Instead of explaining this example in terms of the fair sample defeaters of section 4, we explain it in terms of the diminishers just discussed. Given that $C \Rightarrow A$, it follows that $[(Cx \ \& \ Cy \ \& \ Ay) \Rightarrow Ax]$. This is a kind of limiting case of nonindependence. Then as in (5.6) we get:

(5.8) $\text{prob}\Big(\text{freq}[A/X] \neq 1 \ / \ \#X = n \ \& \ X \subseteq B \ \& \ X \subseteq C \ \&$

$$\text{prob}(A/B) = p \ \& \ [(Cx \ \& \ Cy \ \& \ Ay) \Rightarrow Ax] \Big)$$
$$= 1 - p.$$

This too constitutes a subproperty defeater, but this time the defeater is not (usually) defeated by (DD) because $\ulcorner B_0 \subseteq C \ \&$ $[(Cx \ \& \ Cy \ \& \ Ay) \Rightarrow Ax] \urcorner$ is not a deductive consequence of the argument warranting $\ulcorner freq[A/B_0] = 1 \urcorner$.[6] A probability of $1-p$ will not generate likelihood ratios small enough to allow the statistical induction argument to proceed, so we have a defeater for it and hence for the enumerative induction argument.

It was assumed that we know that $\sim(B \Rightarrow C)$. If we do not know this, then as our sample of B's consists exclusively of C's, it gives us inductive reason for believing that $B \Rightarrow C$. This still provides a defeater for the confirmation of $\ulcorner B \Rightarrow A \urcorner$, but that is now beside the point, because if we know that $C \Rightarrow A$ and we have undefeated confirmation for $\ulcorner B \Rightarrow C \urcorner$, it follows that we are warranted in believing $\ulcorner B \Rightarrow A \urcorner$ on that basis. To summarize, we have the following defeater for enumerative induction:

(5.9) If C is projectible with respect to B then $\ulcorner B_0 \subseteq C \ \& \ (C \Rightarrow A) \ \& \ \sim(B \Rightarrow C) \urcorner$ is an undercutting defeater for the confirmation of $\ulcorner B \Rightarrow A \urcorner$ by the sample B_0.

The preceding reasoning can be generalized in an interesting way that throws further light on the way in which diversity within the sample affects the degree of confirmation. Suppose that instead of knowing that $B_0 \subseteq C$, B_0 can be divided into two disjoint subsets B_1 and B_2, and there are two properties C_1 and C_2 each projectible with respect to B, such that $B_1 \subseteq C_1$ and $B_2 \subseteq C_2$ and $C_1 \Rightarrow A$ and $C_2 \Rightarrow A$. Reasoning as above, we obtain the result that:

(5.10) $prob\big(freq[A/X] \neq 1 \ / \ \#X = n \ \& \ X = X_1 \cup X_2 \ \&$
  $X_1 \cap X_2 = \emptyset \ \& \ X \subseteq B \ \& \ X_1 \subseteq C_1 \ \& \ X_2 \subseteq C_2 \ \&$
  $prob(A/B) = p \ \& \ [(C_1 x \ \& \ C_1 y \ \& \ Ay) \Rightarrow Ax] \ \&$
  $[(C_2 x \ \& \ C_2 y \ \& \ Ay) \Rightarrow Ax]\big)$

$= 1 - p^2.$

---

[6] In some special cases, this defeater is still defeated by (DD). For example, if we let $C = A$ then $\ulcorner B_0 \subseteq C \ \& \ [(Cx \ \& \ Cy \ \& \ Ay) \Rightarrow Ax] \urcorner$ is a deductive consequence of the argument.

This is the same probability that we would get from a two-element sample, and hence the degree of confirmation for $\ulcorner B \Rightarrow A \urcorner$ that results from having this sample is the same as the degree of confirmation that would result from a two-element sample not subject to diversity problems. More generally, if our sample can be divided into n disjoint subsets B_i such that for each i there is a C_i projectible with respect to B for which $B_i \subseteq C_i$ and $C_i \Rightarrow A$, then the degree of confirmation accruing to $\ulcorner B \Rightarrow A \urcorner$ as a result of observation of this sample is the same as the degree of confirmation that would result from observation of an n-membered sample.

6. The Paradox of the Ravens

One of the best known problems for theories of induction is the paradox of the ravens.[7] Any adequate theory of enumerative induction must enable us to resolve this paradox in some congenial way. The paradox results from the fact that although we can confirm that all ravens are black by observing black ravens, it appears that we cannot equally confirm it by observing nonblack nonravens. This is paradoxical because it seems that observation of nonblack nonravens should confirm the generalization that all nonblack things are nonravens, and the latter entails that all ravens are black. Although this reasoning seems persuasive in the abstract, it is clear that observation of green plastic garbage cans and orange pussy cats does not confirm that all ravens are black.

It is tempting to try to resolve the paradox of the ravens by appealing to projectibility and maintaining that 'nonraven' is not projectible with respect to 'nonblack'.[8] Unfortunately, it seems that 'nonraven' *is* projectible with respect to 'nonblack'. If we went about it in the right way, we really could confirm that all nonblack things are nonravens by observing nonblack nonravens. For example, suppose we had a list of all the items in the

---

[7] This is due to Carl Hempel [1943] and [1945].

[8] See, for example, Quine [1969].

universe, and we could run through it and randomly select things to see whether they were nonblack and if so whether they were ravens. If our sampling were sufficiently random, it would provide us with inductive confirmation for the generalization.

My resolution of the paradox of the ravens consists of denying, in a sense, that there is any paradox. As I have just argued, we *can* confirm that all ravens are black by observing nonblack nonravens. Our temptation to deny this turns upon the supposition that certain clearly illegitimate procedures would be legitimate if it were possible to confirm that all ravens are black by examining nonblack nonravens. There are lots of foolish things we might do that would result in the generalization not being confirmed despite our observing a sample of nonblack nonravens. For example, we could not confirm the generalization just by going to a pet store and examining orange pussy cats. Neither could we confirm it by going to a factory manufacturing green plastic garbage cans and observing its output. But this should not be surprising. Such a sampling procedure is biased against ever finding any nonblack ravens, and as such the prima facie reason supplied by such sampling is defeated by a fair sample defeater of the form of (5.7). There is nothing paradoxical about this.

Suppose we do not do anything so blatantly foolish as examining only orange pussy cats or examining only green plastic garbage cans. It might still be hard to get much confirmation for the generalization that all ravens are black by examining nonblack things. This is largely a reflection of a natural lack of imagination. Suppose that instead of deliberately choosing a biased sample, you just stroll around randomly examining nonblack things. You still have to focus your attention on particular nonblack things, and when you do this you are likely to focus on many nonblack things of a few specific kinds. For example, on strolling by the garbage can factory, you might take note of a whole stack of green plastic garbage cans, and when you look in the window of the pet store you might notice an entire litter of orange pussy cats. Then when you pass the paint shop you might attend to a bunch of brightly painted cars. The problem with this procedure is that it produces a sample lacking adequate diversity. If you examine a bunch of green garbage cans, a bunch of orange pussy cats, and a bunch of brightly painted automobiles, you do have a large sample of nonblack things, but as we saw in the last section, the confirmation accruing to 'All ravens are black' from that sample is the same as if you had examined just one thing of each kind. You will get confirmation this way but it is going to accrue with painful slowness, and most of your observations are going to be redundant. I think this is probably the main phenomenon responsible for the intuition that you cannot confirm that all ravens are black by observing nonblack nonravens. You can confirm it this way, but it is very hard to do so efficiently. This is also why it seems that it would be much easier to confirm the generalization if we were supplied with a list of all the things in the universe and we could just dip into it at random. That would make our sampling procedure easier because avoiding an overly homogenous sample would not require so much imagination.

7. Summary

1. Projectibility constraints were originally noted in connection with induction, but inductive projectibility seems to be the same property as the kind of projectibility required by the statistical syllogism.
2. Fair sample defeaters for both statistical and enumerative induction can be derived from the reconstruction of inductive reasoning in terms of the statistical and enumerative induction arguments.
3. We can also make sense of the intuitive requirement that our inductive sample should be as inhomogeneous as possible.
4. The paradox of the ravens can be resolved by noting that normal ways of collecting samples of nonblack nonravens results in overly homogeneous samples.

BIBLIOGRAPHY

Barnard, G. A.
 1949 Statistical inference. *Journal of the Royal Statistical Society B, II*, 115–149.
 1966 The use of the likelihood function in statistical practice. *Proceedings v Berkeley Symposium on Mathematical Statistics and Probability I*, 27–40.
 1967 The Bayesian controversy in statistical inference. *Journal of the Institute of Actuaries 93*, 229–269.
Bennett, Jonathan
 1984 Counterfactuals and temporal direction. *Philosophical Review 93*, 57–91.
Bernoulli, Jacques
 1713 *Ars Conjectandi*. Basle.
Bernstein, Allen R., and Frank Wattenberg
 1969 Nonstandard measure theory. In *Applications of Model Theory to Algebra, Analysis, and Probability*, ed. W. A. J. Luxemburg, 171–185. New York: Holt, Rinehart and Winston.
Bertrand, Joseph
 1889 *Calcul des Probabilites*. Paris.
Birnbaum, Allan
 1962 On the foundations of statistical inference. *Journal of the American Statistical Association 57*, 269–326.
 1969 Concepts of statistical inference. In *Philosophy, Science, and Method*, ed. S. Morganbesser, P. Suppes, and M. White, 112–143. New York: St. Martin's Press.
Borel, Emile
 1909 *Elements de la theorie des probabilities*. Paris.
Braithwaite, R. B.
 1953 *Scientific Explanation*. Cambridge: Cambridge University Press.
Bruckner, A. M., and Jack Ceder
 1975 On improving Lebesgue measure. *Nordisk Matematisk Tidskrift 23*, 59–68.
Burnor, Richard N.
 1984 What's the matter with *The Matter of Chance*? *Philosophical Studies 46*, 349–366.

Butler, Bishop Joseph
1736   *The Analogy of Religion.*   London: George Bell and Sons.

Carnap, Rudolph
1950   *The Logical Foundations of Probability.*   Chicago: University of Chicago Press.
1952   *The Continuum of Inductive Methods.*   Chicago: University of Chicago Press.
1962   The aim of inductive logic. In *Logic, Methodology, and the Philosophy of Science*, ed. Ernest Nagel, Patrick Suppes, and Alfred Tarski, 303–318. Stanford, Calif.: Stanford University Press.

Cartwright, Nancy
1979   Causal laws and effective strategies. *Nous 13*, 419–438.

Chisholm, Roderick
1957   *Perceiving.* Ithaca, N.Y.: Cornell University Press.
1966   *Theory of Knowledge*, 1st edition.   Englewood Cliffs, N.J.: Prentice-Hall.
1977   *Theory of Knowledge*, 2nd edition.   Englewood Cliffs, N.J.: Prentice-Hall.
1981   A version of foundationalism.   *Midwest Studies in Philosophy V*, 543–564.

Church, Alonzo
1940   On the concept of a random sequence. *Bulletin of the American Mathematical Society 46*, 130–135.

Cox, D. R.
1958   Some problems connected with statistical inference. *Annals of Mathematical Statistics 29*, 357–363.

Dempster, A. P.
1964   On the difficulties inherent in Fisher's fiducial argument. *Journal of the American Statistical Association 59*, 56–66.
1966   Review of *The Logic of Statistical Inference. Journal of the American Statistical Association 61*, 1233–1235.

Doyle, J.
1979   A truth maintenance system. *Artificial Intelligence 12*, 231–272.

Edwards, A. W. F.
1972   *Likelihood.* Cambridge: Cambridge University Press.

Eells, Ellery
  1983   Objective probability theory theory.   *Synthese 57*,
          387–442.
Etherington, D. W.
  1983   *Formalizing Nonmonotonic Reasoning Systems*. Com-
          puter Science Technical Report No. 83-1, University of
          British Columbia, Vancouver.
  1987   Formalizing nonmonotonic reasoning systems. *Artificial
          Intelligence 31*, 41–86.
Etherington, D., R. Mercer, and R. Reiter
  1985   On the adequacy of predicate circumscription for
          closed-world reasoning. *Computational Intelligence I*,
          11–15.
Etherington, D., and R. Reiter
  1983   On inheritance hierarchies with exceptions. *Proceedings
          of the AAAI83*, 104–108.
Fetzer, James
  1971   Dispositional probabilities.   *Boston Studies in the
          Philosophy of Science 8*, 473–482.
  1977   Reichenbach, reference classes, and single case 'proba-
          bilities'. *Synthese 34*, 185–217.
  1981   *Scientific Knowledge*. Volume 69 of *Boston Studies in
          the Philosophy of Science*. Dordrecht: Reidel.
Fisher, R. A.
  1921   On the 'probable error' of a coefficient of correlation
          deduced from a small sample. *Metron I*, part 4, 3–32.
  1922   On the mathematical foundations of theoretical
          statistics. *Philosophical Transactions of the Royal Society
          A, 222*, 309–368.
Giere, Ronald N.
  1973   Objective single case probabilities and the foundations
          of statistics. In *Logic, Methodology, and Philosophy of
          Science IV*, ed. Patrick Suppes, Leon Henkin, Athanase
          Joja, and GR. C. Moisil, 467–484. Amsterdam: North
          Holland.
  1973a   Review of Mellor's *The Matter of Chance*.   *Ratio 15*,
          149–155.

1976   A Laplacean formal semantics for single-case propensities. *Journal of Philosophical Logic 5*, 321–353.

Gillies, D. A.
1973   *An Objective Theory of Probability.* London: Methuen.

Ginsberg, M.
1987   *Readings in Nonmonotonic Reasoning.* Los Altos, Calif.: Morgan Kaufman.

Goodman, Nelson
1955   *Fact, Fiction, and Forecast.* Cambridge, Mass.: Harvard University Press.

Hacking, Ian
1965   *Logic of Statistical Inference.* Cambridge: Cambridge University Press.
1975   *The Emergence of Probability.* Cambridge: Cambridge University Press.

Hanks, S., and D. McDermott
1985   *Temporal reasoning and default logics.* Computer Science Research Report No. 430, Yale University, New Haven, Conn.
1987   Nonmonotonic logic and temporal projection. *Artificial Intelligence 13*, 379–412.

Harman, Gilbert
1980   Reasoning and explanatory coherence. *American Philosophical Quarterly 17*, 151–158.

Harper, William
1983   Kyburg on direct inference. In *Profiles: Kyburg and Levi*, ed. R. J. Bogdan.   Dordrecht: Reidel.

Hausdorff, Felix
1949   *Grundzuge der Mengenlehre.* New York: Chelsea.

Hempel, Carl
1943   A purely syntactic definition of confirmation. *Journal of Symbolic Logic 8*, 122–143.
1945   Studies in the logic of confirmation. *Mind 54*, 1–26, 97–121.

Hintikka, Jaakko
1966   A two-dimensional continuum of inductive methods. In *Aspects of Inductive Logic*, ed. J. Hintikka and P. Suppes, 113–132. Amsterdam: North Holland.

Horty, J., R. Thomason, and D. Touretzky
  1987   A skeptical theory of inheritance in nonmonotonic
          semantic networks. *Proceedings of the AAAI87*,
          358–363.
Jackson, Frank, and John Pargetter
  1973   Indefinite probability statements. *Synthese 26*, 205–215.
Keynes, John Maynard
  1921   *A Treatise on Probability*. London: Macmillan.
Kneale, William
  1949   *Probability and Induction*. Oxford: Oxford University
          Press.
Kolmogoroff, A. N.
  1933   *Grundbegriffe der Wahrscheinlichkeitsrechnung*. Berlin.
Konolige, K.
  1987   On the relation between default theories and autoepis-
          temic logic. In Ginsberg [1987].
Kries, J. von
  1871   *Principien der Wahrscheinlichkeitsrechnung*.
Kyburg, Henry, Jr.
  1961   *Probability and the Logic of Rational Belief*. Mid-
          dletown, Conn.: Wesleyan University Press.
  1970   Conjunctivitis. In *Induction, Acceptance, and Rational
          Belief*, ed. Marshall Swain. Dordrecht: Reidel.
  1974   *The Logical Foundations of Statistical Inference*.
          Dordrecht: Reidel.
  1974a   Propensities and probabilities. *British Journal for the
          Philosophy of Science 25*, 321–353.
  1976   Chance. *Journal of Philosophical Logic 5*, 355–393.
  1977   Randomness and the right reference class. *Journal of
          Philosophy 74*, 501–521.
  1980   Conditionalization. *Journal of Philosophy 77*, 98–114.
  1982   Replies. In *Profiles: Kyburg and Levi*, ed. R. J. Bogdan.
          Dordrecht: Reidel.
  1983   The reference class. *Philosophy of Science 50*, 374–397.
Laplace, Pierre Simon Marquis de
  1795   *Essaie philosophique sur les probabilites*. Paris: V.
          Courcier.
  1820   *Theorie analytique des probabilites*. Volume 7 of [1878].

1878   *Oeuvres completes.* 14 volumes. Paris: Gauthier-Villars.
1951   *Essays on Probability.* Translation of [1795]. New York: Dover.

Lehrer, Keith
1981   Coherence and the racehorse paradox. *Midwest Studies in Philosophy*, vol. V, ed. Peter French, Theodore Uehling, Jr., and Howard Wettstein, 183–192.

Levi, Isaac
1973   ... but fair to chance. *Journal of Philosophy 70*, 52–55.
1977   Direct inference. *Journal of Philosophy 74*, 5–29.
1978   Confirmational conditionalization. *Journal of Philosophy 75*, 730–737.
1980   *The Enterprise of Knowledge.* Cambridge, Mass.: MIT Press.
1981   Direct inference and confirmational conditionalization. *Philosophy of Science 48*, 532–552.

Lewis, David
1973   *Counterfactuals.* Cambridge, Mass.: Harvard University Press.
1979   Counterfactual dependence and time's arrow. *Nous 13*, 455–476.

Lifschitz, V.
1984   Some results on circumscription. *Proceedings of the Nonmonotonic Reasoning Workshop.* New Paltz, N.Y.
1985   Closed-world databases and circumscription. *Artificial Intelligence 27*, 229–235.
1987   Pointwise circumscription. In Ginsberg [1987].

Loui, R. P.
1987   Defeat among arguments: A system of defeasible inference. *Computational Intelligence 3*, 100–106.

Makinson, D. C.
1965   The paradox of the preface. *Analysis 25*, 205–207.

Martin-Löf, P.
1966   The definition of random sequences. *Information and Control 9*, 602–619.
1969   Literature on von Mises' *Kollektivs Revisited. Theoria 35*, 12–37.

McCarthy, J.
1980   Circumscription–a form of nonmonotonic reasoning.
        Artificial Intelligence 13, 27–39.
1984   Applications of circumscription to formalizing common
        sense knowledge. *Proceedings of the Nonmonotonic
        Reasoning Workshop.* New Paltz, N.Y.
McDermott, D. V.
1982   Nonmonotonic logic II: Nonmonotonic modal theories.
        Journal of the Association for Computing Machinery 29
        (1), 33–57.
McDermott, D. V., and J. Doyle
1980   Nonmonotonic logic I. *Artificial Intelligence 13*, 41–72.
Mellor, D. H.
1969   Chance. *Proceedings of the Aristotelian Society*, Sup-
        plementary volume, p. 26.
1971   *The Matter of Chance.* Cambridge: Cambridge Univer-
        sity Press.
Mises, Richard von
1957   *Probability, Statistics, and Truth.* New York: Macmillan.
        (Original German edition 1928)
Moore, R.
1983   Semantical considerations on non-monotonic logic.
        Proceedings of the IJCAI83.
Neyman, J.
1967   *A Selection of Early Statistical Papers of J. Neyman.*
        Berkeley: University of California Press.
Neyman, J., and E. S. Pearson
1967   *Joint Statistical Papers.* Berkeley: University of Califor-
        nia Press.
Nute, D.
1986   *LDR: A logic for defeasible reasoning.* ACMC Research
        Report No. 01–0013. Athens: University of Georgia
        Press.
1987   Defeasible reasoning. *Proceedings of the Twentieth
        Annual Hawaii International Conference on Systems
        Sciences*, University of Hawaii, 470–477.

1988   Defeasible reasoning: A philosophical analysis in Prolog.
           In *Aspects of AI*, ed. J. Fetzer, 251-288. Dordrecht:
           Reidel.
Pearson, E. S.
1966   *The Selected Papers of E. S. Pearson*.   Berkeley:
           University of California Press.
Poincare, Henri
1912   *Calcul des Probabilites*. Paris.
Pollock, John
1967   Criteria and our knowledge of the material world.
           Philosophical Review 76, 28-60.
1968   What is an epistemological problem?   *American
           Philosophical Quarterly 5*, 183-190.
1971   Perceptual knowledge.   *Philosophical Review 80*,
           287-319.
1972   The logic of projectibility.   *Philosophy of Science 39*,
           302-314.
1974   *Knowledge and Justification*. Princeton, N.J.: Princeton
           University Press.
1974a   Subjunctive generalizations. *Synthese 28*, 199-214.
1976   *Subjunctive Reasoning*. Dordrecht: Reidel.
1981   A refined theory of counterfactuals.   *Journal of
           Philosophical Logic 10*, 239-266.
1983   A theory of direct inference. *Theory and Decision 15*,
           29-96.
1983a   Epistemology and probability. *Synthese 55*, 231-252.
1983b   How do you maximize expectation value? *Nous 17*,
           409-422.
1984   *Foundations of Philosophical Semantics*. Princeton, N.J.:
           Princeton University Press.
1984a   Nomic probability. *Midwest Studies in Philosophy 9*,
           177-204.
1984b   A solution to the problem of induction. *Nous 18*,
           423-462.
1984c   Reliability and justified belief. *Canadian Journal of
           Philosophy 14*, 103-114.
1984d   Foundations for direct inference. *Theory and Decision
           17*, 221-256.

1986   *Contemporary Theories of Knowledge.* Totowa, N.J.: Rowman and Littlefield.

1986b   The paradox of the preface. *Philosophy of Science 53,* 246–258.

1987   Defeasible reasoning. *Cognitive Science 11,* 481–518.

1987a   Epistemic norms. *Synthese 71,* 61–96.

1987b   Probability and proportions. In *Theory and Decision: Essays in Honor of Werner Leinfellner,* ed. H. Berghel and G. Eberlein. Dordrecht: Reidel.

1988   OSCAR: A General Theory of Reasoning. In preparation.

1989   The theory of nomic probability. *Synthese,* forthcoming.

1989a   Philosophy and AI. *Philosophical Perspectives 3,* forthcoming.

1989b   *How to Build a Person.* Cambridge, Mass.: Bradford Books/MIT Press.

Poole, D.

1985   On the comparison of theories: Preferring the most specific explanation. *Proceedings of the International Joint Conference on AI,* Los Angeles.

Popper, Karl

1938   A set of independent axioms for probability. *Mind 47,* 275ff.

1956   The propensity interpretation of probability. *British Journal for the Philosophy of Science 10,* 25–42.

1957   The propensity interpretation of the calculus of probability, and the quantum theory. In *Observation and Interpretation,* ed. S. Körner, 65–70. New York: Academic Press.

1959   *The Logic of Scientific Discovery.* New York: Basic Books.

1960   The propensity interpretation of probability. *British Journal for the Philosophy of Science 11,* 25–42.

Pratt, John W.

1961   Review of E. L. Lehmann's *Testing Statistical Hypotheses. Journal of the American Statistical Association 56,* 166.

Quine, W. V. O.
1969   Natural kinds. In *Ontological Relativity, and Other Essays*, 114–138. New York: Columbia University Press.

Reichenbach, Hans
1949   *A Theory of Probability*. Berkeley: University of California Press. (Original German edition 1935)

Reiter, R.
1978   On closed world data bases. In *Logic and Data Bases*, ed. H. Gallaire and J. Minker. New York: Plenum Press.
1980   A logic for default reasoning. *Artificial Intelligence 13*, 81–132.

Renyi, Alfred
1955   On a new axiomatic theory of probability. *Acta Mathematica Academiae Scientiarum Hungaricae 6*, 285–333.

Russell, Bertrand
1948   *Human Knowledge: Its Scope and Limits*. New York: Simon and Schuster.

Salmon, Wesley
1977   Objectively homogeneous references classes. *Synthese 36*, 399–414.

Schnorr, C. P.
1971   A unified approach to the definition of random sequences. *Mathematical Systems Theory 5*, 246–258.

Seidenfeld, Teddy
1978   Direct inference and inverse inference. *Journal of Philosophy 75*, 709–730.
1979   *Philosophical Problems of Statistical Inference*. Dordrecht: Reidel.

Shoham, Y.
1987   Non-monotonic logics. In Ginsberg [1987].

Sklar, Lawrence
1970   Is propensity a dispositional concept? *Journal of Philosophy 67*, 355–366.
1973   Unfair to frequencies. *Journal of Philosophy 70*, 41–52.
1974   Review of Mellor's *The Matter of Chance*. *Journal of Philosophy 71*, 418–423.

Suppes, Patrick
1973   New foundations for objective probability: Axioms for propensities. In *Logic, Methodology, and Philosophy of Science IV*, ed. Patrick Suppes, Leon Henkin, Athanase Joja, and GR. C. Moisil, 515–529. Amsterdam: North Holland.

Touretzky, D.
1984   Implicit ordering of defaults in inheritance systems. *Proceedings of the AAAI84.*

van Fraassen, Bas
1981   *The Scientific Image.* Oxford: Oxford University Press.

Venn, John
1888   *The Logic of Chance,* 3rd ed. London.

INDEX

A1, 85, 141, 253 ff.
A2, 99, 155, 161, 253 ff.
A3, 102, 161, 187, 253 ff.
A3*, 105
A4, 104, 165
A5, 170, 173
acceptance class, 164, 169
acceptance interval, 175
acceptance rules, 38, 75 ff., 244 ff.
addition principle, 64
AGREE, 72, 141, 147, 189, 203, 212 ff.
associative principle, 62

Barnard, G. A., 161, 177
Bennett, Jonathan, 41
Bernoulli, Jacques, 6, 17
Bernoulli's theorem, 16, 150
Bernstein, Allen R., 51
Bertrand, Joseph, 8
Birnbaum, Alan, 154, 155, 161
Boolean theory of proportions, 48 ff., 53
Borel, Emile, 37
Braithwaite, R. B., 10, 14, 37
Bruckner, A. M., 51
Burnor, Richard, 28
Butler, Bishop, 4, 108

Carnap, Rudolph, 9, 22
Cartwright, Nancy, 22
Cavalieri spaces, 205
CDI, 126, 138
Ceder, Jack, 51
Chisholm, Roderick, 78
Church, Alonzo, 14
collective defeat, 80 ff., 226 ff., 232 ff., 240
concatenation principle, 63
conceptual roles of concepts, 35 ff.
constancy principle, 205
countable additivity, 50
counterfactual conditionals, 40 ff.

counterfactual probability, 25, 267 ff.
counterlegal properties, 42, 54, 172
Cox, D. R., 155
cross product principle, 60
CSD, 126, 138
curve fitting, 193

D1, 86, 103, 142, 253 ff.
D2, 103, 253 ff.
D3, 103, 187, 253 ff.
DD, 256
default logic, 238
default rules, 238
normal, 241
defeasible inference, 78 ff., 220 ff.
defeat, analysis of, 225
defeaters, 78 ff., 221
rebutting, 79, 112, 221
subproperty, 86, 103
subset, 112 ff.
undercutting, 79, 112, 222
definite probabilities, 22, 108 ff.
Dempster, A. P., 155
denseness principle, 52
DI, 128, 140, 190
DI*, 141
direct inference, 23, 37, 108 ff., 190, 267 ff.
classical, 109, 110 ff.
nonclassical, 109, 127 ff.
rule of, 111
diversity of the sample, 315
DOM, 259
DOM-D, 261
domination defeaters, 256 ff., 282 ff.
Doyle, Jon, 79, 238, 241

E-DOM, 265
Edwards, A. W. F., 161, 177
Eells, Ellory, 26
Ellis, R. L., 10

empirical theories, 20 ff.
enumerative induction argument,
    179 ff.
epistemic conditionalization, 137
epistemic basis, 222
epistemic justification, 87
epistemic probability, 4
Etherington, D., 79, 238, 239,
    240, 241
existential generalization principle,
    207
exotic principles, 68 ff., 203 ff.
extended domination defeaters,
    291 ff.
extended frequency principle, 49

fair sample defeaters
    for enumerative induction,
        312 ff.
    for statistical induction,
        307 ff.
Fetzer, James, 26, 27 ff., 270 ff.
Fisher, R. A., 17, 158, 161, 177
frequency principle, 48
frequency theories, 10 ff.
    finite, 12
    hypothetical, 15 ff.
    limiting, 12

gambler's fallacy, 245 ff.
Giere, Ronald, 26 ff., 270 ff.
Gillies, D. A., 170, 177
Ginsberg, Matt, 239
Goodman, Nelson, 149, 196, 296,
    297, 302

Hacking, Ian, 4, 5, 6, 25, 26, 161,
    177
Hanks, S., 79, 238, 239
hard propensities, 270 ff.
Harman, Gilbert, 89
Harper, William, 137
Hausdorff, Felix, 204
Hempel, Carl, 320
Hintikka, Jaakko, 9
Horty, John, 240

IND, 66, 189, 263, 294
indefinite probabilities, 21
indifference, principle of, 6
indirect arguments, 223
induction, 149 ff., 190, 296 ff.
    enumerative, 150, 177 ff.
    material, 179 ff.
    statistical, 150, 229
inference to the best explanation,
    195
insufficient reason, principle of, 6,
    128

Jackson, Frank, 21
joint defeat, 235 ff.
joint subset defeaters, 278 ff.

Keynes, John Maynard, 6, 9
Kneale, William, 10, 14
Kolmogoroff, A. N., 46
Konolige, Kurt, 238
Kyburg, Henry, x, 8, 10 ff., 17, 26,
    78, 80, 115 ff., 123, 137, 155,
    196, 278, 294, 304

Laplace, Pierre, 4, 5 ff.
legal conservatism 41
Lehrer, Keith, 83
Leibniz, G. W., 6
level n argument, 90, 225
Levi, Isaac, 25, 113, 137
Lewis, David, 40, 41
Lifschitz, Vladimir, 79, 238
likelihood 158
likelihood principle, 155, 161, 170
likelihood ratio, 160
    maximum acceptable, 167
linear argument, 223
linear model of arguments, 92
logical probability, 8 ff.
lottery paradox, 80, 227
Loui, Ron, 79, 87, 238

Makinson, D. C., 244
Martin-Löf, P., 14
material probability, 16

material statistical induction argument, 181 ff.
McCarthy, John, 79, 238, 239, 241
McDermott, D., 79, 238, 239, 241
Mellor, D. H., 26, 27 ff.
Mercer, R., , 238, 241
Moore, R., 79, 238, 241
multiplicative axiom, 56, 57
multiple extension problem, 239

Neyman, J., 154, 160
nomic generalizations, 32 ff., 42 ff.
nomic probabilities, 25, 32 ff.
nomically rigid designators, 70
nonmonotonic logic, 79, 238 ff.
Nute, Donald, 79, 87, 238

OSCAR, 187

paradox of the preface, 244, 250 ff.
paradox of the ravens, 320 ff.
paradoxes of defeasible reasoning, 227 ff.
Pargetter, Robert, 21
Pearson, E. S., 154, 160
Peirce, C. S., 10, 17
PFREQ, 70, 106, 120, 181, 189, 203, 211 ff., 306
physical necessity, 42
physical possibility, 42
physical probability, 4
Poincare, Henri, 8, 37
Poole, D., 79, 87, 238
Popper, Karl, 12, 17, 24, 25, 51
Popper functions, 51
PPROB, 54, 59, 172, 180, 189, 313
Pratt, Ian, 294
Pratt, John, 155
principle of agreement, 71 ff., 212 ff.
product defeaters, 294
projectibility, 81 ff., 123 ff., 171, 194, 196 ff.
  inductive, 296
projection defeaters, 294
propensities, 17, 24 ff., 226 ff.

properties, 40
proportions, 45 ff.
Putnam, Hillary, 304

Quine, W. V., 320

reasons,
  conclusive, 221
  defeasible, 78, 221
  non-primitive, 93
  prima facie, 78 ff., 221
  primitive, 93
rectangle, 206
Reichenbach, Hans, 7, 12, 14, 50, 110 ff.
Reichenbach's rules, 110
Reiter, R., 79, 238 ff., 241
rejection class, 164, 168
Renyi, Alfred, 51, 205
reversibility principle, 208
rigid designator, 58
rules of inference, 223
Russell, Bertrand, 10, 14

Salmon, Wesley, 14
Schnorr, C. P., 14
SD, 131, 140, 190
Seidenfeld, Teddy, 137, 155
self-defeating argument, 230
Shoham, Yoav, 238
Sklar, Lawrence, 10, 28, 32
SS, 41
states of affairs, 39
statistical induction, 37, 149
statistical induction argument, 155, 170 ff.
statistical inference, 192
statistical syllogism, 75 ff.
  material, 77, 104 ff.
strong indefinite propensities, 276 ff.
subjunctive generalizations, 33
subjunctive indefinite probability, 33
subproperty, 72, 127
  strict, 72

subsample problem, 303 ff.
subset defeat, 111
    rule of, 111
subsidiary arguments, 223
Suppes, Patrick, 26

Thomason, Richmond, 241
Touretzky, D., 87, 241
translation invariance, principle of,
    204
    vertical and horizontal, 205

ultimately undefeated arguments,
    89 ff., 225

undercutting, 41

van Fraassen, Bas, 17, 19, 26, 189,
    206
Venn, John, 10, 12
von Kries, J., 6
von Mises, Richard, 7, 8, 12, 13, 17

warrant, 87, 222 ff.
Wattenberg, Frank, 51
weak indefinite propensities, 273 ff.

APPENDIX
MAIN PRINCIPLES

The principles listed here are cited from other chapters.

NAMED PRINCIPLES

Probability Calculus:

(*PPROB*) If r is a rigid designator of a real number and $\Diamond[\exists G \;\&\; \text{prob}(F/G) = r]$ then
$$\text{prob}\left(F \;/\; G \;\&\; \text{prob}(F/G) = r\right) = r.$$

(*PFREQ*) If r is a nomically rigid designator of a real number and $\Diamond_p[\exists G \;\&\; \text{freq}[F/G] = r]$ then
$$\text{prob}\left(F \;/\; G \;\&\; \text{freq}[F/G] = r\right) = r.$$

(*IND*) $\quad \text{prob}(Axy \;/\; Rxy \;\&\; y=b) = \text{prob}(Axb/Rxb).$

(*AGREE*) If F and G are properties and there are infinitely many physically possible G's and $\text{prob}(F/G) = p$ (where p is a nomically rigid designator) then for every $\delta > 0$:
$$\text{prob}\left(\text{prob}(F/X) \approx_\delta p \;/\; X \text{ is a strict subproperty of } G\right) = 1.$$

Acceptance Rules:

(*A1*) $\quad$ If F is projectible with respect to G and $r > .5$ then $\lceil Gc \;\&\; \text{prob}(F/G) \geq r \rceil$ is a prima facie reason for believing $\lceil Fc \rceil$, the strength of the reason depending upon the value of r.

(*D1*)   If F is projectible with respect to H then $\ulcorner Hc$ &
          $\text{prob}(F/G\&H) \neq \text{prob}(F/G)\urcorner$ is an undercutting defeater
          for $\ulcorner Gc$ & $\text{prob}(F/G) \geq r\urcorner$ as a prima facie reason for
          $\ulcorner Fc\urcorner$.

(*A2*)   If F is projectible with respect to G and $r > .5$ then
          $\ulcorner \text{prob}(F/G) \geq r$ & $\sim Fc\urcorner$ is a prima facie reason for
          $\ulcorner \sim Gc\urcorner$, the strength of the reason depending upon the
          value of r.

(*D2*)   If F is projectible with respect to H then $\ulcorner Hc$ &
          $\text{prob}(F/G\&H) \neq \text{prob}(F/G)\urcorner$ is an undercutting defeater
          for $\ulcorner \sim Fc$ & $\text{prob}(F/G) \geq r\urcorner$ as a prima facie reason for
          $\ulcorner \sim Gc\urcorner$.

(*A3*)   If F is projectible with respect to G and $r > .5$ then
          $\ulcorner \text{prob}(F/G) \geq r\urcorner$ is a prima facie reason for the
          conditional $\ulcorner Gc \supset Fc\urcorner$, the strength of the reason
          depending upon the value of r.

(*D3*)   If F is projectible with respect to H then $\ulcorner Hc$ &
          $\text{prob}(F/G\&H) \neq \text{prob}(F/G)\urcorner$ is an undercutting defeater
          for $\ulcorner \text{prob}(F/G) \geq r\urcorner$ as a prima facie reason for $\ulcorner Gc \supset$
          $Fc\urcorner$.

(*DD*) $\ulcorner$'$\sim Fc$' is warranted, and 'Hc' is a deductive
          consequence of propositions supported by the argument
          on the basis of which '$\sim Fc$' is warranted$\urcorner$ is an
          undercutting defeater for each of (*D1*), (*D2*), and (*D3*).

(*A3**) If F is projectible with respect to G and $r > .5$ then
          $\ulcorner \text{freq}[F/G] \geq r\urcorner$ is a prima facie reason for $\ulcorner Gc \supset Fc\urcorner$,
          the strength of this reason being precisely the same as
          the strength of the reason $\ulcorner \text{prob}(F/G) \geq r\urcorner$ provides
          for $\ulcorner Gc \supset Fc\urcorner$ by (*A3*).

(*A4*)   If F is projectible with respect to G and $r > .5$ then $\ulcorner \sim Fc$ & $(\forall x)[Hx \supset \text{prob}_x(Fy/Gxy) \geq r]\urcorner$ is a prima facie reason for $\ulcorner (\forall x)(Hx \supset \sim Gxc)\urcorner$.

(*A5*)   If F is projectible with respect to B and X is a set such that:
   (a)   we are warranted in believing that $(\exists x)[x \in X$ & $Bxc]$;
   (b)   we are warranted in believing $\ulcorner \sim Fc$ and for each b in X, if 'Bbc' is true then $\text{prob}(Fy/Bby) \geq r\urcorner$; and
   (c)   the only defeaters for the instances of (*A2*) and (*A4*) that result from (b) are those "internal" ones that result from those instances conflicting with one another due to (a);
then we are warranted to degree α in believing that $(\exists x)[x \in \mathbf{A}_\alpha(X)$ & $Bxc]$ and $(\forall x)[x \in \mathbf{R}_\alpha(X) \supset \sim Bxc]$.

(*DOM*)   If A is projectible with respect to both B and D then $\ulcorner \text{prob}(A/B) = \text{prob}(A/C)$ & $C \preceq B\urcorner$ is an undercutting defeater for each of the following inferences:
   (a) from $\ulcorner Cc$ & $\text{prob}(A/C) \geq r\urcorner$ to $\ulcorner Ac\urcorner$ by (*A1*);
   (b) from $\ulcorner \sim Ac$ & $\text{prob}(A/C) \geq r\urcorner$ to $\ulcorner \sim Cc\urcorner$ by (*A2*);
   (c) from $\ulcorner \text{prob}(A/C) \geq r\urcorner$ to $\ulcorner Cc \supset Ac\urcorner$ by (*A3*).

(*DOM-D*) If A is projectible with respect to E then $\ulcorner C \preceq E \preceq B$ & $\text{prob}(A/E) \neq \text{prob}(A/B)\urcorner$ is a defeater for (*DOM*).

(*E-DOM*) Consider the following:
   (a) $Bb_1 \ldots, b_n$ & $\text{prob}(A/B) \geq r$;
   (b) $B^*c_1 \ldots c_k$ & $\text{prob}(A^*/B^*) \geq r$;
   If P is a domination defeater for (a) as a prima facie reason for $\ulcorner Ab_1 \ldots b_n\urcorner$, then
   (c) f is a nomic transform such that
   $f(\langle A,B,\langle b_1,\ldots,b_n\rangle\rangle) = \langle A^*,B^*,\langle c_1,\ldots,c_k\rangle\rangle$, and
   P & $[Bb_1\ldots b_n \Rightarrow (Ab_1\ldots b_n \equiv A^*c_1\ldots c_k)]$ & $(Bb_1\ldots b_n$
   $\Leftrightarrow B^*c_1\ldots c_k)$

is a defeater for (b) as a prima facie reason for $\ulcorner A^*c_1 \dots c_k \urcorner$. Defeaters of the form $(DOM\text{-}D)$ for P are also defeaters for (c).

Direct Inference:

(CDI)   If F is projectible with respect to G then $\ulcorner \text{prob}(F/G) = r \ \& \ \mathbf{W}(Gc) \ \& \ \mathbf{W}(P \equiv Fc) \urcorner$ is a prima facie reason for $\ulcorner \text{PROB}(P) = r \urcorner$.

(CSD)   If F is projectible with respect to H then $\ulcorner \text{prob}(F/H) \neq \text{prob}(F/G) \ \& \ \mathbf{W}(Hc) \ \& \ \Box \forall (H \supset G) \urcorner$ is an under-cutting defeater for (CDI).

(DI)   If F is projectible with respect to G then $\ulcorner H \preccurlyeq G \ \& \ \text{prob}(F/G) = r \urcorner$ is a prima facie reason for $\ulcorner \text{prob}(F/H) = r \urcorner$;

(SD)   If F is projectible with respect to J then $\ulcorner H \preccurlyeq J \preccurlyeq G \ \& \ \text{prob}(F/J) \neq \text{prob}(F/G) \urcorner$ is an undercutting defeater for (DI).

Chapter 2

(3.3) $(F \Rightarrow G)$ iff $\diamond_p \exists F > \Box_p \forall (F \supset G)$.

(3.17) $[\exists F \ \& \ \Box_p \forall (F \supset G)] \supset (F \Rightarrow G)$.

(5.2)   If X and Y are finite and Y is nonempty then $\wp(X/Y) = \text{freq}[X/Y]$.

(5.6)   If $Z \neq \varnothing$ and $Z \cap X \cap Y = \varnothing$ then $\wp(X \cup Y/Z) = \wp(X/Z) + \wp(Y/Z)$.

(5.12)   If $X \subseteq Y$ and $r < \wp(X/Y)$ and $(\exists Z)[Z \subseteq Y \,\&\, \wp(Z/Y) = r]$ then $(\exists Z)[Z \subseteq X \subseteq Y \,\&\, \wp(Z/Y) = r]$.

(5.13)   If $X \subseteq Y$ and $r > \wp(X/Y)$ and $(\exists Z)[Z \subseteq Y \,\&\, \wp(Z/Y) = r]$ then $(\exists Z)[X \subseteq Z \subseteq Y \,\&\, \wp(Z/Y) = r]$.

(6.13)   If $\ulcorner \diamond_p \exists (G \,\&\, F_1 \,\&\, \ldots \,\&\, F_n) \urcorner$ then $\text{prob}(F_1 \,\&\, \ldots \,\&\, F_n / G) = \text{prob}(F_1/F_2 \,\&\, \ldots \,\&\, F_n \,\&\, G) \cdot \text{prob}(F_2/F_3 \,\&\, \ldots \,\&\, F_n \,\&\, G) \cdot \ldots \cdot \text{prob}(F_n/G)$.

(6.18)   If $\text{prob}(H/G) = 1$ then $\text{prob}(F/G \,\&\, H) = \text{prob}(F/G)$.

(6.20)   If r is a rigid designator of real number and $\diamond \exists G$ and $[\diamond_p \exists G > \text{prob}(F/G) = r]$ then $\text{prob}(F/G) = r$.

(6.21) $(\diamond_p \exists G \,\&\, \Box_p P) \supset \text{prob}(F/G \,\&\, P) = \text{prob}(F/G)$.

(7.2)   If $C \neq \varnothing$ and $D \neq \varnothing$ then $\wp(A \mathrm{x} B / C \mathrm{x} D) = \wp(A/C) \cdot \wp(B/D)$.

(7.9)   If $A \subseteq R$, $B \subseteq S$, $R \neq \varnothing$, $S \neq \varnothing$, and either R and S are both relations or one is a relation and the other is a nonrelational set then $\wp(A \,\hat{}\, B \;/\; R \,\hat{}\, S) = \wp(A/R) \cdot \wp(B/S)$.

(7.11)   If φ and θ are open formulas containing the free variables $x_1, \ldots, x_n$ (ordered alphabetically) then $\wp(\varphi/\theta) = \wp\big(\{\langle x_1, \ldots, x_n \rangle \mid \varphi\} \;/\; \{\langle x_1, \ldots, x_n \rangle \mid \theta\}\big)$.

(7.14)   If B is infinite, $b_1, \ldots, b_n \in B$ and $b_1, \ldots, b_n$ are distinct, $C \subseteq B$ and $b_1, \ldots, b_n \notin C$, then $\wp(b_1, \ldots, b_n \in X \;/\; C \subseteq X \subseteq B \,\&\, X \text{ infinite}) = 1/2^n$.

(8.4) $\wp(X/Y) = \text{prob}(x \in X / x \in Y)$.

(9.2)   For every $\delta, \gamma > 0$, there is an n such that if B is a finite set containing at least n members then

$$\text{freq}\Big[\text{freq}[A/X] \approx_{\delta} \text{freq}[A/B] \;/\; X \subseteq B\Big] > 1\text{-}\gamma.$$

Chapter 3

(2.3)   If A and B are projectible with respect to C, then $(A\&B)$ is projectible with respect to C.

(2.4)   If A is projectible with respect to both B and C, then A is projectible with respect to $(B\&C)$.

(3.3)   If Q provides a reason for R then $(P \supset Q)$ provides a reason for $(P \supset R)$. If S is a defeater for the argument from Q to R, then $(P \supset S)$ is a defeater for the argument from $(P \supset Q)$ to $(P \supset R)$.

(3.6)   If P and Q both provide reasons for R, so does $(P \lor Q)$.

(3.7)   If $\ulcorner Fx \urcorner$ is a reason schema for $\ulcorner Gx \urcorner$, then $\ulcorner (\forall x)(Hx \supset Fx) \urcorner$ is a reason for $\ulcorner (\forall x)(Hx \supset Gx) \urcorner$. If $\ulcorner Jx \urcorner$ is a defeater for the argument schema from $\ulcorner Fx \urcorner$ to $\ulcorner Gx \urcorner$, then $\ulcorner (\exists x)(Hx \;\&\; Jx) \urcorner$ is a defeater for $\ulcorner (\forall x)(Hx \supset Fx) \urcorner$ as a reason for $\ulcorner (\forall x)(Hx \supset Gx) \urcorner$.

(3.12)   If P is a reason for $\ulcorner Fx \urcorner$ then P is a reason for $\ulcorner (\forall x)Fx \urcorner$.

(3.13)   If P is a reason for Q then $\ulcorner \square_p P \urcorner$ is a reason for $\ulcorner \square_p Q \urcorner$.

Chapter 4

(3.5)   If F is projectible with respect to G then $\ulcorner \square \forall (H \supset G)$ & $(\lozenge_p \exists G \;\rangle\; \lozenge_p \exists H)$ & $\text{prob}(F/G) = r \urcorner$ is a prima facie reason for $\ulcorner \text{prob}(F/H) = r \urcorner$.